Fundamentals of Mean Field Theories of Liquid Crystals

Fundamentals of Mean Field Theories of Liquid Crystals offers a clear and comprehensive introduction to mean field approaches for understanding liquid crystalline systems, with a focus on rod-like molecules and polymeric liquid crystals. These theories form the foundation for describing the rich phase behavior of anisotropic fluids, which underpin soft matter physics and advanced materials science.

Beginning with the classical Onsager and Maier–Saupe theories, the book develops a unified framework connecting microscopic molecular interactions with macroscopic ordering phenomena. It guides readers through the statistical mechanics of orientational order and phase transitions, extending naturally to biaxial nematic and cholesteric phases, field-induced phase transitions, liquid crystalline gels, and self-assembled liquid crystals. This book is primarily intended for senior undergraduate and graduate students, as well as academic researchers in physics, chemistry, and materials science.

Key Features:

- Introduces liquid crystalline phases, orientational order, and phase separations with many useful figures.
- Develops mean field models for uniaxial and biaxial nematic, smectic, cholesteric, and columnar phases, and explores phase transitions under external fields.
- Extends models to polymer liquid crystals and binary mixtures.
- Highlights phase transition phenomena in liquid crystalline gels.
- Examines self-assembled liquid crystals and their theoretical treatment.

Akihiko Matsuyama received his Ph.D. from the Graduate School of Engineering at the University of Tokyo in 1991. He began his career the same year as a Researcher at the Research Institute for Polymers and Textiles, National Institute of Advanced Industrial Science and Technology, in Japan. In 1993, he joined the Faculty of Engineering at Mie University as an Assistant Professor. From 1998 to 1999, he was a Visiting Research Fellow in the Department of Physics at the University of Edinburgh. He moved to the Faculty of Computer Science and Systems Engineering at Kyushu Institute of Technology in 2003 as an Associate Professor and has been a Professor there since 2014. He is currently a Professor at the Kyushu Institute of Technology. He has dedicated his career to the theoretical study of soft matter systems, with a particular focus on mean field theories of liquid crystals, polymeric materials, and related phase transitions. His research covers microscopic modeling, statistical mechanics, and the connection between molecular interactions and macroscopic physical properties.

Fundamentals of Mean Field Theories of Liquid Crystals

Akihiko Matsuyama

CRC Press
Taylor & Francis Group
Boca Raton London New York

CRC Press is an imprint of the
Taylor & Francis Group, an **informa** business

Designed cover image: AdobeStock

First edition published 2026
by CRC Press
2385 NW Executive Center Drive, Suite 320, Boca Raton FL 33431

and by CRC Press
4 Park Square, Milton Park, Abingdon, Oxon, OX14 4RN

CRC Press is an imprint of Taylor & Francis Group, LLC

ISBN: 978-1-041-23660-3 (hbk)
ISBN: 978-1-041-23711-2 (pbk)
ISBN: 978-1-003-73771-1 (ebk)

DOI: 10.1201/9781003737711

Typeset in CMR10 font
by KnowledgeWorks Global Ltd.

Publisher's note: This book has been prepared from camera-ready copy provided by the authors.

To my family and to all who have inspired and supported my journey in science.

Contents

Preface

Liquid crystals are a key class of soft matter systems that exhibit both orientational and positional order. This book, Fundamentals of Mean Field Theories of Liquid Crystals, is intended as an introductory text for undergraduate and graduate students, as well as for newcomers who wish to understand the theoretical foundations of liquid crystal physics. A basic knowledge of thermodynamics and statistical mechanics is assumed, but no prior experience with liquid crystals is required. Mean field theory plays a central role in the study of liquid crystals and other soft matter systems. By reducing complex many-body interactions to average effective fields, it provides a tractable framework that captures the essential physics of phase transitions and ordering phenomena. While it may neglect certain fluctuations, mean field theory offers valuable physical insight and serves as a foundation for developing more refined models and for guiding computational approaches.

Liquid crystals offer a fascinating platform for exploring concepts in statistical mechanics. While many textbooks provide excellent coverage of the general properties of liquid crystals, their various phases, and continuum theoretical descriptions, there is often limited emphasis on molecular-level and mean field approaches. Both statistical continuum theory and molecular mean field theory are complementary, and a comprehensive understanding of liquid crystal behavior benefits from integrating these perspectives. This book aims to address this need by presenting a thorough introduction to mean field theories of liquid crystals. It equips readers with the essential tools to connect molecular interactions to macroscopic phenomena, bridging the gap between microscopic models and observable behavior. Throughout the text, emphasis is placed on clear explanations, illustrative examples, and practical calculations, allowing readers to develop both conceptual understanding and theoretical proficiency. By the end of the book, readers will gain a solid foundation in mean field approaches and be prepared to explore more advanced theoretical models and computational studies in soft matter physics.

The structure of the book is designed to provide a logical and pedagogical progression. Chapter 1 provides an introduction to liquid crystalline molecules and several characteristic phase behaviors that are central to this textbook. Chapter 2 introduces the fundamental concepts of liquid crystal phases, with a particular focus on their classification through symmetry breaking and the corresponding order parameters. Various types of order: orientational, helical, and positional, are discussed, and representative order parameters are introduced as key quantities. Chapter 3 explains the concept of elastic free energy, or continuum theory, associated with director deformations, forming the basis for the theoretical developments in later chapters.

Chapter 4 begins with the Onsager theory for hard rod-like molecules, which explains the isotropic-nematic transition based on excluded volume effects and entropy. It then introduces the classical Maier-Saupe theory, which incorporates mean field attractive interactions to describe orientational ordering. The chapter also covers the Landau-de Gennes expansion of the free energy near phase transitions and introduces theoretical models for the smectic A and columnar phases. These foundational models provide a comprehensive framework for analyzing liquid crystal phases.

Chapter 5 introduces the theoretical framework for cholesteric liquid crystals (also known as chiral nematic phases). It explains how molecular chirality leads to the spontaneous formation of a helical structure in the director field and defines the pitch and helical wavenumber as key order parameters. The chapter also discusses the role of twist elastic energy and outlines how external factors such as temperature or dopants can influence the pitch. This forms the basis for understanding optical phenomena and applications unique to cholesteric phases. Chapter 6 explores the phase transitions of liquid crystals under external fields, focusing on field-induced alignment, the twist-untwist transition, and the heliconical cholesteric phase.

Chapter 7 discusses the theoretical framework for phase separation in mixtures of polymers and liquid crystalline molecules. It examines how the interplay between entropic mixing, liquid crystalline ordering, and molecular interactions leads to rich phase behaviors, including isotropic-nematic demixing and coexistence of anisotropic phases. Mean field models combining Flory-Huggins theory with orientational order parameters are introduced to describe the coupling between composition and alignment.

Chapter 8 discusses volume phase transitions in nematic and cholesteric liquid crystalline gels. These systems combine the anisotropic properties of liquid crystals with the elasticity of polymer networks, leading to rich coupled behavior. The chapter introduces theoretical models that describe how orientational ordering interacts with swelling and mechanical deformation, with emphasis on volume phase transitions triggered by temperature, solvent conditions, and orientational ordering.

Chapter 9 introduces the basic theory of self-assembled liquid crystals, such as those formed by reversible aggregation of amphiphilic molecules or short rod-like units. The coupling between aggregation, orientational ordering, and phase behavior is analyzed using statistical mechanical models.

This book is a substantially revised and expanded version of two previous works: Introduction to Soft Matter Physics (Japanese edition), published by Morikita Publishing in 2022, and lecture notes (Japanese edition) published in EKISHO, the Japanese Liquid Crystal Society, in 2020. It will serve both as a learning resource for students and as a foundation for those wishing to delve deeper into the theory of soft matter and liquid crystals. A list of references is provided at the end of the book to guide readers interested in further study.

Finally, it is my great pleasure and honor to thank my esteemed mentors—Professor Fumihiko Tanaka (Professor Emeritus, Kyoto University, formerly Tokyo University of Agriculture and Technology), Professor Koji Okano (Professor Emeritus, the University of Tokyo), and Professor Michael E. Cates (the University of Cambridge, formerly the University of Edinburgh)—who introduced me to the fascinating world of soft matter physics early in my career.

Akihiko Matsuyama

December 2025
Iizuka, Fukuoka, Japan

1

Introduction

As de Gennes wrote, "Liquid crystals are beautiful and mysterious; I am fond of them for both reasons" [1], a fascination shared by many scientists working on these materials. A wide variety of molecules can form liquid crystalline phases. In this chapter, we introduce some typical examples and their characteristic phase behaviors. The anisotropic shapes and the specific intermolecular interactions of these molecules give rise to various mesophases, including nematic, smectic, and cholesteric phases. Understanding how the molecular structure influences phase behavior is essential for designing new liquid crystalline materials and their applications [2].

1.1 Liquid crystalline molecules

Liquid crystalline materials exhibit a variety of mesophases, including the nematic phase characterized by long-range orientational order, the smectic A phase with one-dimensional layered positional order, the cholesteric phase featuring a helical modulation of the nematic director, and the columnar phase where molecules are arranged into two-dimensional lattices of columns (Fig. 1.1) [3,4]. The details of each phase will be discussed in Chapter 2.

Liquid crystals (LCs) are broadly classified into two categories: thermotropic and lyotropic, according to the physical mechanism that drives their phase transitions. Thermotropic liquid crystals exhibit mesophases as a function of temperature, typically in the absence of solvent molecules. Thermal energy alters the molecular orientation and degree of order of anisotropic molecules, thereby inducing phase transitions. Lyotropic liquid crystals, on the other hand, form mesophases in solution, where the concentration of anisotropic

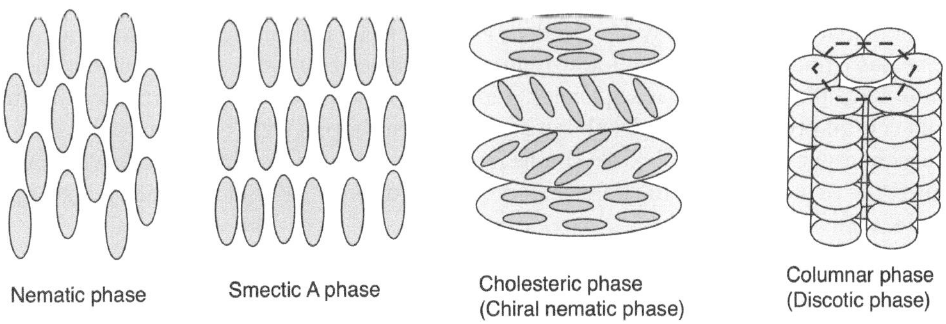

FIGURE 1.1
Schematic illustration of liquid crystalline phases.

DOI: 10.1201/9781003737711-1

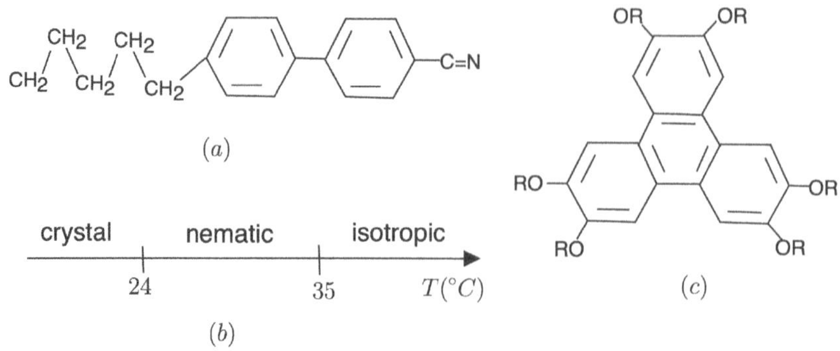

FIGURE 1.2
(a) Low-molecular-weight liquid crystal: 5CB. (b) Phase transition temperatures of a thermotropic LC. (c) Disc-shaped liquid crystalline molecule.

solutes is the key control parameter. Their phase behaviors depend not only on temperature but also critically on solute concentration.

A characteristic feature of molecules that exhibit these liquid crystalline phases is their anisotropic shape and relatively rigid structure. Because of these features, they are often referred to as rod-like molecules. Figure 1.2(a) shows a representative low-molecular-weight liquid crystalline molecule, 5CB (4-Cyano-4'-pentylbiphenyl). It consists of two benzene rings connected linearly, with a five-carbon alkyl chain and a cyano group attached at opposite ends, forming a rigid and elongated molecular structure.

Figure 1.2(b) shows the phase transition temperatures of thermotropic 5CB. As the temperature decreases, the material undergoes a transition from the isotropic phase to the nematic phase at 35°C, and then to the crystal phase at 24°C. The transition temperatures vary depending on the length of the alkyl chain attached to the molecule. By coarse-graining the molecular structure, such liquid crystalline molecules can be effectively regarded as rod-like particles. A typical representation is a rod with a diameter of $D \simeq 0.4$ nm and a length of $L \simeq 2$ nm, resulting in an axial ratio of $L/D \simeq 5$. In the case of low-molecular-weight liquid crystals, the axial ratio of the molecules is generally of order one, reflecting their relatively compact shape compared to rod-like polymers. Most liquid crystalline molecules used in liquid crystal displays (LCDs) are indeed rod-like in shape.

Figure 1.2(c) shows an example of a disc-shaped liquid crystalline molecule. As the temperature decreases, the system undergoes a phase transition from the isotropic phase to the nematic phase, and at even lower temperatures, to a columnar phase. Various molecules exhibit such phase transition sequences, depending on their molecular structure. To display liquid crystal properties, a molecule typically requires both a rigid core (known as mesogens, such as benzene rings) and flexible chains (spacer units, such as alkyl chains). Without alkyl chains, the nematic phase does not emerge; instead, only the isotropic and crystal phases are observed.

In addition to rod-like and disc-like molecules, bent-shaped (or banana-shaped) liquid crystalline molecules have attracted considerable interest [5]. These molecules possess a bent or kinked structure, typically due to the presence of an angular core or asymmetric substitution in the rigid part of the molecule. The unique geometry of bent-shaped molecules can give rise to novel phases such as the biaxial nematic phase, ferroelectric phases, and antiferroelectric smectic phases, which are not typically observed in systems composed of rod-like molecules.

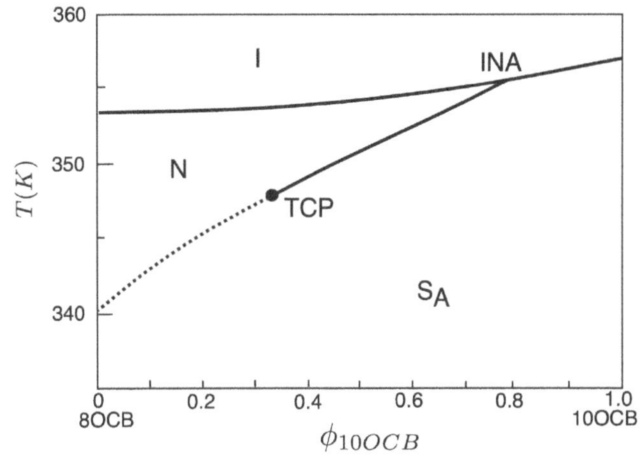

FIGURE 1.3
Phase diagram for a mixture of two low-molecular-weight liquid crystals.

Figure 1.3 shows the phase diagram for a mixture of two low-molecular-weight liquid crystals: 8OCB (4-Cyano-4'-n-octyloxybiphenyl) and 10OCB (4'-Cyano-4-decyloxybiphenyl) [6]. The vertical axis represents temperature, while the horizontal axis denotes the mole fraction of 10OCB. The solid line indicates a first-order phase transition, and the dotted line represents a second-order transition. By varying temperature and composition, three distinct liquid crystalline phases: isotropic (I), nematic (N), and smectic A (S_A), are observed. As the concentration of 10OCB increases, the temperature range of the nematic phase becomes narrower. For mole fractions above 0.8, the nematic phase disappears entirely, and a direct transition occurs from the I to the S_A phase. The point where the three phases: I, N, and S_A, coexist is labeled INA, and the intersection of the first-order and second-order transition curves is known as the tricritical point (TCP). At high temperatures, thermal motion dominates, stabilizing the isotropic phase. Although pure 10OCB does not exhibit a nematic phase on its own, mixing it with 8OCB induces nematic ordering. This example highlights how molecular mixtures can give rise to complex and tunable phase behavior, a principle that plays a key role in the design of functional liquid crystalline materials. Similar phase diagrams have been reported for many other combinations of liquid crystalline molecules.

1.2 Liquid crystalline polymers and gels

A liquid crystalline polymer (LCP) is a high-molecular-weight material in which rod-like molecules (mesogens) and flexible chains (spacers) are covalently connected in an alternating fashion [7]. As illustrated in Fig. 1.4, when mesogens and spacers are linked alternately along the backbone, the resulting structure is referred to as a main-chain liquid crystalline polymer. In contrast, when mesogenic units are attached as pendant side groups to a flexible main chain, the material is called a side-chain liquid crystalline polymer. In addition to these, various architectures exist depending on how the mesogens and spacers are incorporated. These materials combine the anisotropic properties of liquid crystals with the

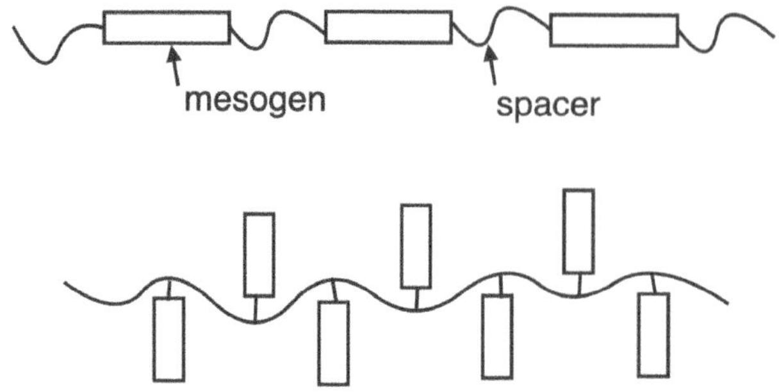

FIGURE 1.4
Main-chain liquid crystalline polymer and side-chain LCP.

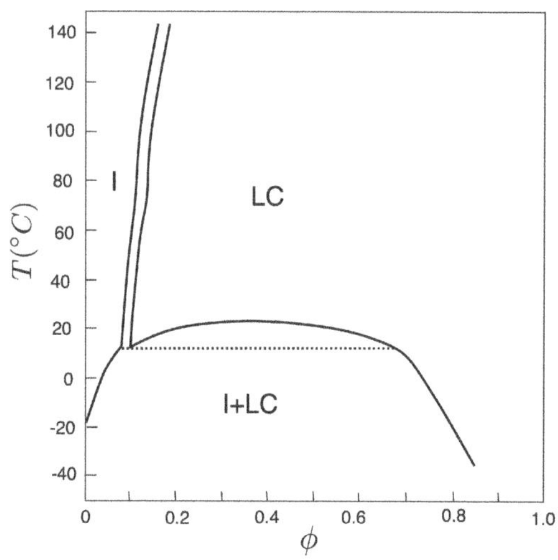

FIGURE 1.5
Phase diagram of a binary mixture of an LCP (PBLG) and a solvent (DMF).

mechanical robustness of polymers, leading to practical applications as high-performance fibers in products such as tennis rackets, skis, and automobile components. Moreover, certain biopolymers—including DNA, polypeptides, and the tobacco mosaic virus—also exhibit liquid crystalline behavior. Similarly, carbon nanotubes (CNT) and other long, rigid, rod-like molecules can form liquid crystalline phases and are often studied within the same theoretical frameworks, such as the Onsager theory and the Maier-Saupe model for nematic ordering.

Various phase behaviors emerge when LCPs are mixed with solvent molecules. Figure 1.5 presents the typical phase diagram of a binary mixture composed of a LCP, poly-γ-benzyl-L-glutamate (PBLG), and a solvent molecule, dimethylformamide (DMF) [7–9]. The vertical axis indicates temperature, while the horizontal axis shows the concentration of the LCP. The solid line represents the coexistence curve. At low polymer concentrations, the system remains in the isotropic (I) phase. As the concentration increases, a liquid crystal (LC) phase emerges, which includes nematic and cholesteric ordering. The region labeled I+LC corresponds to the two-phase coexistence between I and LC phases. On the

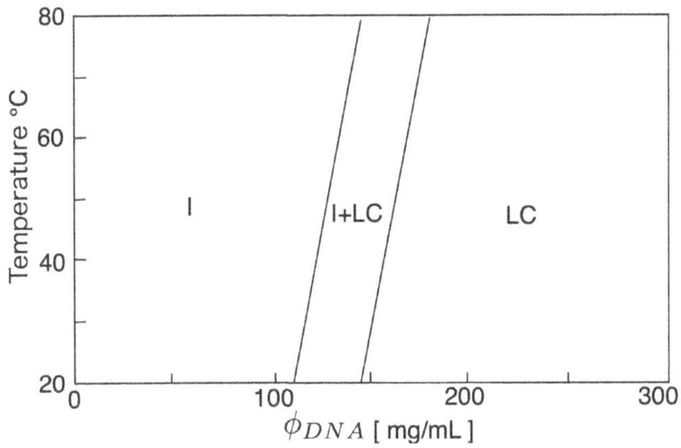

FIGURE 1.6
Coexistence curve of a DNA (rod-like molecule) aqueous solution.

high-temperature side, the phase separation I+LC exhibits a chimney-type phase separation, where the concentration difference between the coexisting phases is relatively small. In contrast, low-temperature phase separation occurs with a larger concentration difference. Within the LC region, an upper critical solution temperature (UCST) is observed, below which the system undergoes LC+LC phase separation into two distinct liquid crystalline phases. Around 15 °C, a three-phase equilibrium (I+LC+LC) appears, in which the I phase and two LC phases coexist simultaneously.

Figure 1.6 shows the typical coexistence curve of an aqueous DNA solution, where DNA acts as a rod-like molecule [10]. As the DNA concentration increases, the system undergoes a phase transition from the I phase to the LC phase. A chimney-shaped I+LC phase separation appears in the low-concentration region, similar to that shown in Fig.1.5. As the temperature increases, the coexistence curve becomes slightly tilted. In solutions of rod-like molecules, the excluded volume interactions universally give rise to a chimney-like two-phase region (I+LC), especially on the high-temperature side.

Various types of phase separation phenomena have also been observed in mixtures of polymers and liquid crystalline molecules. Due to the differences in molecular size, shape, and intermolecular interactions, these systems often exhibit rich and complex phase behavior, including isotropic-nematic, isotropic smectic, and polymer-rich/liquid-crystal-rich phase coexistence. Of particular interest is the formation of polymer-dispersed liquid crystals (PDLCs), where microdroplets of liquid crystal become phase-separated and dispersed within a polymer matrix [11,12]. PDLCs are widely studied and utilized as LCDs and light-modulating materials because electric fields can tune their optical properties. The interplay between polymer compatibility, liquid crystalline ordering, and external stimuli makes these systems highly versatile for both fundamental research and technological applications.

Figure 1.7 shows the phase diagrams of mixtures of 8CB and polystyrene (PS): (a) for PS with molecular weight $M_w = 4750$; (b) for $M_w = 65000$ [8]. Different symbols indicate the experimental transition temperatures determined by optical microscopy. The horizontal axis represents the weight fraction of 8CB. The phase diagrams demonstrate a strong dependence of the phase behavior on the molecular weight of the polymer. For the low molecular weight polymer (Fig. 1.7(a)), no isotropic-isotropic (I+I) phase separation is observed. In contrast, for the high molecular weight polymer ($M_w = 65000$) (Fig. 1.7(b)), an I+I biphasic region appears, exhibiting a UCST and an asymmetric shape. The critical point

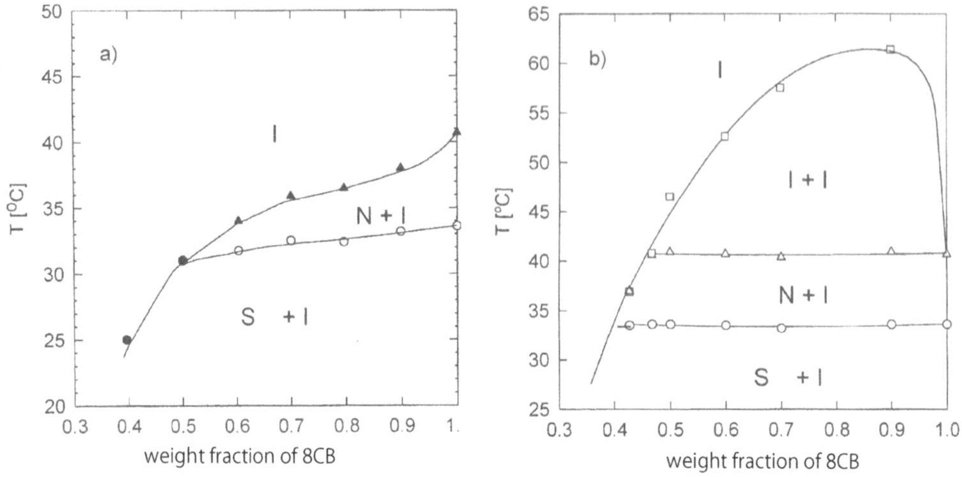

FIGURE 1.7
Equilibrium phase diagrams of 8CB/PS mixtures: (a) for PS of molecular weight $M_w = 4750$; (b) $M_w = 65000$. Reprinted with permission from Ref. [8].

is located at a weight fraction of 85 % 8CB and a temperature of $T_c = 62°C$. In the upper region of the phase diagram, a homogeneous isotropic phase (I) is observed, where 8CB and PS form a single-phase solution. In the lower region, multiple types of phase separation appear, including I+I, nematic-isotropic (N+I), and smectic-isotropic (S+I) coexistence. Similar phase behaviors have also been reported in other mixtures of polymer and LC molecules [13–15]. The phase behaviors introduced in this chapter will be discussed in detail theoretically in Chapters 4 and 5. In addition, the effects of external fields, such as electric and magnetic fields, will be examined in Chapter 6.

LC gels, which incorporate liquid crystalline structures into polymer networks, exhibit volume phase transitions and anisotropic swelling that are strongly coupled to the orientational order of the embedded liquid crystalline components [16–18]. These transitions can be triggered by external stimuli such as temperature, solvent composition, pH, or electric fields. The orientational ordering of liquid crystals gives rise to unique phase behaviors that are not observed in conventional isotropic gels. Figure 1.8 shows the temperature dependence of the degree of swelling (Q) for a nematic gel immersed in a nematic solvent, as observed by Urayama et al. [16]. Two distinct nematic-isotropic transition (NIT) temperatures are identified. One is that of the pure surrounding LC solvent (T_{NI}^S), below which the solvent molecules form a nematic phase. The other is that of the LC gel itself (T_{NI}^G), below which the gel becomes nematic. When $T > T_{NI}^G$, the gel is in a swollen isotropic state, surrounded by isotropic solvent. As the temperature decreases, the gel undergoes a discontinuous volume change at T_{NI}^G, where it transitions from the isotropic to the nematic state. Upon further cooling, Q increases continuously at the solvent NIT temperature T_{NI}^S, leading to a reentrant volume phase transition. Such LC gel systems provide a valuable platform for studying the interplay between phase transitions and elasticity, and they hold promise for applications in soft actuators, sensors, and other stimuli-responsive materials.

The cholesteric ordering in polymer networks exhibits unique properties due to the coupling between rubber elasticity and liquid crystalline ordering. As a result, external fields and temperature directly affect the liquid crystalline structure [19, 20]. Kim and Finkelmann [17] observed that a side-chain cholesteric elastomer undergoes an oblate chain

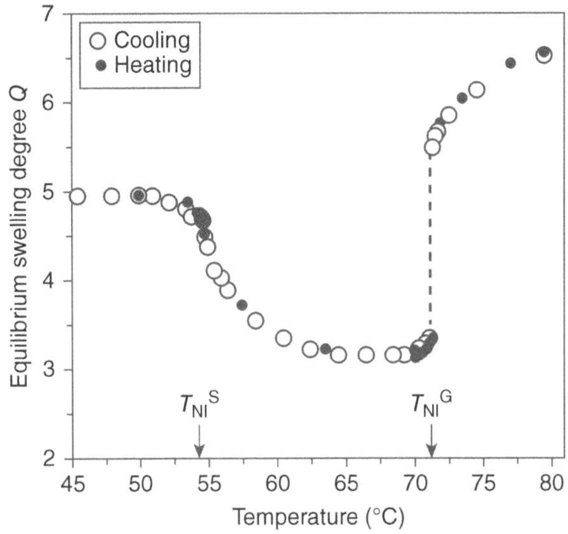

FIGURE 1.8
Temperature (T) dependence of the degree of swelling (Q) for a nematic gel in a nematic solvent. Reprinted with permission from Ref. [16].

conformational change at the isotropic-cholesteric phase transition temperature (67 °C). Specifically, the elastomer shows spontaneous compression along the z direction, parallel to the cholesteric pitch axis, and simultaneous elongation in the x and y directions along the director. Figure 1.9 shows the temperature dependence of the dimensions of the cholesteric elastomer in the x and y directions. L_x and L_y are scaled by the isotropic network length L_i. At the isotropic-cholesteric transition, the network spontaneously elongates equally in both lateral directions.

Liquid crystalline ordering can therefore play a major role in driving shape changes in liquid crystalline elastomers and gels. The phase transitions of LC gels and elastomers will be discussed in Chapter 8.

1.3 Self-assembled liquid crystals

Self-assembled liquid crystals represent a fascinating class of soft materials in which mesoscopic order emerges spontaneously through non-covalent interactions, without the need for external alignment fields. Typical systems include amphiphilic molecules, block copolymers, peptides, and nucleic acid-based structures. These building blocks organize into various liquid crystalline phases, such as nematic, smectic, or columnar, driven by intermolecular forces including hydrogen bonding, π-π stacking, hydrophobic interactions, and electrostatic interactions.

Various liquid crystalline phases are observed in aqueous solutions of amphiphilic molecules, in a manner similar to low-molecular-weight liquid crystals and liquid crystalline polymers. A key feature is that the shape and size of the molecular aggregates of amphiphiles, or surfactant molecules, vary with temperature and concentration [1, 21, 22]. Because the intermolecular forces involved, such as hydrogen bonding and hydrophobic

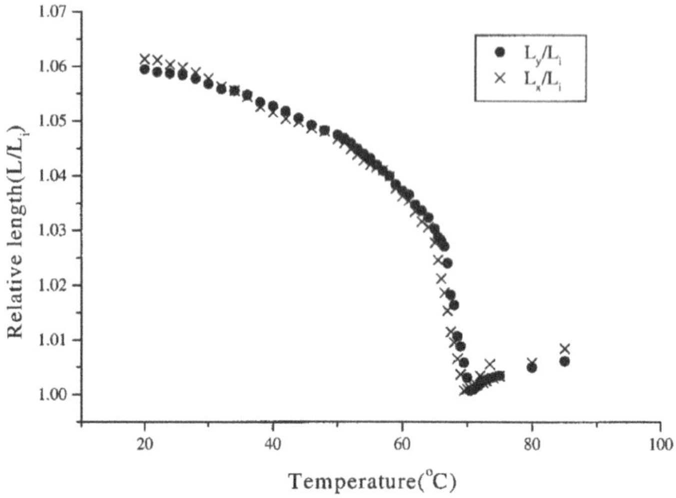

FIGURE 1.9
Dimensions of the cholesteric elastomer in x- and y-direction as a function of temperature. Reprinted with permission from Ref. [17].

interactions, are relatively weak, the assemblies can form and dissociate reversibly with temperature changes. Such physical bonding underlies the thermoreversible nature of self-assembly, in contrast to chemical bonding based on covalent interactions, which are essentially irreversible. Amphiphilic molecules, possessing both hydrophilic and hydrophobic moieties, self-assemble in water into a wide variety of supramolecular aggregates. When the concentration of amphiphiles exceeds the critical micelle concentration (CMC), they typically form spherical micelles. With a further increase in concentration, the morphology evolves into rod-like micelles, which can then pack into ordered liquid crystalline phases.

Figure 1.10 schematically illustrates a typical phase diagram of an aqueous solution of amphiphilic molecules as a function of temperature and amphiphile concentration. With increasing amphiphile concentration, spherical micelles first transform into rod-like micelles. At higher concentrations, these rod-like aggregates organize into liquid crystalline phases such as the hexagonal columnar and lamellar (smectic) phases. In the lamellar phase, amphiphilic molecules arrange into stacked, membrane-like bilayers, which are often regarded as a model system for biological membranes. In regions of reduced water content, the aggregates may invert their curvature and form inverse micelles.

Another important class of self-assembled liquid crystals formed through physical bonding is chromonic liquid crystals, in which disc-like or plank-like aromatic molecules aggregate in aqueous solutions to form liquid crystalline phases [23–25]. The aggregation process typically proceeds through face-to-face stacking of the planar aromatic molecules, resembling a column of coins, which subsequently grow into elongated rod-like assemblies. These structures exhibit lyotropic mesophases, such as nematic and columnar phases, driven primarily by molecular concentration rather than temperature. Such features have stimulated interest in applications ranging from biosensing and drug delivery to dyeing processes and environmentally friendly alignment layers [26]. Figure 1.11 shows the experimentally observed phase diagram of aqueous solutions of the chromonic dye Sunset Yellow (SSY) [23]. Three distinct mesophases are observed: isotropic (I), nematic (N), and columnar (C), accompanied by broad biphasic coexistence regions (I+N and N+C). Self-assembled liquid crystals

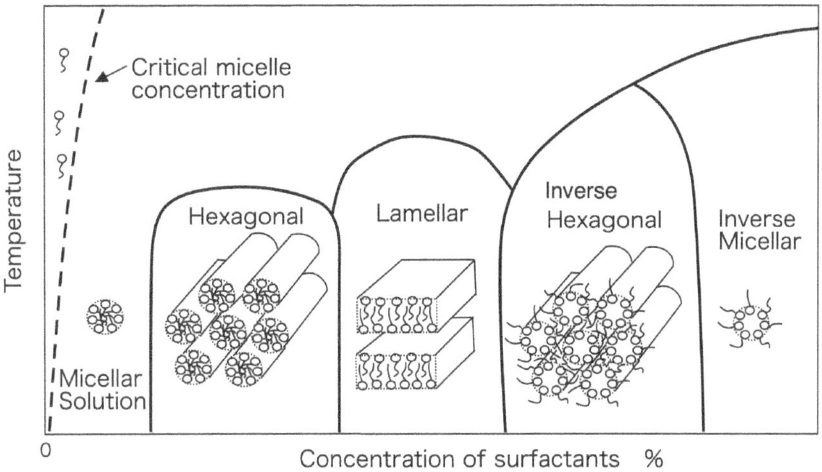

FIGURE 1.10
Schematic phase diagram of amphiphilic molecules in water.

significantly broaden the landscape of liquid crystalline materials, offering new design principles and functionalities beyond those of traditional rod-like mesogens. The molecular mechanisms underlying these aggregation processes will be discussed in Chapter 9.

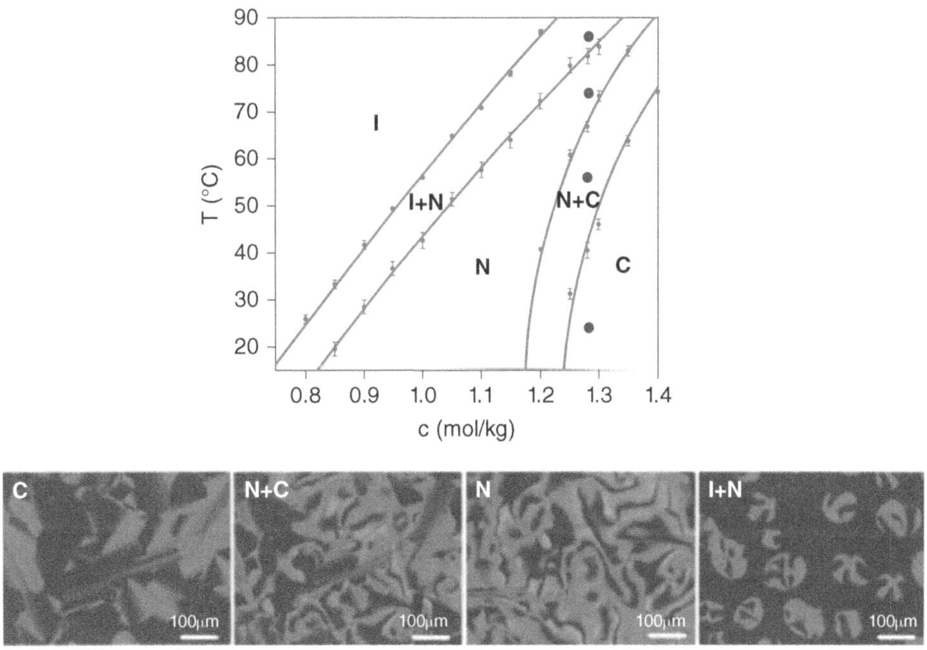

FIGURE 1.11
Phase diagram and polarizing microscope textures of SSY water solutions. The phase diagram shows an isotropic-nematic (I+N) and nematic-columnar (N+C) phase separation. Reprinted with permission from Ref. [23].

By reading all nine chapters, readers will gain a solid foundation in the mean field theories of liquid crystalline phase transitions and phase separation introduced in this chapter. The author hopes that this also serves as a basis for readers to explore further and develop new theoretical models.

2

Order Parameters of Liquid Crystals

Order parameters provide a quantitative measure of the degree of orientational and positional order in liquid crystalline phases, and they play a central role in linking microscopic molecular arrangements with macroscopic physical properties. In this chapter, we introduce the most widely used scalar and tensor order parameters, with particular emphasis on those characterizing nematic, cholesteric, columnar, and smectic phases. We also discuss a tensor order parameter relevant to biaxial nematic phases. These order parameters provide the foundation for constructing the free energy of liquid-crystalline phases and for developing theoretical descriptions of their phase behavior.

2.1 Scalar order parameters

The three classical states of matter are gas, liquid, and solid (crystal). However, many materials do not fall neatly into these categories. In everyday life, substances such as water and beer are found in the isotropic phase, where the positions and orientations of the constituent molecules are randomly distributed. In contrast, a crystal phase—as in the case of ice—exhibits a well-defined three-dimensional long-range order in the arrangement of the molecular centers of mass. Common crystal structures include simple cubic, face-centered cubic, and body-centered cubic lattices. Liquid crystalline phases represent an intermediate state between the isotropic liquid and crystal phases, exhibiting characteristics of both fluidity and orientational or positional order. When a system with a certain symmetry spontaneously transitions to a state of lower symmetry, the process is called spontaneous symmetry breaking. The isotropic phase possesses the highest degree of symmetry, characterized by continuous rotational symmetry. As the system transitions from the isotropic to the crystal phase, symmetries are progressively broken. Liquid crystalline phases are defined by the nature and extent of this symmetry breaking [1, 3].

2.1.1 Three states of matter

How can we define the three classical states (phases) of matter? Figure 2.1 shows a schematic representation of the three phases of a single type of molecule. The primary distinction between a gas and a liquid lies in their densities. For example, the density (ρ) of water under normal conditions is approximately $\rho \sim 1$ g/cm^3. However, when water is heated and transitions to the gaseous phase, its density decreases by roughly a factor of 10^3. This three orders of magnitude difference in density provides a clear physical distinction between the gas and liquid phases. What about the density difference between the liquid and crystal phases? Both phases typically have densities close to 1 g/cm^3, making it difficult to distinguish between them based on density alone; though precise density measurements can still detect small differences. For instance, rubber and water both have densities near

DOI: 10.1201/9781003737711-2

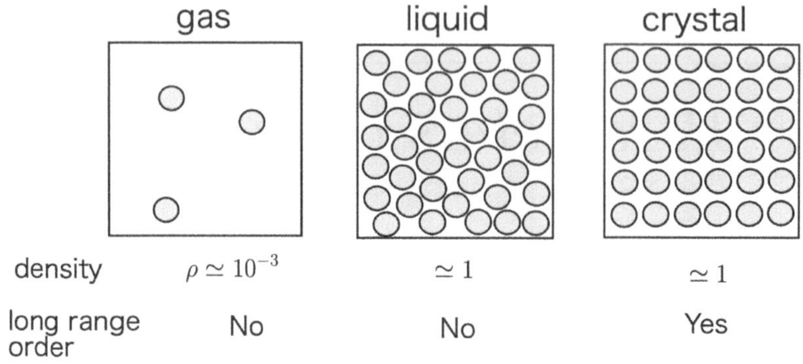

FIGURE 2.1
Three states (phases) of matter.

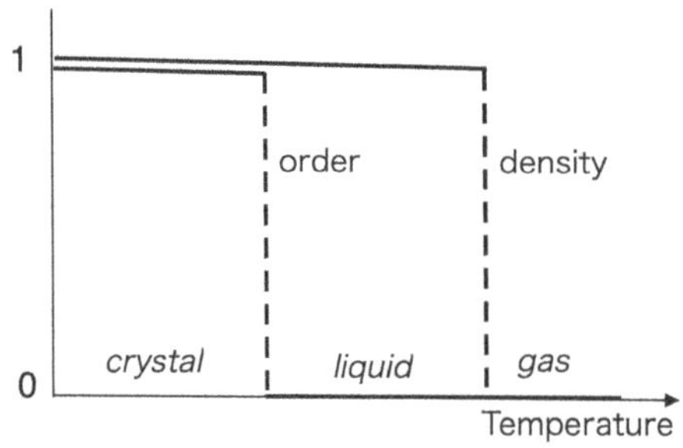

FIGURE 2.2
Three states defined by density and long-range order.

1 g/cm^3, yet they are fundamentally different materials. Thus, while density is a useful criterion to differentiate gases from liquids or solids clearly, it is insufficient to distinguish between liquid and crystal phases. This limitation highlights the need for an additional concept: long-range translational order, which provides a more definitive characterization of the crystal phase.

In the gas and liquid phases, the positions of the molecules' centers of mass are random and disordered, making these phases isotropic and lacking translational order. On the other hand, in the crystal phase, the positions of the molecules' centers of mass form a regular, ordered arrangement known as a crystal structure. By examining the state of matter through the lens of order (order parameter), the difference between liquid (or gas) and crystal becomes apparent. Using two physical quantities: density and order, we can define the three phases of matter as follows:

Gas: A state with sufficiently low density $\rho \sim 10^{-3}$ g/cm^3 and no long-range order.

Liquid: A state with density around 1 g/cm^3 and no long-range translational order.

Crystal: A state with density around 1 g/cm^3 but possessing long-range translational order.

This definition is illustrated in Fig. 2.2. Alternatively, these phases can also be characterized using dynamic physical quantities such as viscosity, elasticity, and density, as well as by their

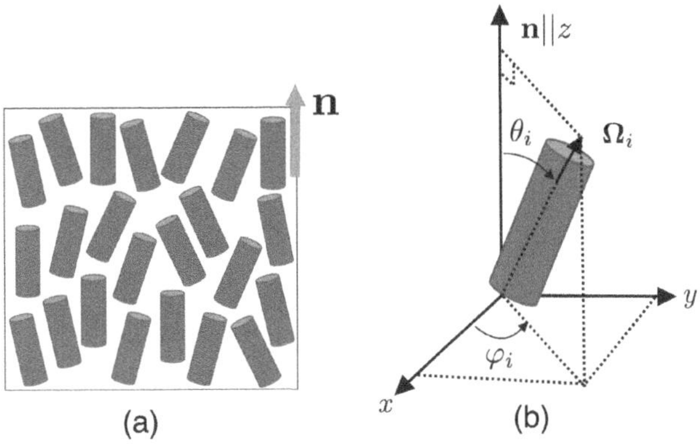

FIGURE 2.3
(a) The director **n** of a nematic liquid crystal composed of rod-like molecules. (b) The relationship between the long axis direction $\mathbf{\Omega}_i$ of a rod-like molecule and the director **n**.

associated order parameters. In the following sections, we introduce representative liquid crystalline phases and define their corresponding order parameters.

2.1.2 Nematic phase

Molecules that exhibit liquid crystalline phases typically possess anisotropic shapes, such as rod-like or disk-like geometries. This anisotropy in molecular shape plays a crucial role in the formation of liquid crystalline phases. The nematic phase exhibits several distinctive properties that set it apart from both the isotropic liquid and crystal phases. The nematic phase is characterized by the average alignment of molecules along a common preferred direction, known as the director, denoted by **n** (Fig. 2.3 (a)). Its key characteristics are summarized below:

Lack of Positional Order: The centers of mass of individual molecules are randomly distributed, exhibiting no long-range translational order. As a result, the nematic phase behaves as a fluid, similar to an ordinary liquid.

Orientational Order: The nematic phase is invariant under inversion of the director, i.e., $\mathbf{n} \rightarrow -\mathbf{n}$. This head-tail symmetry implies that the physical state remains unchanged when the alignment direction is reversed.

Uniaxial Symmetry: Since the system remains unchanged when the orientation vector **n** is reversed to $-\mathbf{n}$, the two states (**n** and $-\mathbf{n}$) are indistinguishable. This symmetry defines the uniaxial nematic phase, in which the system is symmetric with respect to a single preferred axis. When the two states, **n** and $-\mathbf{n}$, are distinguishable, the system is said to be in a polar nematic phase.

Fluidity: Nematic liquid crystals can flow like conventional liquids. This property is crucial for applications such as liquid crystal displays, where electric fields can control the alignment of molecules.

As shown in Fig. 2.3 (b), a convenient measure of this orientational order is given by the angle θ_i between the director **n** and the unit vector $\mathbf{\Omega}_i$, which represents the orientation of the long axis of the i-th molecule. Without loss of generality, we can choose the director to lie along the z-axis, i.e., $\mathbf{n} = (0, 0, 1)$. In this coordinate system, the orientation of molecule i can be described using spherical coordinates by the polar angle θ_i $(0 \le \theta_i \le \pi)$ and the azimuthal angle φ_i $(0 \le \varphi_i \le 2\pi)$. The unit vector $\mathbf{\Omega}_i$ is then expressed as

$$\mathbf{\Omega}_i = (\sin\theta_i \cos\varphi_i, \sin\theta_i \sin\varphi_i, \cos\theta_i). \tag{2.1}$$

The orientational state of rod-like molecules can be described by the orientational distribution function $f_d(\theta, \varphi)$. In systems with uniaxial symmetry, each molecule maintains a fixed polar angle θ while freely rotating about the director $\mathbf{n}$. As a result, the distribution function is independent of the azimuthal angle φ, and can be written as a function of θ alone: $f_d(\theta, \varphi) = f_d(\theta)$. Moreover, since the orientations corresponding to $\mathbf{n}$ and $-\mathbf{n}$ are physically indistinguishable, the following symmetry condition must be satisfied: $f_d(\theta) = f_d(\pi - \theta)$. In the isotropic phase, $f_d(\theta)$ is a constant, reflecting the uniform distribution of orientations. In contrast, in the nematic phase, $f_d(\theta)$ exhibits an angular dependence, indicating a preference for alignment along the director. The function $f_d(\theta)$ can be interpreted as a probability distribution representing the fraction of molecules (among the total number N) that are oriented at an angle θ with respect to the director.

Let's consider the solid angle $d\mathbf{\Omega} = \sin\theta d\theta d\varphi$ of the unit hemisphere with a radius of 1. Since the orientational distribution function does not depend on the azimuthal angle φ, integrating with respect to φ gives $d\mathbf{\Omega} = 2\pi \sin\theta d\theta$. The number of molecules with a polar angle θ within this solid angle is $N f_d(\theta) d\mathbf{\Omega}$. Since the total number N of molecules is conserved when integrated over the entire solid angle, the following equation holds:

$$\int N f_d(\theta) d\mathbf{\Omega} = N. \tag{2.2}$$

Dividing both sides of this equation by N, the distribution function satisfies the normalization condition:

$$\int f_d(\theta) d\mathbf{\Omega} = 2\pi \int_0^{\pi} f_d(\theta) \sin\theta d\theta = 1. \tag{2.3}$$

To define the nematic phase quantitatively, we consider the average of the inner product between the director $\mathbf{n}$ and the orientation vector $\mathbf{\Omega}_i$ of the i-th molecule. As noted earlier, the nematic phase exhibits head-tail symmetry: the states $\mathbf{n}$ and $-\mathbf{n}$ are physically indistinguishable. This symmetry implies that any quantity linear in $\mathbf{\Omega}_i$ averages to zero in the nematic phase. Specifically, taking the average over all molecules, we obtain

$$\langle \cos\theta \rangle = \frac{1}{N} \sum_{i=1}^{N} (\mathbf{n} \cdot \mathbf{\Omega}_i) = 0, \tag{2.4}$$

where $\langle \cdots \rangle$ denotes the ensemble average over all N molecules. This result shows that $\langle \cos\theta \rangle$ is not a suitable order parameter for describing the nematic phase, as it vanishes due to the symmetry $\mathbf{\Omega}_i \to -\mathbf{\Omega}_i$.[1]

In the isotropic phase, the orientations of molecules are uniformly distributed in all directions. Therefore, the orientational distribution function $f_{\text{iso}}(\theta)$ does not depend on the polar angle θ or the azimuthal angle φ, and is simply a constant. To determine this constant, we require that the distribution function be normalized over the unit sphere:

$$\int f_{\text{iso}}(\theta) d\mathbf{\Omega} = f_{\text{iso}}(\theta) \int_0^{\pi} \sin\theta d\theta \int_0^{2\pi} d\varphi. \tag{2.5}$$

[1]For a polar nematic phase, in which the head-tail symmetry is broken, $\langle \cos\theta \rangle$ serves as a polar order parameter.

Calculating the integrals, we get

$$\int f_{\text{iso}}(\theta)d\mathbf{\Omega} = 4\pi f_{\text{iso}}(\theta) = 1. \tag{2.6}$$

Thus, the normalized orientational distribution function in the isotropic phase is

$$f_{\text{iso}}(\theta) = \frac{1}{4\pi}. \tag{2.7}$$

Next, let us consider the square of the inner product between the director $\mathbf{n}$ and the orientation vector $\mathbf{\Omega}_i$. This quantity, $\cos^2\theta_i = (\mathbf{n}\cdot\mathbf{\Omega}_i)^2$, is invariant under the transformation $\mathbf{\Omega}_i \to -\mathbf{\Omega}_i$, making it suitable for describing systems with head-tail symmetry, such as the nematic phase. In the isotropic phase, where molecular orientations are uniformly distributed, the average of $\cos^2\theta$ is given by

$$\begin{aligned}\langle\cos^2\theta\rangle &= \frac{1}{N}\sum_{i=1}^{N}(\mathbf{n}\cdot\mathbf{\Omega}_i)^2 \\ &= \int f_{\text{iso}}(\theta)\cos^2\theta d\mathbf{\Omega}.\end{aligned} \tag{2.8}$$

Substituting $f_{\text{iso}}(\theta) = 1/(4\pi)$, we find:

$$\langle\cos^2\theta\rangle = \frac{1}{3}, \tag{2.9}$$

in the isotropic phase. This result reflects the fact that, under uniform angular distribution, the average squared projection of molecular orientation onto any fixed direction is 1/3, consistent with spherical symmetry.

In contrast to the isotropic phase, the nematic phase exhibits orientational order, which is quantified by the average value of $\cos^2\theta$. This average differs from the isotropic value of 1/3 and serves as the basis for defining the nematic order parameter. The orientational order parameter S is defined as the ensemble average of the second-order Legendre polynomial $P_2(\cos\theta)$ given by:

$$\begin{aligned}S = \langle P_2(\cos\theta)\rangle &= \frac{1}{N}\sum_{i=1}^{N}\frac{3}{2}\left[(\mathbf{n}\cdot\mathbf{\Omega}_i)^2 - \frac{1}{3}\right] \\ &= \int P_2(\cos\theta)f_d(\theta)d\mathbf{\Omega},\end{aligned} \tag{2.10}$$

where the second-order Legendre polynomial is

$$P_2(\cos\theta) = \frac{3}{2}\left(\cos^2\theta - \frac{1}{3}\right). \tag{2.11}$$

The quantity S characterizes the degree of molecular alignment along the director $\mathbf{n}$, with typical values ranging from $S = 0$ (isotropic) to $S \neq 0$ in nematic phases.[2] Equation (2.10) can be interpreted as the weighted average of $P_2(\cos\theta)$ over all molecular orientations, using the distribution function $f_d(\theta)$.

[2]Note that S should not be confused with entropy, often denoted by similar symbols.

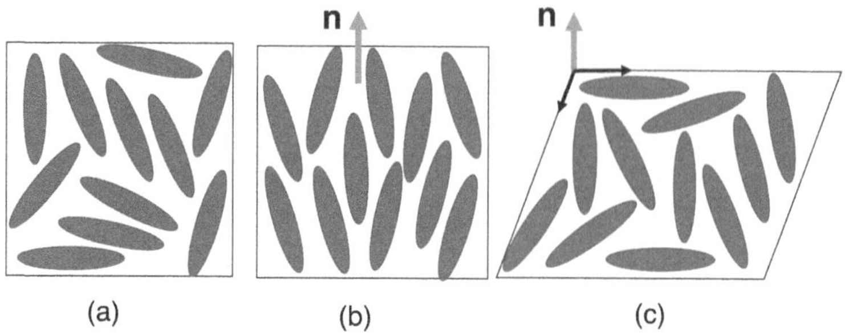

FIGURE 2.4
(a) Isotropic phase $S = 0$, (b) nematic phase $S > 0$, and (c) $S = -0.5$.

The orientational order parameter S quantifies the degree of alignment of molecules in the nematic phase and takes values in the range:

$$-\frac{1}{2} \leq S \leq 1. \tag{2.12}$$

As shown in Fig. 2.4, the value $S = 0$ corresponds to the isotropic phase, in which molecular orientations are completely random and there is no preferred direction. $S = 1$ represents the ideal case of perfect alignment, where all molecular axes are exactly parallel to the director $\mathbf{n}$. The value of S can also be negative. For instance, $S = -1/2$ arises when all molecules are oriented at $\theta = \pi/2$, i.e., perpendicular to the director (see Fig. 2.4(c)). In this configuration, molecular orientations are uniformly distributed within a plane normal to the director. Such negative values of S can occur in systems composed of disk-like (oblate) molecules, where the preferred orientation is perpendicular to the director rather than parallel to it. This behavior contrasts with that of rod-like (prolate) molecules, which typically yield positive values of S due to their tendency to align along the director.

In most conventional nematic liquid crystals, the orientational order is uniaxial, meaning there is a single preferred alignment direction, represented by the director $\mathbf{n}$. In such cases, the orientational distribution is axially symmetric around $\mathbf{n}$, and the phase can be fully characterized by the scalar order parameter S. However, under certain conditions, such as in materials composed of non-cylindrically symmetric molecules or at high concentrations, a higher degree of orientational organization can emerge, leading to a biaxial nematic phase (we discuss in the next section).

2.1.3 Cholesteric phase

In the nematic phase, the director $\mathbf{n}$, representing the average molecular orientation, is uniform throughout macroscopic regions of the system. This uniformity defines the uniaxial nature of nematic order. In contrast, the cholesteric phase, also known as the chiral nematic phase and denoted by N^*, exhibits a spatially varying director field. In this phase, the director rotates helically around a fixed axis, forming a periodic structure known as a helix, as illustrated in Fig. 2.5. The pitch of the helix, typically on the order of hundreds of nanometers to micrometres, is the distance over which the director rotates by 2π. The cholesteric phase arises due to the intrinsic chirality of the constituent molecules. A chiral molecule cannot be superimposed on its mirror image, analogous to the asymmetry between the right and left hands. This lack of mirror symmetry gives rise to a twist in the preferred molecular alignment, leading to the helical structure of the phase. By contrast,

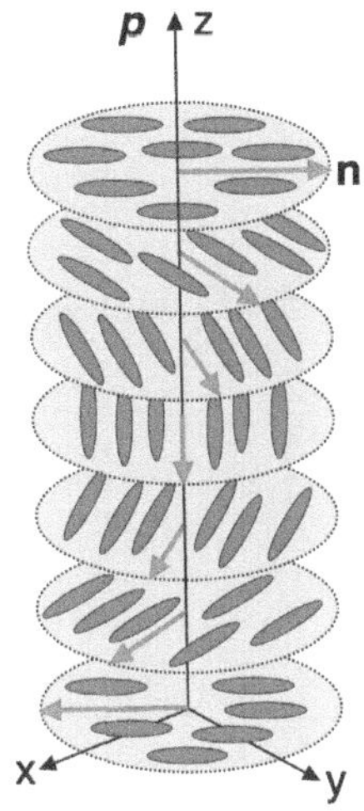

FIGURE 2.5
Director ($\mathbf{n}$) in the cholesteric phase. The director rotates along the pitch axis (the z-axis), forming a right-handed helix.

molecules that are superimposable on their mirror images are called achiral molecules. Common examples include rod-like mesogens, which possess mirror symmetry and typically form standard uniaxial nematic phases without helical modulation. The spontaneous formation of a cholesteric helix is a striking manifestation of molecular chirality at the macroscopic scale. The handedness of the helix (left- or right-handed) directly reflects the chirality of the constituent molecules.

In the cholesteric phase, the distance along the helical axis ($\mathbf{p}$-axis) over which the director completes a full 2π rotation is called the pitch length (p). Bragg scattering occurs when this pitch length is comparable to the wavelength of visible light (a few hundred nanometers), producing vivid colors. Changing the pitch length using external factors, such as temperature, electric fields, or the addition of impurities, makes it possible to induce color changes in the cholesteric phase. This property underlies liquid crystal displays, optical sensors, color-changing films, and e-paper technologies [26]. Beyond technological applications, the cholesteric phase plays a crucial role in biological systems. Many biological macromolecules, such as DNA, filamentous viruses, and collagen, exhibit cholesteric ordering due to their intrinsic chirality and anisotropic shapes. These structures are essential in understanding the self-organization of complex biological materials [27].

As illustrated in Fig. 2.5, the director field $\mathbf{n}$ in the cholesteric phase exhibits a continuous helical rotation around the pitch axis $\mathbf{p}$. If we choose the pitch axis to lie along the z-axis, the spatial variation of the director can be expressed as:

$$\mathbf{n} = (\cos qz, \sin qz, 0), \tag{2.13}$$

where q is the pitch wavenumber, which characterizes the spatial periodicity of the helical twist. The pitch length p, defined as the distance over which the director rotates by 2π, is related to q by $p = 2\pi/|q|$. The sign of q determines the handedness of the helical structure, distinguishing between right-handed ($q > 0$) and left-handed ($q < 0$) helices. In the case of a standard nematic phase, $q = 0$ (which corresponds to $p = \infty$), and the director $\mathbf{n}$ becomes parallel to the x-axis, independent of position. The wavenumber q serves as an order parameter that characterizes the unique helical structure of the cholesteric phase. While

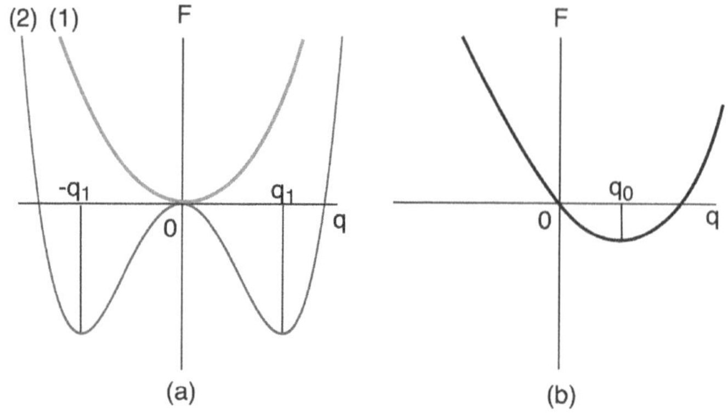

FIGURE 2.6
The wavenumber (q) dependence of the free energy (F) for systems with various types of twist is illustrated as follows: (a) System where right-handed and left-handed helices coexist, (b) System where right-handed and left-handed helices are distinguishable.

the scalar nematic order parameter $S \neq 0$ reflects the presence of orientational order, the additional condition $q \neq 0$ signifies chiral modulation, a hallmark of the cholesteric phase. Thus, the cholesteric phase can be viewed as a modulated nematic phase, characterized by both orientational alignment (through S) and spatial twist (through q).

To understand the cholesteric phase, we now consider its description in terms of the free energy F. The free energy of the cholesteric phase depends on the twist angle $\theta(z) = qz$ or, equivalently, on the pitch wavenumber $q = \partial\theta/\partial z$. As shown in Fig. 2.6, two scenarios can be considered to analyze the free energy behavior:

Case (a): In the case where there is no distinction between right-handed and left-handed helices, the free energy satisfies the relation $F(q) = F(-q)$, as shown in Fig. 2.6(a).

Curve (1): The free energy has its minimum at $q = 0$, corresponding to the nematic phase with no twist. This indicates that the system prefers a uniform director alignment without any helical structure.

Curve (2): The free energy is symmetric for $q = 0$ and exhibits two minima at $q = \pm q_1$. This implies that the system can spontaneously form a right-handed ($q = q_1$) or a left-handed ($q = -q_1$) helical structure with equal probability. This behavior is possible in systems composed of achiral molecules, such as dumbbell-shaped molecules, which are often used as analogies in strength training equipment.

Case (b): In systems with chiral molecules, the free energy $F(q)$ becomes asymmetric concerning the wavenumber $q = 0$. For example, as shown in Fig. (2.6(b)), the free energy has a minimum at $q = q_0$. This indicates that the system prefers a helical structure with a specific handedness. In this case, the cholesteric phase (N*) with a right-handed helix is realised.

These two cases highlight the role of chirality in cholesteric liquid crystals. In achiral systems, the right-handed and left-handed twists are equivalent, resulting in a symmetric free energy. In contrast, chiral systems exhibit asymmetry, leading to the spontaneous formation of a helical structure.

The essential distinction between the nematic and cholesteric phases lies in the behavior of the director field $\mathbf{n}$. In the nematic phase, the director remains uniform throughout space, whereas in the cholesteric phase, it exhibits a helical twist along a fixed axis. Experimentally,

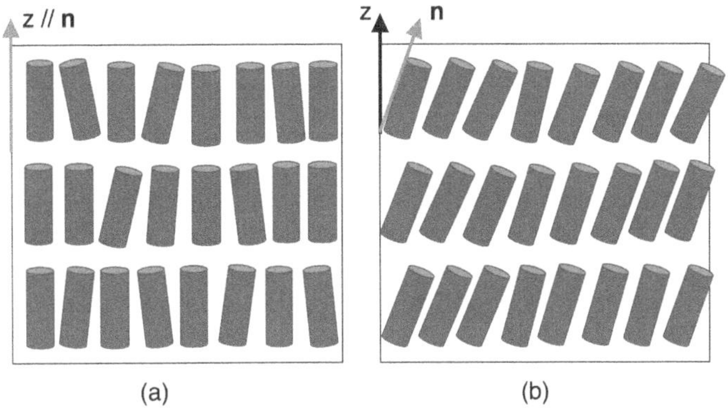

FIGURE 2.7
(a) Smectic A phase(S_A) and (b) smectic C phase (S_C).

a cholesteric (or chiral nematic) phase is often induced by adding a small concentration of chiral dopant to an otherwise achiral nematic host. The chiral dopant molecules break the mirror symmetry of the system and introduce a preferred sense of twist, either right-handed or left-handed, depending on their molecular chirality [28].

2.1.4 Smectic phase

The smectic phase is a class of liquid crystal phases characterized by a layered structure. Smectic phases generally occur at lower temperatures than the nematic phase, reflecting a higher degree of molecular ordering. To date, various types of smectic phases have been identified [29]. Based on the director's orientation and the positional order of the molecular centers of mass, smectic phases are classified into types ranging from smectic A, B, and C, $\cdots$. We now introduce two representative smectic phases: smectic A (S_A) and smectic C (S_C), as illustrated in Fig. 2.7. These phases share a layered structure but differ in the orientation of the director relative to the layer normal. The key characteristics of smectic phases are as follows:

One-dimensional positional order : Molecules are oriented along the normal of the layers (the z-axis), and the positions of their centers of mass exhibit periodic variation along this direction, resulting in one-dimensional positional order. Therefore, the smectic phase is more ordered than the nematic phase.

Two-dimensional liquid : The mass centers of the molecules are distributed randomly within each layer (in the $x-y$ plane), and each layer can be regarded as a two-dimensional liquid. When viewed from above, the positions of molecular centers appear random. Membrane-like layered structures formed by surfactant molecules in water, or by biological membranes, are known as lyotropic lamellar phases, which are also classified as smectic phases. Incidentally, "smectic" is derived from the Greek word for "soap".

Director orientation : The director $\mathbf{n}$ is aligned along the normal layer in the S_A phase. However, in the S_C phase, the director tilts from the normal direction by an angle ϵ_c. Furthermore, if the director in S_C is twisted along the layer normal direction (typically the z-axis), the resulting phase is called the chiral smectic C phase, denoted $S_C{}^*$. This phase exhibits ferroelectricity, where a spontaneous polarization arises from the combination of molecular chirality and tilt, a feature of great interest in electro-optical applications.

The one-dimensional positional order is captured by a spatially varying density $\rho(z)$, which can be expressed as:

$$\rho(z) = \rho\big(1 + \sigma_A \cos(2\pi z/L_s)\big), \tag{2.14}$$

where ρ is the average molecular density, L_s is the layer spacing (the distance between adjacent smectic layers), and σ_A is the smectic A order parameter, which characterizes the amplitude of the density modulation. To characterize the smectic A phase more rigorously, we incorporate both molecular position and orientation into the order parameter. Specifically, we define the order parameter of smectic A phase σ_A as the ensemble average:

$$\begin{aligned} \sigma_A &= \langle \cos(2\pi z/L_s) P_2(\cos\theta) \rangle \\ &= \iint \cos(2\pi z/L_s) P_2(\cos\theta) f_d(z,\theta) dz d\mathbf{\Omega}, \end{aligned} \tag{2.15}$$

where θ is the angle between the molecular axis and the layer normal, and $f_d(z, \theta)$ is the distribution function of molecules, dependent on both the spatial coordinate z and the orientation θ. When $\sigma_A = 0$, the density is uniform in the z-axis, and there is no smectic layering, corresponding to an isotropic or nematic phase, depending on whether $S = 0$ or $S \neq 0$. When $\sigma_A > 0$, periodic modulation in molecular density appears along the z-axis, indicating the emergence of smectic A ordering. Thus, the condition for the smectic A phase is: $S \neq 0$ and $\sigma_A \neq 0$.

In the smectic C phase, the director $\mathbf{n}$ is tilted from the layer normal (z-axis) by an angle ϵ_c. The director $\mathbf{n}$ is expressed as:

$$\mathbf{n} = (\sin\epsilon_c, 0\ \cos\epsilon_c), \tag{2.16}$$

where the tilt angle ϵ_c is the order parameter for the smectic C phase. The smectic C phase is characterized by the following conditions: $S \neq 0$, $\sigma_A \neq 0$, $\epsilon_c \neq 0$.

2.1.5 Columnar phase

A phase with lower symmetry than the smectic phase is the columnar phase, which exhibits two-dimensional positional order concerning the molecules' mass centers, as shown in Fig. 2.8(a). In this phase, the positions $\mathbf{r}$ of the molecular centers of mass form a two-dimensional periodic structure in the x and y directions while being random along the z-axis. For a rectangular columnar phase, the molecular density $\rho(x, y)$ is given by

$$\rho(x,y) = \rho\big(1 + \sigma_{cl} \cos(2\pi x/L_s)\cos(2\pi y/L_s)\big), \tag{2.17}$$

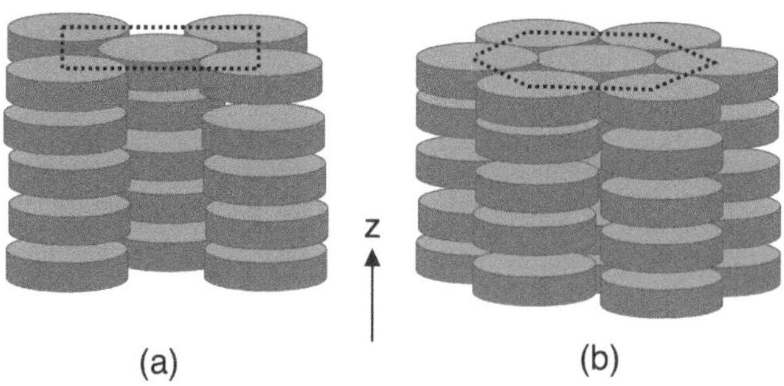

FIGURE 2.8
(a) Rectangular columnar phase and (b) hexagonal columnar phase.

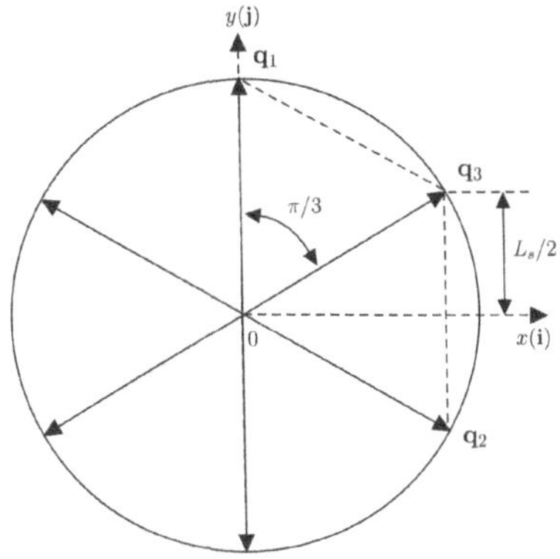

FIGURE 2.9
Three wavevectors: $\mathbf{q}_1$, $\mathbf{q}_2$, $\mathbf{q}_3$, corresponding to the symmetry of the hexagonal lattice.

where ρ is the average molecular density, σ_{cl} is the amplitude of the density modulation, representing the degree of order in the columnar phase, and L_s is the lattice constant in the x and y directions. Taking molecular orientation into account, the order parameter σ_{cl} of the columnar phase is defined as the ensemble average:

$$\sigma_{cl} = \langle \cos(2\pi x/L_s)\cos(2\pi y/L_s)P_2(\cos\theta)\rangle. \tag{2.18}$$

When $\sigma_{cl} = 0$, the molecular density becomes constant. However, when $\sigma_{cl} > 0$, a two-dimensional periodic structure appears in the molecular density on the $x-y$ plane; therefore, the conditions for the columnar phase are $S \neq 0$ and $\sigma_{cl} \neq 0$.

The molecular density in the hexagonal columnar phase exhibits a periodic structure in two dimensions, typically represented by three wavevectors: $\mathbf{q}_1$, $\mathbf{q}_2$, $\mathbf{q}_3$, corresponding to the symmetry of the hexagonal lattice, as shown in Fig. 2.8(b). These three wavevectors are arranged in a way that reflects the sixfold symmetry of the hexagonal arrangement, as shown in Fig. 2.9: $\mathbf{q}_1 = (4\pi/(\sqrt{3}L_s))\mathbf{j}$, $\mathbf{q}_2 = (4\pi/(\sqrt{3}L_s))((\sqrt{3}/2)\mathbf{i} - (1/2)\mathbf{j})$, $\mathbf{q}_3 = \mathbf{q}_1 + \mathbf{q}_2$, and the density is given by

$$\rho(x,y) = \rho\big(1 + \sigma_{hcl}\big[\cos(\mathbf{q}_1\cdot\mathbf{r}) + \cos(\mathbf{q}_2\cdot\mathbf{r}) + \cos(\mathbf{q}_3\cdot\mathbf{r})\big]\big), \tag{2.19}$$

where the symbols $\mathbf{i}$, $\mathbf{j}$, and $\mathbf{k}$ represent the basis vectors along the three coordinate axes in a three-dimensional Cartesian coordinate system. Then, the order parameter of the hexagonal columnar phase is given by

$$\begin{aligned}\sigma_{hcl} &= \frac{1}{3}\langle\big(\cos(\mathbf{q}_1\cdot\mathbf{r}) + \cos(\mathbf{q}_2\cdot\mathbf{r}) + \cos(\mathbf{q}_3\cdot\mathbf{r})\big)P_2(\cos\theta)\rangle \\ &= \frac{1}{3}\iiint\big(\cos(\mathbf{q}_1\cdot\mathbf{r}) + \cos(\mathbf{q}_2\cdot\mathbf{r}) + \cos(\mathbf{q}_3\cdot\mathbf{r})\big) \\ &\quad \times P_2(\cos\theta)f_d(x,y,\theta)dxdyd\mathbf{\Omega}. \end{aligned}\tag{2.20}$$

The conditions for the hexagonal columnar phase are given by $S \neq 0$ and $\sigma_{hcl} \neq 0$. Disc-shaped molecules primarily form this type of columnar phase [30]. Additionally, long, cylindrical micelles can be formed by self-assembly of amphiphilic molecules in water. These micelles can arrange themselves into a hexagonal columnar structure, as shown in Fig. 1.10.

2.1.6 Crystal phase

As discussed in Section 2.1.1, defining a phase requires order parameters related to the molecules' translational and orientational order. In general, order parameters can be defined as the amplitude of the modes when the molecular concentration (or density) is expanded in terms of a mode expansion: [31]

$$\rho(\mathbf{r}) = \rho \sum_{q_x, q_y, q_z = -\infty}^{\infty} \sigma_{q_x q_y q_z} \exp(i\mathbf{q} \cdot \mathbf{r}), \tag{2.21}$$

$$\mathbf{q} = \frac{2\pi}{L_s}(q_x \mathbf{i} + q_y \mathbf{j} + q_z \mathbf{k}), \tag{2.22}$$

where we use Euler's formula $\exp(i\psi) = \cos\psi + i\sin\psi$. The vector $\mathbf{q}$ represents the wavevector, and the q_x, q_y, and q_z are the components of the wavevector along the x, y, and z axes, respectively. In Eq. (2.21), the periodicity of the ordered phase is taken as $L_s = L_x = L_y = L_z$. For example, if we consider the condition $|\mathbf{q}| \leq (2\pi/L_s)\sqrt{3}$, this means that the wavevector magnitude $|\mathbf{q}|$ is constrained within a specific range. Using Eq. (2.22), the magnitude of the wavevector is related to the components q_x , q_y, and q_z by the following expression: $|\mathbf{q}| \leq (2\pi/L_s)\sqrt{q_x^2 + q_y^2 + q_z^2}$. This implies that the allowed values of q_x, q_y, and q_z are integer combinations satisfying this inequality. For example, the allowed values of (q_x, q_y, q_z) could include $(1,0,0), (1,1,0), (1,1,1), \cdots$.

For a given symmetry, expanding Eq. (2.21) in terms of the basis vectors that represent the symmetry allows us to derive the coefficients that correspond to the order parameters (specifically, the translational order parameter) of the system. For example, expanding Eq. (2.21) with $q_x = q_y = 0$ and $q_z = 1$ results in the density distribution of the smectic phase (as described by Eq. (2.14)). We assume that $\sigma_{000} = 1$ and $\sigma_{001} = \sigma_A$, where σ_A represents the order parameter characterising the density modulation along the z-axis. The molecular density exhibits a two-dimensional periodic structure in the x and y directions in the columnar phase. Therefore, expanding equation (refmodel) with $q_z = 0$, $q_x = \pm 1$, and $q_y = \pm 1$ results in the density distribution of the columnar phase (as described by Eq. (2.17)). This reflects the periodic arrangement of molecules in a columnar lattice with two-dimensional ordering in the x-y plane.

Similarly, in the face-centered cubic (FCC) crystal phase, characterized by three-dimensional long-range order, the molecular density exhibits a periodic structure along the x-, y-, and z-axes. Expanding the density in terms of wavevector components with $q_x = q_y = q_z = \pm 1$, the molecular density can be expressed as

$$\rho(\mathbf{r}) = \rho\big(1 + 8\sigma_{fcc} \cos(2\pi x/L_s)\cos(2\pi y/L_s)\cos(2\pi z/L_s)\big), \tag{2.23}$$

where the order parameter σ_{fcc} of the FCC phase is defined as

$$\sigma_{fcc} = \langle \big[\cos(2\pi x/L_s)\cos(2\pi y/L_s)\cos(2\pi z/L_s)\big] P_2(\cos\theta) \rangle. \tag{2.24}$$

The order parameters of crystal phases such as body-centered cubic (BCC) and simple cubic (SC) lattices can be defined using the same approach as [32]

$$\begin{aligned} \sigma_{bcc} \quad = \quad & \langle \big[\cos(2\pi x/L_s)\cos(2\pi y/L_s) + \cos(2\pi y/L_s)\cos(2\pi z/L_s) \\ & + \cos(2\pi z/L_s)\cos(2\pi x/L_s)\big] P_2(\cos\theta) \rangle, \end{aligned} \tag{2.25}$$

and

$$\sigma_{sc} = \langle \big[\cos(2\pi x/L_s) + \cos(2\pi y/L_s) + \cos(2\pi z/L_s)\big] P_2(\cos\theta) \rangle. \tag{2.26}$$

Using the scalar order parameters introduced in this section, we will discuss the free energies of the nematic, smectic A, and cholesteric phases in Chapters 4 and 5.

2.2 Tensor order parameter

The director and the orientational order parameters characterizing a liquid crystal phase are spatial variables of different natures. The director is a hydrodynamic variable that describes the collective orientation of the material as a continuum, varying slowly in space and time compared to molecular scales. If the director $\mathbf{n}$ of the nematic phase is uniform and position-independent, the bulk free energy of the system remains unchanged regardless of the direction of $\mathbf{n}$, and no restoring force acts on it. Such a slow collective rotation of the director, for which the restoring force vanishes in the long-wavelength (zero wavenumber) limit, is referred to as a Nambu-Goldstone mode.

On the other hand, the orientational order parameter S indicates the degree of molecular alignment, and it is not a hydrodynamic variable, since it does not correspond to a conserved or symmetry-breaking mode but relaxes quickly to equilibrium. In the continuum theory of liquid crystals, the director $\mathbf{n}$ plays a central role. In contrast, in mean field theory, the orientational order parameter is the key variable. In this section, we introduce the tensor order parameter, which allows for the simultaneous treatment of the director and orientational order parameters. The physical meaning of the tensor order parameter corresponds to the deviation of the magnetic permeability tensor or the dielectric tensor of liquid crystalline molecules from that in the isotropic phase [1]. Typically, since the director and the scalar order parameter are functions of position $\mathbf{r}$, it would be more precise to write them as $\mathbf{n}(\mathbf{r})$ and $S(\mathbf{r})$. However, for simplicity, we will denote them as $\mathbf{n}$ and S in the following discussion.

2.2.1 Tensor order parameter for uniaxial nematic phases

As shown in Fig. 2.3, let the director $\mathbf{n}$ be aligned parallel to the z-axis. Each component of the unit vector $\mathbf{\Omega}$, which represents the orientation of the long axis of a rod-like molecule, is denoted as Ω_α ($\alpha, \beta = x, y, z$). Consider the value of Ω_α averaged using the orientational distribution function $f_d(\theta)$:

$$\langle \Omega_\alpha \rangle = \int \Omega_\alpha f_d(\theta) d\mathbf{\Omega}. \tag{2.27}$$

In the nematic phase, there is no distinction between the states $\mathbf{\Omega}$ and $-\mathbf{\Omega}$, meaning that the distribution function satisfies $f_d(\mathbf{\Omega}) = f_d(-\mathbf{\Omega})$. As a result, the average value is always $\langle \Omega_\alpha \rangle = 0$.

Next, let us consider $\langle \Omega_\alpha \Omega_\beta \rangle$. The system exhibits spherical symmetry in the isotropic phase, meaning that swapping the x, y, z axes does not change it. As a result, the following equipartition relations hold:

$$\langle \Omega_x^2 \rangle = \langle \Omega_y^2 \rangle = \langle \Omega_z^2 \rangle, \tag{2.28}$$

$$\langle \Omega_x \Omega_y \rangle = \langle \Omega_y \Omega_z \rangle = \langle \Omega_z \Omega_x \rangle = 0. \tag{2.29}$$

We also have

$$\Omega_x^2 + \Omega_y^2 + \Omega_z^2 = 1, \tag{2.30}$$

and then[3]

$$\langle \Omega_\alpha \Omega_\beta \rangle = \frac{1}{3} \delta_{\alpha\beta}, \tag{2.31}$$

where the symbol $\delta_{\alpha\beta}$ represents the Kronecker delta function:

$$\delta_{\alpha\beta} = \begin{cases} 1 & \text{if } \alpha = \beta, \\ 0 & \text{if } \alpha \neq \beta. \end{cases} \tag{2.32}$$

[3] Eq. (2.31) is a reformulation of $\langle \Omega_x \Omega_x \rangle = \langle \Omega_y \Omega_y \rangle = \langle \Omega_z \Omega_z \rangle = 1/3$ using the Kronecker delta function.

Therefore, to describe the nematic phase, we define a second-rank tensor using the deviation from the isotropic phase as follows:

$$Q_{\alpha\beta} = \frac{3}{2}\langle \Omega_\alpha \Omega_\beta - \frac{1}{3}\delta_{\alpha\beta} \rangle. \tag{2.33}$$

Let each component of the director **n** be denoted as $n_\alpha (\alpha, \beta = x, y, z)$. By multiplying both sides of Eq. (2.33) by $n_\alpha n_\beta$, we obtain[4]

$$\begin{aligned} n_\alpha n_\beta Q_{\alpha\beta} &= \frac{3}{2}\langle n_\alpha n_\beta \Omega_\alpha \Omega_\beta - \frac{1}{3} n_\alpha n_\beta \delta_{\alpha\beta} \rangle \\ &= \frac{3}{2}\langle (\mathbf{n} \cdot \mathbf{\Omega})^2 - \frac{1}{3} \rangle = S, \end{aligned} \tag{2.34}$$

where S is the orientational order parameter defined in Eq. (2.10). Multiplying both sides of Eq. (2.34) by $(\frac{3}{2} n_\alpha n_\beta - \frac{1}{2}\delta_{\alpha\beta})$, we get

$$n_\alpha n_\beta \left(\frac{3}{2} n_\alpha n_\beta - \frac{1}{2}\delta_{\alpha\beta} \right) Q_{\alpha\beta} = S \left(\frac{3}{2} n_\alpha n_\beta - \frac{1}{2}\delta_{\alpha\beta} \right). \tag{2.35}$$

Calculating the left-hand side, we have

$$\begin{aligned} n_\alpha n_\beta \left(\frac{3}{2} n_\alpha n_\beta - \frac{1}{2}\delta_{\alpha\beta} \right) &= \frac{3}{2} n_\alpha n_\alpha n_\beta n_\beta - \frac{1}{2} n_\alpha n_\beta \delta_{\alpha\beta} \\ &= \frac{3}{2} - \frac{1}{2} n_\alpha n_\alpha = 1, \end{aligned} \tag{2.36}$$

and substituting this into Eq. (2.35), the tensor order parameter at a position **r** is given by

$$Q_{\alpha\beta} = S \left(\frac{3}{2} n_\alpha n_\beta - \frac{1}{2}\delta_{\alpha\beta} \right), \tag{2.37}$$

where since α and β can each be x, y, or z, the total number of degrees of freedom is nine (corresponding to Q_{xx}, Q_{xy}, Q_{xz}, Q_{yx}, $\ldots$, Q_{zz}). However, since $Q_{\alpha\beta}$ is a symmetric tensor ($Q_{\alpha\beta} = Q_{\beta\alpha}$) and traceless ($Q_{\alpha\alpha} = 0$), the number of independent degrees of freedom is reduced to five.

Equation (2.37) represents the tensor order parameter for the uniaxial nematic phase. The tensor order parameter simultaneously describes both the scalar orientational order parameter S, which indicates how the molecules are orientationally aligned around the director **n**, and the spatial distortion of the director **n**. The continuum theory of liquid crystals accounts for spatial variations in the tensor order parameter to describe the three fundamental elastic deformations: splay, twist, and bend (as discussed in Chapter 3). However, in most continuum theories, the scalar order parameter $S(\mathbf{r})$ is typically assumed to be spatially uniform.

2.2.2 Tensor order parameter for biaxial nematic phases

The conventional nematic phase, as shown in Fig. 2.3, has uniaxial symmetry, where the direction of the molecular long axis ($\mathbf{\Omega}$) rotates 360 degrees around a single director **n**.

[4] The notation such as $Q_{\alpha\alpha}$ is referred to as Einstein summation convention. According to this rule, when the same index appears twice in a term, the summation is implied over that index. For example, $Q_{\alpha\alpha}$ means $Q_{\alpha\alpha} = Q_{xx} + Q_{yy} + Q_{zz}$. The inner product of the vectors **n** and $\mathbf{\Omega}$ is written as $n_\alpha \Omega_\alpha = n_x \Omega_x + n_y \Omega_y + n_z \Omega_z$. Therefore, $n_\alpha n_\beta \Omega_\alpha \Omega_\beta = (n_\alpha \Omega_\alpha)(n_\beta \Omega_\beta) = (\mathbf{n} \cdot \mathbf{\Omega})^2$. The director **n** is a unit vector, satisfying $n_\alpha n_\alpha = n_x^2 + n_y^2 + n_z^2 = 1$. The same description applies to the directors **l** and **m** used in Eq. (2.55).

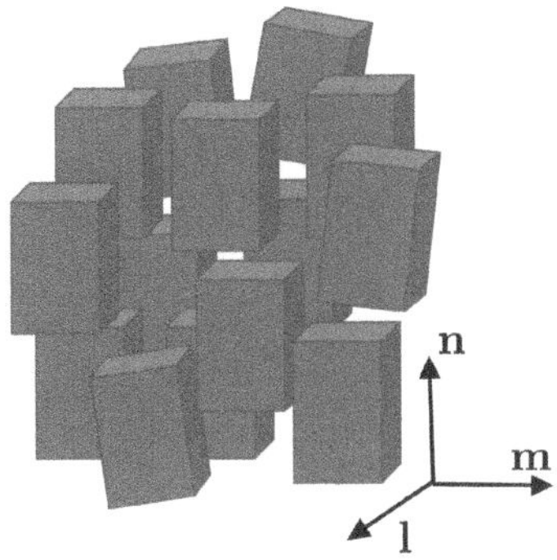

FIGURE 2.10
Biaxial nematic phase of a rectangular molecule that itself possesses biaxiality.

On the other hand, in the biaxial nematic phase, as shown in Fig. 2.10, the molecules are typically aligned along two perpendicular directors (the first director $\mathbf{n}$ and the second director $\mathbf{m}$ or $\mathbf{l}$), with the average molecular alignment in these two directions. The sum of the uniaxial and biaxial terms gives the tensor order parameter for the biaxial nematic phase:

$$Q_{\alpha\beta} = S_n\left(\frac{3}{2}n_\alpha n_\beta - \frac{1}{2}\delta_{\alpha\beta}\right) + \frac{1}{2}B\left(m_\alpha m_\beta - l_\alpha l_\beta\right). \tag{2.38}$$

The first term, S_n, represents the scalar order parameter for the uniaxial nematic phase, while the second term, B, corresponds to the scalar order parameter for the biaxial nematic phase. When $B = 0$, the tensor order parameter reduces to that of the uniaxial nematic phase. The biaxial nematic phase is expected for molecular systems with intrinsic biaxiality, such as rectangular or bent-shaped molecules. To understand the biaxial nematic phase, as shown in Fig. 2.11(a), three Euler angles (ϕ, θ, ψ) are required to describe the molecular coordinate axes in the laboratory frame, which are defined by three directors $\mathbf{l}$, $\mathbf{m}$, and $\mathbf{n}$.

In the following, we consider the tensor order parameters using the bent-shaped molecule (also known as a banana-shaped molecule) as an example, which is a biaxial molecule [5]. "Four" order parameters and Eq. (2.38) can be derived rigorously [33].

Consider a bent-shaped molecule with a bending angle θ_B, as shown in Fig. 2.11(b). We define three orthogonal molecular coordinate systems $\{\mathbf{a}, \mathbf{b}, \mathbf{c}\}$ fixed to the molecule, and the unit vectors $\mathbf{u}_A$ and $\mathbf{u}_B$ along the two mesogen groups (referred to here as arm vectors). When $\theta_B = 180°$, there is no distinction between the molecular axes $\mathbf{a}$ and $\mathbf{b}$, but once the molecule bends, $\mathbf{a}$ and $\mathbf{b}$ become distinguishable. This refers to the molecular biaxiality inherent in rectangular or bent-shaped molecules. It should be distinguished from the biaxial phase of the liquid crystal phase. The angle between the arm vector $\mathbf{u}_A$ and $\mathbf{c}$-axis is χ_B: $\chi_B = (\pi - \theta_B)/2$. Then, the arm vectors are given by

$$\mathbf{u}_A = \mathbf{b}\sin\chi_B + \mathbf{c}\cos\chi_B, \tag{2.39}$$

$$\mathbf{u}_B = \mathbf{b}\sin\chi_B - \mathbf{c}\cos\chi_B. \tag{2.40}$$

As shown in Fig. 2.11(a), we define three orthogonal director coordinate systems $\{\mathbf{l}, \mathbf{m}, \mathbf{n}\}$. The two coordinate systems are related by the Euler angles (ϕ, θ, ψ):

$$\begin{aligned} \mathbf{a} \;=\; & (\cos\psi\cos\phi - \cos\theta\sin\phi\sin\psi)\mathbf{l} \\ & +(\cos\psi\sin\phi + \cos\theta\cos\phi\sin\psi)\mathbf{m} \\ & +(\sin\psi\sin\theta)\mathbf{n}, \end{aligned} \tag{2.41}$$

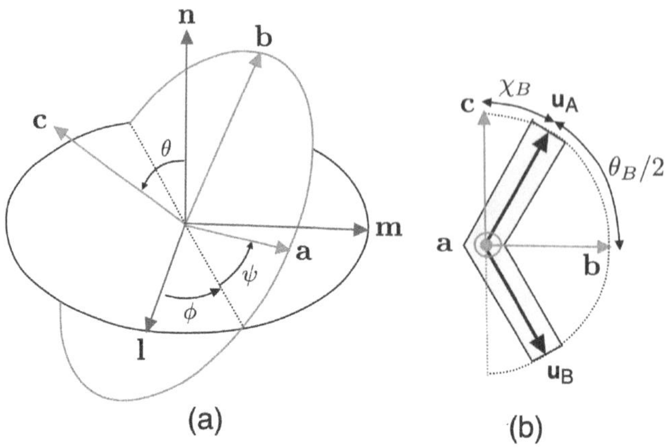

FIGURE 2.11
(a) Director coordinate system $\{\mathbf{l}, \mathbf{m}, \mathbf{n}\}$ and Euler angles (ϕ, θ, ψ). (b) Molecular coordinate system $\{\mathbf{a}, \mathbf{b}, \mathbf{c}\}$ of a bent-shaped molecule with a bending angle θ_B. The bending angle θ_B satisfies the relationship $\sin\chi_B = \cos(\theta_B/2)$, $\cos\chi_B = \sin(\theta_B/2)$ with $\theta_B + 2\chi_B = \pi$. Reprinted with permission from Ref. [33].

$$\begin{aligned} \mathbf{b} &= -(\sin\psi\cos\phi + \cos\theta\sin\phi\cos\psi)\mathbf{l} \\ &\quad -(\sin\psi\sin\phi - \cos\theta\cos\phi\cos\psi)\mathbf{m} \\ &\quad +(\cos\psi\sin\theta)\mathbf{n}, \end{aligned} \tag{2.42}$$

$$\mathbf{c} = (\sin\theta\sin\phi)\mathbf{l} - (\sin\theta\cos\phi)\mathbf{m} + (\cos\theta)\mathbf{n}, \tag{2.43}$$

where $0 \leq \theta \leq \pi$ and $0 \leq \phi, \psi \leq 2\pi$.

To describe the orientation of a bent-shaped molecule, three orientational order parameters are needed, which are formed between the arm vector $\mathbf{u}_A$ (or $\mathbf{u}_B$) and each director. If we neglect the polarity of the molecule, the orientational order parameter between the arm vector $\mathbf{u}_A$ and the director $\mathbf{n}$ is given by

$$S_n = \frac{1}{2}\langle P_2(\mathbf{u}_A \cdot \mathbf{n}) + P_2(\mathbf{u}_B \cdot \mathbf{n})\rangle. \tag{2.44}$$

The inner product between the arm vector $\mathbf{u}_A$ and the director $\mathbf{n}$ can be decomposed into the inner products $\mathbf{b}\cdot\mathbf{n}$ and $\mathbf{c}\cdot\mathbf{n}$ by using Eq. (2.39). ($\mathbf{u}_B$ and $\mathbf{n}$ are decomposed similarly). The orientational order parameter between the arm vector $\mathbf{u}_A$ and the director $\mathbf{m}$ is given by

$$S_m = \frac{1}{2}\langle P_2(\mathbf{u}_A \cdot \mathbf{m}) + P_2(\mathbf{u}_B \cdot \mathbf{m})\rangle, \tag{2.45}$$

and the orientational order parameter between the arm vector $\mathbf{u}_A$ and the director $\mathbf{l}$ is given by

$$S_l = \frac{1}{2}\langle P_2(\mathbf{u}_A \cdot \mathbf{l}) + P_2(\mathbf{u}_B \cdot \mathbf{l})\rangle. \tag{2.46}$$

We also have

$$P_2(\mathbf{a}\cdot\mathbf{i}) + P_2(\mathbf{b}\cdot\mathbf{i}) + P_2(\mathbf{c}\cdot\mathbf{i}) = 0, \tag{2.47}$$

$(\mathbf{i} = \mathbf{l}, \mathbf{m}, \mathbf{n})$ and

$$P_2(\mathbf{k}\cdot\mathbf{l}) + P_2(\mathbf{k}\cdot\mathbf{m}) + P_2(\mathbf{k}\cdot\mathbf{n}) = 0, \tag{2.48}$$

$(\mathbf{k} = \mathbf{a}, \mathbf{b}, \mathbf{c})$.

Using Eq. (2.39), we obtain

$$P_2(\mathbf{u}_A \cdot \mathbf{n}) = \sin^2\chi_B P_2(\mathbf{b}\cdot\mathbf{n}) + \cos^2\chi_B P_2(\mathbf{c}\cdot\mathbf{n}) + \frac{3}{2}\sin 2\chi_B P_1(\mathbf{b}\cdot\mathbf{n})P_1(\mathbf{c}\cdot\mathbf{n}), \tag{2.49}$$

and

$$P_2(\mathbf{u}_B \cdot \mathbf{n}) = \sin^2\chi_B P_2(\mathbf{b}\cdot\mathbf{n}) + \cos^2\chi_B P_2(\mathbf{c}\cdot\mathbf{n}) - \frac{3}{2}\sin 2\chi_B P_1(\mathbf{b}\cdot\mathbf{n})P_1(\mathbf{c}\cdot\mathbf{n}), \tag{2.50}$$

where $P_1(x) = x$ is the first-order Legendre polynomial and we have $P_1(\mathbf{b}\cdot\mathbf{n})P_1(\mathbf{c}\cdot\mathbf{n}) = (1/2)\cos\psi\sin 2\theta$.

Then, Eq. (2.44) is given by

$$S_n = \sin^2\chi_B P_2(\mathbf{b}\cdot\mathbf{n}) + \cos^2\chi_B P_2(\mathbf{c}\cdot\mathbf{n}). \tag{2.51}$$

Similarly, we obtain

$$S_l = \sin^2\chi_B P_2(\mathbf{b}\cdot\mathbf{l}) + \cos^2\chi_B P_2(\mathbf{c}\cdot\mathbf{l}), \tag{2.52}$$

$$S_m = \sin^2\chi_B P_2(\mathbf{b}\cdot\mathbf{m}) + \cos^2\chi_B P_2(\mathbf{c}\cdot\mathbf{m}). \tag{2.53}$$

The sum of the three order parameters is given by

$$S_l + S_m + S_n = 0, \tag{2.54}$$

and using these three eigenvalues, the tensorial order parameter can be defined as

$$Q_{\alpha\beta} = S_l l_\alpha l_\beta + S_m m_\alpha m_\beta + S_n n_\alpha n_\beta. \tag{2.55}$$

It can be easily confirmed that $Q_{\alpha\beta}$ is a symmetric tensor ($Q_{\alpha\beta} = Q_{\beta\alpha}$) and traceless ($Q_{\alpha\alpha} = 0$). Since the director axis $\mathbf{l}$ is defined as the cross product of $\mathbf{m}$ and $\mathbf{n}$: $\mathbf{l} = \mathbf{m}\times\mathbf{n}$, we can express this relation using the Levi-Civita symbol:

$$l_\alpha = \varepsilon_{\alpha\beta\gamma} m_\beta n_\gamma, \tag{2.56}$$

where

$$\varepsilon_{\alpha\beta\gamma} = \begin{cases} 1 & \text{if } \alpha,\beta,\gamma \text{ are unequal and in cyclic order,} \\ -1 & \text{if } \alpha,\beta,\gamma \text{ are unequal and in non-cyclic order,} \\ 0 & \text{if any two } \alpha,\beta, \text{or } \gamma \text{ are equal.} \end{cases} \tag{2.57}$$

We then have

$$l_\alpha l_\beta = \epsilon_{\alpha\gamma\delta} m_\gamma n_\delta \epsilon_{\beta\rho\sigma} m_\rho n_\sigma = \delta_{\alpha\beta} - m_\alpha m_\beta - n_\alpha n_\beta. \tag{2.58}$$

Using Eqs. (2.54) and (2.58), tensor order parameter (Eq. (2.55) can be rewritten by $n_\alpha n_\beta$ and $l_\alpha l_\beta$:

$$Q_{\alpha\beta} = S_l l_\alpha l_\beta + S_m\big(\delta_{\alpha\beta} - l_\alpha l_\beta - n_\alpha n_\beta\big) + S_n n_\alpha n_\beta = \big(S_n - \frac{1}{3}B\big)\big(\frac{3}{2}n_\alpha n_\beta - \frac{1}{2}\delta_{\alpha\beta}\big) - B\big(l_\alpha l_\beta - \frac{1}{3}\delta_{\alpha\beta}\big), \tag{2.59}$$

where we define the order parameter of the biaxial nematic phase:

$$B = S_m - S_l, \tag{2.60}$$

and we have $S_l = -(B + S_n)/2$ and $S_m = (B - S_n)/2$. In the uniaxial nematic phase: $S_m = S_l = 0$, we have $B = 0$. Substituting Eq. (2.58) into Eq. (2.59), and eliminating the $\delta_{\alpha\beta}$ function, we obtain Eq. (2.38).

In the following discussion, we show that S_n and B each contain two distinct order parameters. Using, Eqs. (2.42) and (2.43), the order parameter S_n (Eq. (2.51)) for the uniaxial nematic phase is given by

$$S_n = SP_2(\cos\chi_B) + \frac{3}{4}U\sin^2\chi_B, \tag{2.61}$$

where we define

$$S = \langle P_2(\mathbf{c}\cdot\mathbf{n})\rangle = \langle \frac{3}{2}\cos^2\theta - \frac{1}{2}\rangle, \tag{2.62}$$

and

$$U = \frac{2}{3}\langle P_2(\mathbf{b}\cdot\mathbf{n}) - P_2(\mathbf{a}\cdot\mathbf{n})\rangle = \langle \sin^2\theta\cos 2\psi\rangle. \tag{2.63}$$

The order parameter S represents the degree of alignment of the molecular axis $\mathbf{c}$ relative to the director $\mathbf{n}$. In contrast, the order parameter U indicates the difference in the alignment of the molecular axes $\mathbf{a}$ and $\mathbf{b}$ concerning the director $\mathbf{n}$. In the case of uniaxial rod-like molecules, there is no distinction between $\mathbf{a}$ and $\mathbf{b}$, meaning that $\mathbf{a}$ and $\mathbf{b}$ can be interchanged. As a result, $U = 0$ always holds. Similarly, when the two molecular axes satisfy $\mathbf{a}\cdot\mathbf{n} = \mathbf{b}\cdot\mathbf{n}$ (which corresponds to $\psi = 45°$), $U = 0$ also holds. For typical rod-like molecules where $\chi_B = 0$, Eq. (2.61) becomes $S_n = S$, which is identical to the orientational order parameter defined in Eq. (2.10). Using these two orientational order parameters S and U, the orientational order parameter of the molecular axis $\mathbf{a}$ relative to the director $\mathbf{n}$ is given by $\langle P_2(\mathbf{a}\cdot\mathbf{n})\rangle = -(1/2)(S + 3U/2)$. Similarly, the orientational order parameter of the molecular axis $\mathbf{b}$ relative to the director $\mathbf{n}$ is $\langle P_2(\mathbf{b}\cdot\mathbf{n})\rangle = -(1/2)(S - 3U/2)$.

We discuss the three orientational order parameters: $\langle P_2(\mathbf{a}\cdot\mathbf{n})\rangle$, $\langle P_2(\mathbf{b}\cdot\mathbf{n})\rangle$, and $\langle P_2(\mathbf{c}\cdot\mathbf{n})\rangle$, using a bent-shaped molecule as an example. This provides a comprehensive description of the molecular orientation.

Figure 2.12 shows the schematic representation of three uniaxial ((a), (b), (c)) and three biaxial ((d), (e), (f)) nematic phases. Each bent-shaped molecule has a body-fixed orthonormal coordinate system with unit vectors $\{\mathbf{a}, \mathbf{b}, \mathbf{c}\}$ under a local reference frame $\{\mathbf{l}, \mathbf{m}, \mathbf{n}\}$. The molecular axis $\mathbf{a}$ is defined as perpendicular to the two arms of the banana-shaped molecule, the vector $\mathbf{b}$ is parallel to the twofold rotation axis, and $\mathbf{c} = \mathbf{a}\times\mathbf{b}$. In the uniaxial nematic phase N_U^a (Fig. 2.12(a)), the molecular axis $\mathbf{a}$ rotates around the director $\mathbf{n}$ with $\langle P_2(\mathbf{a}\cdot\mathbf{n})\rangle > 0$. This corresponds to a disc-like molecule with a bending angle $\theta_B \simeq 90°$. In the uniaxial nematic phase N_U^b (Fig. 2.12(b)), the molecular axis $\mathbf{b}$ rotates around the director $\mathbf{n}$ with $\langle P_2(\mathbf{b}\cdot\mathbf{n})\rangle > 0$. This corresponds to a V-shaped molecule with a small bending angle $\theta_B < 90°$. In the uniaxial nematic phase N_U^c (Fig. 2.12(c)), the molecular axis $\mathbf{c}$ rotates around the director $\mathbf{n}$ with $\langle P_2(\mathbf{c}\cdot\mathbf{n})\rangle > 0$, corresponding to rod-like molecules with a bending angle $\theta_B > 90°$. The N_U^b (or N_U^c) and N_U^a phases are commonly referred to as the N_U^+ and N_U^- phases, respectively [34]. The biaxial nematic phases, N_B^a (Figure 2.12(d)), N_B^b(Figure 2.12(e)), and N_B^c (Figure 2.12(f)), are derived from the uniaxial nematic phases N_U^a, N_U^b, and N_U^c, respectively, by imposing a twofold rotational symmetry in the plane perpendicular to the director $\mathbf{n}$. These biaxial phases are also referred to as the N+2 phase, following the terminology introduced by Lubensky and Radzihovsky [35].

Using Eq. (2.52) and Eq. (2.53), the biaxial order parameter B of Eq. (2.59) is given by

$$\begin{aligned} B \;=\; & \sin^2\chi_B\langle P_2(\mathbf{b}\cdot\mathbf{m}) - P_2(\mathbf{b}\cdot\mathbf{l})\rangle \\ & + \cos^2\chi_B\langle P_2(\mathbf{c}\cdot\mathbf{m}) - P_2(\mathbf{c}\cdot\mathbf{l})\rangle. \end{aligned} \tag{2.64}$$

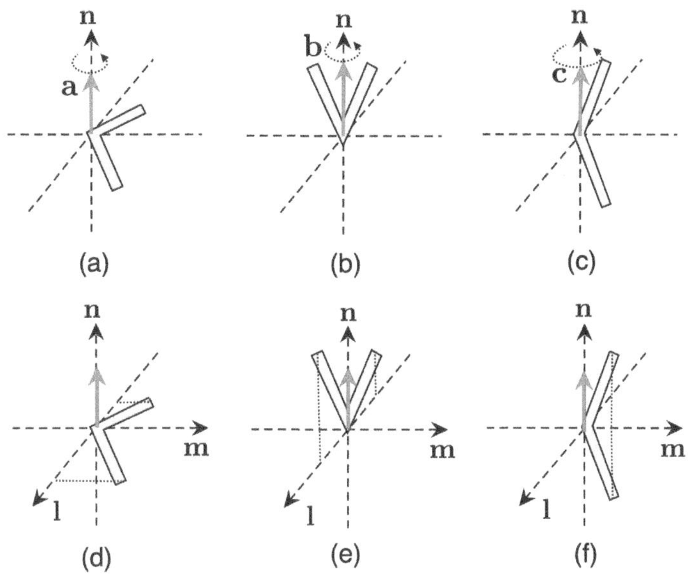

FIGURE 2.12
Schematic representation of uniaxial and biaxial nematic phases. There are three uniaxial nematic phases, N_U^a (a), N_U^b (b), and N_U^c (c), characterized by the predominant alignment of the molecular axes **a**, **b**, and **c** along the director **n**, respectively. The corresponding biaxial nematic phases, N_B^a (d), N_B^b (e), and N_B^c (f), are derived from the uniaxial phases (a), (b), and (c), respectively, by restricting molecular rotations in the plane perpendicular to **n** to have two-fold symmetry. Reprinted with permission from Ref. [33].

Substituting Eqs. (2.42) and (2.43) into Eq. (2.64), we obtain

$$B = \frac{3}{2}\Big(B_1 P_2(\cos\chi_B) + B_2 \sin^2\chi_B\Big), \tag{2.65}$$

where two biaxial order parameters are given by

$$\begin{aligned} B_1 &= \frac{2}{3}\langle P_2(\mathbf{c}\cdot\mathbf{m}) - P_2(\mathbf{c}\cdot\mathbf{l})\rangle \\ &= \langle \sin^2\theta\cos 2\phi\rangle, \end{aligned} \tag{2.66}$$

and

$$\begin{aligned} B_2 &= \frac{1}{3}\langle P_2(\mathbf{b}\cdot\mathbf{m}) - P_2(\mathbf{b}\cdot\mathbf{l}) - P_2(\mathbf{a}\cdot\mathbf{m}) + P_2(\mathbf{a}\cdot\mathbf{l})\rangle \\ &= \langle \frac{1}{2}(1+\cos^2\theta)\cos 2\phi\cos 2\psi - \cos\theta\sin 2\phi\sin 2\psi\rangle. \end{aligned} \tag{2.67}$$

The order parameter B_1 represents the orientation order of the molecular axis **c** in the $\mathbf{l}-\mathbf{m}$ plane, while B_2 represents the orientation order of the **a**-axis and **b**-axis in the same plane. For example, considering a bent-shaped molecule with a large bending angle, as the uniaxial order parameter S increases, the **c**-axis becomes parallel to the director **n**. When $S = 1$, $\mathbf{c}\cdot\mathbf{m} = \mathbf{c}\cdot\mathbf{n} = 0$, and thus $B_1 = 0$. Therefore, the order parameter B_1 in highly ordered states does not reflect true biaxial order. The true biaxial nematic order parameter is B_2. The four scalar order parameters, S, U, B_1, and B_2, introduced in this section were first proposed by Straley [36].

Using the tensor order parameter (Eq. (2.59)), the free energy F of a bent-shaped molecular system can be constructed. (The detailed form of the free energy will be discussed in Section 4.5.) Regardless of the specific form, the transition from the isotropic phase to the uniaxial nematic phase is expected to be a first-order phase transition, since the states with $S_n > 0$ and $S_n < 0$ are not equivalent: $F(S_n) \neq F(-S_n)$. A Landau expansion of the free energy in terms of the order parameter S_n yields $F = aS_n^2 - bS_n^3 + cS_n^4$, where a, b, and c are phenomenological constants. On the other hand, in the biaxial nematic phase, exchanging the secondary directors $\mathbf{l}$ and $\mathbf{m}$ does not alter the system's free energy, implying that $F(B) = F(-B)$ (see Eq. (2.64)). As a result, the free energy can be expressed as a Landau expansion in terms of the biaxial order parameter B as $F = -aB^2 + cB^4$, where the cubic term B^3 is absent due to symmetry considerations. This form of the free energy suggests that the phase transition from the uniaxial to the biaxial nematic phase is of second order. The tensor order parameter (Eq. (2.59)) thus provides a unified framework for describing a wide range of liquid crystal phases in systems composed of bent-core molecules.

3

Elastic Energy of Liquid Crystals

We now introduce the elastic energy associated with spatial distortions of the scalar order parameter S and the director field $\mathbf{n}$. For the director $\mathbf{n}$, three fundamental types of deformations can arise: splay, twist, and bend. These distortions are described by the Frank elastic energy, which forms the basis of the continuum theory of liquid crystals and is widely used to analyze various liquid crystal systems. In this chapter, we also discuss distinctive features of the nematic phase, including topological defects such as disclinations, Freedericksz transition, and orientational fluctuations.

3.1 Elastic distortions and the elastic free energy

The scalar order parameter S quantifies the degree of orientational alignment of liquid crystalline molecules. In contrast, the director $\mathbf{n}$ specifies the average direction along which the molecules are aligned. To describe the state of the nematic phase, it is necessary to simultaneously consider the scalar order parameter $S(\mathbf{r})$ and the director $\mathbf{n}(\mathbf{r})$ at a given position $\mathbf{r}$. For this purpose, we use a second-rank tensor order parameter (Eq. (2.37)). When the director $\mathbf{n}$ and the scalar order parameter S are not spatially uniform, the elastic free energy density f_{dis} associated with distortions can be expressed by expanding the spatial derivatives of the tensor order parameter $Q_{\alpha\beta}(\mathbf{r})$ up to second order. This leads to the following form [37]:

$$f_{dis} = \frac{1}{2}L_1(\partial_\gamma Q_{\alpha\beta}) \cdot (\partial_\gamma Q_{\alpha\beta}) + \frac{1}{2}L_2(\partial_\alpha Q_{\alpha\gamma}) \cdot (\partial_\beta Q_{\beta\gamma}), \tag{3.1}$$

where L_1 and L_2 are constants, $\partial_\alpha = \partial/\partial\alpha$ denotes the partial derivative with respect to the spatial coordinate α (with $\alpha, \beta, \gamma = x, y, z$), and the Einstein summation is assumed for repeated indices.

Noting that

$$\partial_\gamma Q_{\alpha\beta} = (\frac{3}{2}n_\alpha n_\beta - \frac{1}{2}\delta_{\alpha\beta})\partial_\gamma S + \frac{3}{2}S\partial_\gamma(n_\alpha n_\beta), \tag{3.2}$$

$$n_\alpha(\mathbf{r})\partial_\beta n_\alpha(\mathbf{r}) = \frac{1}{2}\partial_\beta\left(n_x^2(\mathbf{r}) + n_y^2(\mathbf{r}) + n_z^2(\mathbf{r})\right) = \frac{1}{2}\partial_\beta(1) = 0, \tag{3.3}$$

$$\begin{aligned} n_\alpha\partial_\alpha n_\gamma &= n_x\partial_x n_k + n_y\partial_y n_k + n_z\partial_z n_k = (\mathbf{n}\cdot\nabla)\mathbf{n} \\ &= -\mathbf{n}\times\nabla\times\mathbf{n}, \end{aligned} \tag{3.4}$$

$$n_\gamma\partial_\alpha n_\alpha = n_\gamma\nabla\cdot\mathbf{n} = \mathbf{n}(\nabla\cdot\mathbf{n}), \tag{3.5}$$

$$n_\alpha\partial_\alpha S = n_x\partial_x S + n_y\partial_y S + n_z\partial_z S = \mathbf{n}\cdot(\nabla S), \tag{3.6}$$

$$(\partial_\gamma n_\alpha n_\beta)^2 = (\partial_\gamma n_\alpha)^2 + 2(n_\alpha\partial_\gamma n_\alpha)(n_\beta\partial_\gamma n_\beta) + (\partial_\gamma n_\alpha)^2 = 2(\partial_\gamma n_\beta)^2, \tag{3.7}$$

DOI: 10.1201/9781003737711-3

we then obtain the first term of Eq. (3.1):

$$\text{first term} = \frac{3}{4}L_1(\nabla S)^2 + \frac{9}{4}L_1 S^2(\partial_\alpha n_\beta)(\partial_\alpha n_\beta), \tag{3.8}$$

and the second term:

$$\begin{aligned}\text{second term} &= \frac{9}{8}L_2 S^2\big((\mathbf{n}\times\nabla\times\mathbf{n})^2 + (\nabla\cdot\mathbf{n})^2\big) + \frac{3}{8}L_2(\mathbf{n}\cdot\nabla S)^2 \\ &\quad + \frac{1}{8}L_2(\nabla S)^2 + \frac{3}{2}L_2 S(\nabla\cdot\mathbf{n})(\mathbf{n}\cdot\nabla S) \\ &\quad + \frac{3}{4}L_2 S(\nabla S)\cdot(\mathbf{n}\times\nabla\times\mathbf{n}).\end{aligned} \tag{3.9}$$

The second term in Eq. (3.8) is expressed as

$$\begin{aligned}\big(\partial_\alpha n_\beta(\mathbf{r})\big)\big(\partial_\alpha n_\beta(\mathbf{r})\big) &= \big(\nabla\cdot\mathbf{n}(\mathbf{r})\big)^2 + \big(\mathbf{n}(\mathbf{r})\cdot(\nabla\times\mathbf{n}(\mathbf{r}))\big)^2 + \big(\mathbf{n}(\mathbf{r})\times(\nabla\times\mathbf{n}(\mathbf{r}))\big)^2 \\ &\quad -\nabla\cdot\big[\mathbf{n}(\mathbf{r})(\nabla\cdot\mathbf{n}(\mathbf{r})) + \mathbf{n}(\mathbf{r})\times(\nabla\times\mathbf{n}(\mathbf{r}))\big],\end{aligned} \tag{3.10}$$

and, substituting this, the distortion free energy becomes

$$\begin{aligned}f_{dis} &= \frac{3}{4}\Big(L_1 + \frac{1}{6}L_2\Big)(\nabla S)^2 + \frac{3}{8}L_2(\mathbf{n}\cdot\nabla S)^2 \\ &\quad + \frac{9}{4}S^2\left[\Big(L_1 + \frac{1}{2}L_2\Big)(\nabla\cdot\mathbf{n})^2 + L_1(\mathbf{n}\cdot\nabla\times\mathbf{n})^2 + \Big(L_1 + \frac{1}{2}L_2\Big)(\mathbf{n}\times\nabla\times\mathbf{n})^2\right] \\ &\quad + \frac{3}{2}L_2 S(\nabla\cdot\mathbf{n})(\mathbf{n}\cdot\nabla S) + \frac{3}{4}L_2 S(\nabla S)\cdot(\mathbf{n}\times\nabla\times\mathbf{n}).\end{aligned} \tag{3.11}$$

The last term in Eq. (3.10) can be neglected because it represents a contribution from the surface.[1]

The first two terms in Eq. (3.11) represent the spatial variation of the scalar order parameter $S(\mathbf{r})$. The second line describes the spatial variation of the orientation field $\mathbf{n}(\mathbf{r})$, where $(\nabla\cdot\mathbf{n})^2$, $(\mathbf{n}\cdot\nabla\times\mathbf{n})^2$, and $(\mathbf{n}\times\nabla\times\mathbf{n})^2$ correspond to the elastic free energy contributions from splay, twist, and bend deformations, respectively. The last two terms represent coupling between the spatial variations of the scalar order parameter $S(\mathbf{r})$ and the director $\mathbf{n}(\mathbf{r})$. It is straightforward to verify that the distortion free energy density f_{dis} is a scalar quantity.[2]

3.1.1 Frank elastic constants

When the orientational order parameter (S) is spatially uniform: $\nabla S = 0$, Eq. (3.11) simplifies to

$$f_F = \frac{1}{2}K_1(\nabla\cdot\mathbf{n})^2 + \frac{1}{2}K_2(\mathbf{n}\cdot\nabla\times\mathbf{n})^2 + \frac{1}{2}K_3(\mathbf{n}\times\nabla\times\mathbf{n})^2, \tag{3.12}$$

where

$$K_1 = \frac{9}{2}S^2\Big(L_1 + \frac{1}{2}L_2\Big), \tag{3.13}$$

$$K_2 = \frac{9}{2}S^2 L_1, \tag{3.14}$$

[1]For any vector $\mathbf{E}$ leaving a closed surface, the volume integral $\int(\nabla\cdot\mathbf{E})dV$ over the region enclosed by the surface can be transformed into a surface integral $\int(\mathbf{E}\cdot\mathbf{u})dA$ by Gauss's theorem, where $\mathbf{u}$ denotes the outward unit normal vector to the closed surface.

[2]Here, ∇ is a vector differential operator. The dot product $\nabla\cdot\mathbf{n}$ yields a scalar, while the cross product $\nabla\times\mathbf{n}$ and the gradient ∇S yield vector quantities.

and

$$K_3 = \frac{9}{2}S^2\Big(L_1 + \frac{1}{2}L_2\Big). \tag{3.15}$$

This is referred to as Frank elastic free energy. The coefficients K_1, K_2, and K_3 represent the Frank elastic constants, corresponding to the splay, twist, and bend distortions, respectively. In Eq. (3.11), the elastic constants for the distortions are determined by two independent constants, L_1 and L_2. However, in the actual nematic phase, the elastic energy is given by three elastic constants as described by Eq. (3.12). This discrepancy arises from the expansion of the spatial derivatives of $Q_{\alpha\beta}$ only up to the second order in Eq. (3.1). To address this, three independent constants are redefined in Eq. (3.12). Using Eq. (3.12), various distortions in the orientation field, such as the Fredericks transition, will be discussed in the next Section. Although the orientational order parameter S does not explicitly appear in Eq. (3.12), the elastic constants are proportional to S^2.

Equation (3.12) can be directly derived using a more straightforward approach. Considering the spatial distortion of the director $\mathbf{n}$ up to the second order $(\partial_i n_j)^2$ and ignoring surface terms, the spatial derivatives of the director $\mathbf{n}$ can be decomposed using vector operators into $\nabla \cdot \mathbf{n}$ and $\nabla \times \mathbf{n}$. The former is a scalar quantity, while the latter is a vector quantity. Since the free energy is a scalar, scalar quantities are constructed from these terms. The first-order term of $\nabla \cdot \mathbf{n}$ is excluded because, in the nematic phase, the states $\mathbf{n}$ and $-\mathbf{n}$ represent the same physical orientation. Therefore, $(\nabla \cdot \mathbf{n})^2$ contributes to the elastic free energy from splay deformation. The vector quantity $\nabla \times \mathbf{n}$ can be further decomposed into the scalar triple product $\mathbf{n} \cdot \nabla \times \mathbf{n}$ (horizontal component) and the vector triple product $\mathbf{n} \times \nabla \times \mathbf{n}$ (vertical component). The former is a pseudoscalar, while the latter is a vector quantity. The term $\mathbf{n} \cdot \nabla \times \mathbf{n}$ does not contribute to systems with mirror symmetry and is relevant only in chiral phases like the cholesteric phase. Squaring these terms gives $(\mathbf{n} \cdot \nabla \times \mathbf{n})^2$ for twist deformation and $(\mathbf{n} \times \nabla \times \mathbf{n})^2$ for bend deformation. Hence, by assigning coefficients K_i to each type of distortion, the elastic free energy is given as the sum of the three terms.

Let us now examine each distortion term in Eq. (3.12) in more detail. Suppose that the director $\mathbf{n}$ lies in the x-y plane and is nearly aligned along the x-axis, forming a small angle θ with respect to it. Then, assuming spatial uniformity along the z-direction, the director field can be expressed as:

$$\mathbf{n} = (\cos\theta, \sin\theta, 0) \simeq (1, \theta, 0), \tag{3.16}$$

where the approximation on the right-hand side holds for small θ. Substituting Eq. (3.16) into Eq. (3.12), the first term is given by

$$\nabla \cdot \mathbf{n} = \frac{\partial \cos\theta}{\partial x} + \frac{\partial \sin\theta}{\partial y} \simeq \partial_y \theta. \tag{3.17}$$

Assuming the deformation is uniform and letting the right-hand side be a constant c, we obtain $\theta = cy$. Therefore, as shown in Fig. 3.1(a), the orientation angle θ changes along the y-axis. This is referred to as 'splay' deformation.

Similarly, we obtain

$$\nabla \times \mathbf{n} = \begin{vmatrix} \mathbf{e}_x & \mathbf{e}_y & \mathbf{e}_z \\ \partial_x & \partial_y & \partial_z \\ 1 & \theta & 0 \end{vmatrix} = -\partial_z \theta\, \mathbf{e}_x + \partial_x \theta\, \mathbf{e}_z, \tag{3.18}$$

where $\mathbf{e}_i$ is the basis vector and the second term in Eq. (3.12) is given by

$$\mathbf{n} \cdot \nabla \times \mathbf{n} \simeq -\partial_z \theta. \tag{3.19}$$

If the right-hand side is a constant c, we get $\theta = cz$. This means that the orientation angle θ varies along the z-axis, resulting in a rotation of the director along the z-axis, as shown in Fig. 3.1(b). This type of deformation is called 'twist' deformation.

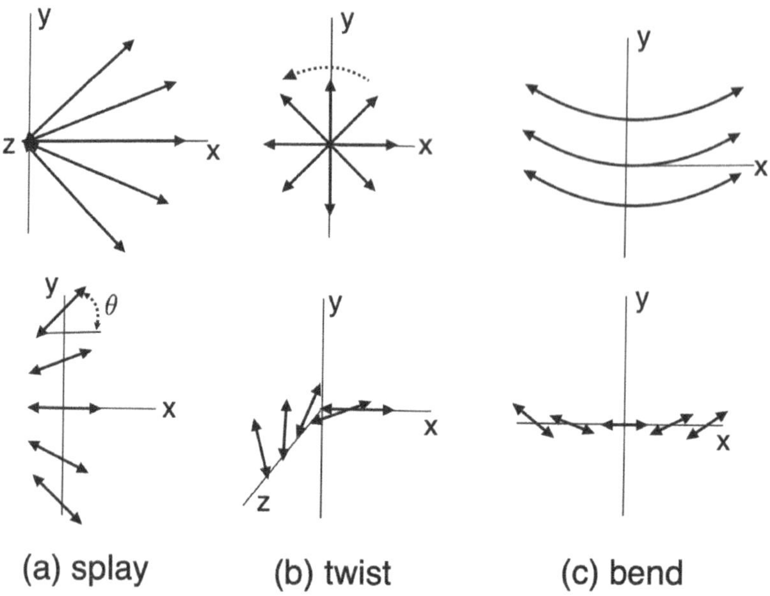

FIGURE 3.1
Three (splay, twist, bend) distortions of the director. The arrows represent the direction of the director.

The third term of Eq. (3.12) is given by

$$\mathbf{n} \times \nabla \times \mathbf{n} = \begin{vmatrix} \mathbf{e}_x & \mathbf{e}_y & \mathbf{e}_z \\ 1 & \theta & 0 \\ -\partial_z \theta & 0 & \partial_x \theta \end{vmatrix} = \theta \partial_x \theta \, \mathbf{e}_x - \partial_x \theta \, \mathbf{e}_y + \theta \partial_z \theta \, \mathbf{e}_z, \tag{3.20}$$

and

$$|\mathbf{n} \times \nabla \times \mathbf{n}|^2 = (\partial_x \theta)^2 + \theta^2 (\partial_x \theta)^2 + \theta^2 (\partial_z \theta)^2 \simeq (\partial_x \theta)^2. \tag{3.21}$$

We then obtain $\partial_x \theta = c$: $\theta = cx$. This means that the orientation angle θ varies along the x-axis, resulting in a tilt of the director along the x-axis, as shown in Fig. 3.1(c). This type of deformation is called 'bend' deformation.

3.1.2 Director near an interface

Let us examine the expansion coefficients L_1 and L_2 in Eq. (3.11). When the director $\mathbf{n}$ is spatially uniform, such that $\nabla \cdot \mathbf{n} = \nabla \times \mathbf{n} = 0$, Eq. (3.11) simplifies to

$$f_{dis} = \frac{3}{4}\Big(L_1 + \frac{1}{6}L_2\Big)(\nabla S)^2 + \frac{3}{8}L_2(\mathbf{n} \cdot \nabla S)^2, \tag{3.22}$$

indicating that the distortion free energy depends solely on the spatial variation of the scalar order parameter $S(\mathbf{r})$. Both the magnitude and directional gradient of S contribute to the elastic free energy through the two independent terms in Eq. (3.22). For example, at the interface between the isotropic and nematic phases (IN interface), the scalar order parameter S varies continuously from zero in the isotropic phase to a finite value in the nematic phase. In Eq. (3.22), the coefficient L_1 must be positive ($L_1 > 0$) to ensure thermodynamic stability.

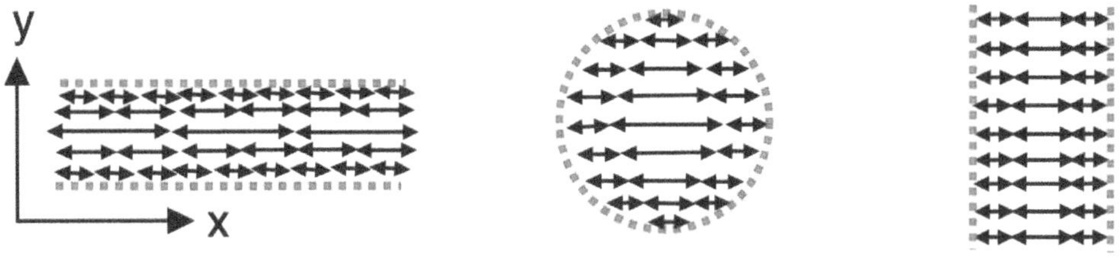

(*a*) $L_2 > 0$ (*b*) $L_2 = 0$ (*c*) $L_2 < 0$

FIGURE 3.2
Illustration of the spatial anisotropy of the director for different values of L_2. The arrows indicate the direction of the director **n** (fixed along the x-axis), and the length of the arrows is proportional to the magnitude of the orientational order parameter S. Outside the IN interface, the system is in the isotropic phase.

Specifically, a positive L_1 penalizes spatial variations in the scalar order parameter via the term $(\nabla S)^2$, thereby preventing unbounded growth in the free energy. If L_1 were negative, such variations would lower the free energy indefinitely, leading to an unstable system. The sign of the coefficient L_2, on the other hand, determines the preferred orientation of the director **n** relative to the IN interface. When $L_2 > 0$, the term $(\mathbf{n} \cdot \nabla S)^2$ is energetically unfavorable, so the director tends to align perpendicular to the gradient of S, i.e., parallel to the interface. Conversely, when $L_2 < 0$, this term becomes favorable, and the director prefers to align along the gradient of S, i.e., perpendicular to the interface. This coupling plays a key role in determining the director configuration at the boundary between isotropic and nematic phases.

We consider a two-dimensional x-y plane ($n_z = 0$) and fix the direction of the director **n** along the x-axis. When the director and the gradient of the scalar order parameter are perpendicular ($\mathbf{n} \perp \nabla S$), we have $(\mathbf{n} \cdot \nabla S) = 0$. The free energy Eq. (3.22) is given by

$$f_{dis}^{\perp} = \frac{3}{4}(L_1 + \frac{1}{6}L_2)(\nabla S)^2. \tag{3.23}$$

If the director and the gradient of the scalar order parameter are parallel ($\mathbf{n} \parallel \nabla S$), we have $(\mathbf{n} \cdot \nabla S)^2 = \mathbf{n}^2(\nabla S)^2 = (\nabla S)^2$. Then, the free energy Eq. (3.22) is given by

$$f_{dis}^{\parallel} = \frac{3}{4}(L_1 + \frac{2}{3}L_2)(\nabla S)^2. \tag{3.24}$$

Comparing Eqs. (3.23) and (3.24), if $L_2 > 0$, we have $f_{dis}^{\parallel} > f_{dis}^{\perp}$, and the state of $\mathbf{n} \perp \nabla S$ becomes stable. In this case, the director (**n**) aligns in the x-axis direction, and the gradient of the orientational order parameter is along the y-axis. That is, the director stabilises parallel to the IN interface and relaxes toward the isotropic phase (Fig. 3.2(a)). Conversely, if $L_2 < 0$, then $f_{dis}^{\parallel} < f_{dis}^{\perp}$, and the configuration with $\mathbf{n} \parallel \nabla S$ becomes stable. In this case, both the director and the gradient of the order parameter align along the x-axis, meaning the director tends to orient perpendicular to the IN interface and relax toward the isotropic phase (Fig. 3.2(c)). When $L_2 = 0$, we have $f_{dis}^{\parallel} = f_{dis}^{\perp}$, and the order parameter relaxes isotropically without directional preference (Fig. 3.2(b)).

3.2 Characteristics of the nematic phase

3.2.1 Disclinations

The line defects of liquid crystal directors are called disclinations. The term "disclination" means the turning of the tilt of the director $\mathbf{n}$. Disclinations are topological defects characterized by singularities or discontinuities in the orientation of the liquid crystalline molecules. It is distinguished from "dislocation", which refers to line defects in the crystal lattice of solids.

Figure 3.3 illustrates four representative types of disclinations. The solid lines represent the directors around the disclination core (closed circle). At the disclination core, the director $\mathbf{n}$ no longer changes continuously, forming a defect. When the orientation vector (arrow) $\mathbf{n}$ is placed on a two-dimensional plane, the angle between $\mathbf{n}$ and the x-axis is denoted as $\psi(\mathbf{r})$, where $\mathbf{r} = (x, y, 0)$. The director field is then given by

$$\mathbf{n}(\mathbf{r}) = (\cos\psi(\mathbf{r}), \sin\psi(\mathbf{r}), 0), \tag{3.25}$$

where $\psi(\mathbf{r})$ describes the angular configuration of the director around the disclination core.

By approximating the elastic constants as a single averaged value $K_{\text{el}} = K_1 = K_2 = K_3$, we can significantly simplify the mathematical treatment of Frank elastic free energy. Substituting the director field Eq. (3.25) into the elastic energy Eq. (3.12), we obtain

$$\begin{aligned} f_F &= \frac{1}{2}K_{el}\Big((\nabla\cdot\mathbf{n})^2 + (\mathbf{n}\times\nabla\times\mathbf{n})^2\Big) \\ &= \frac{1}{2}K_{el}\Big(\big(\frac{d\psi}{dx}\big)^2 + \big(\frac{d\psi}{dy}\big)^2\Big). \end{aligned} \tag{3.26}$$

This one-constant approximation significantly reduces the complexity of the analysis and is often employed for qualitative studies of orientational distortions and topological defects, such as disclinations, in liquid crystal systems. To determine the equilibrium configuration of the director field, we minimize the total Frank free energy: $F = \int f_F(d\psi/dx, d\psi/dy)\, dV$. Applying the Euler-Lagrange equation to this variational problem:

$$\frac{\partial f_F}{\partial\psi} - \frac{\partial}{\partial x}\left(\frac{\partial f_F}{\partial(d\psi/dx)}\right) - \frac{\partial}{\partial y}\left(\frac{\partial f_F}{\partial(d\psi/dy)}\right) = 0, \tag{3.27}$$

and using the form of the free energy density given in Eq. (3.26), we obtain

$$\frac{\partial^2\psi}{\partial x^2} + \frac{\partial^2\psi}{\partial y^2} = \nabla^2\psi = 0, \tag{3.28}$$

which is the Laplace equation. This result implies that, under the one-constant approximation and in the absence of external fields or boundary anchoring forces, the director angle $\psi(x, y)$ behaves as a harmonic function. The Laplace equation thus governs the equilibrium director configuration near defects in this simplified model. As a general solution to Eq. (3.28) in two dimensions, we consider an azimuthal form:

$$\psi(\mathbf{r}) = s\theta + c, \tag{3.29}$$

where $\theta = \tan^{-1}(y/x)$ is the polar angle measured from the x-axis (with $0 < \theta < 2\pi$), s is the topological charge (or strength) of the disclination, and c is a constant.

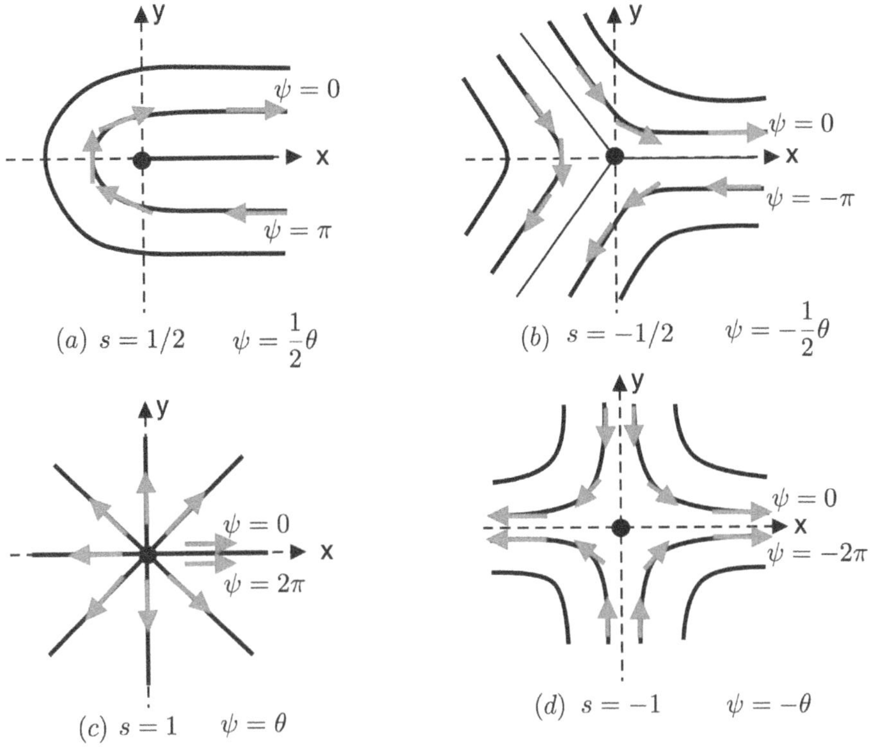

FIGURE 3.3
Schematic representation of disclinations in a nematic phase for (a) $s = 1/2$, (b) $s = -1/2$, (c) $s = 1$, and (d) $s = -1$ with $c = 0$ in Eq. (3.29). The arrows represent the direction of the director.

When $c = 0$, the four typical disclination patterns shown in Fig. 3.3 appear, corresponding to different values of the topological charge s, such as $s = \pm 1$ and $s = \pm\frac{1}{2}$. In the case of $s = 1/2$, if we follow the director around the disclination core, starting at $\psi = 0$ (at $\theta = 0$), and complete a full rotation, the direction of the arrow rotates in a "counterclockwise" direction, returning to the original position. However, the final arrow is at $\psi = \pi$ (at $\theta = 2\pi$). Therefore, we can write $\psi = \theta/2$. For a disclination with $s = -1/2$, if we follow the director around the core, starting at $\psi = 0$ (at $\theta = 0$), and complete a full rotation, the arrow rotates in a "clockwise" direction, returning to the original position. The final arrow is at $\psi = -\pi$ (at $\theta = 2\pi$). Therefore, we can write $\psi = -\theta/2$. In the x-y plane, we define the "counterclockwise" increase of ψ from the first quadrant to the second quadrant as positive. A disclination with $s = 1$ is the simplest to understand, where $\psi = \theta$. This corresponds to the emergence of an electric field from a point charge in the context of electromagnetism. At this point, understanding $s = -1$ also becomes straightforward. When $c \neq 0$, more complex disclinations appear. (practice: consider what happens for $s = 1$ and $c = \pi/2$). Due to the presence of disclinations and other defects, liquid crystals observed under a polarising microscope exhibit a variety of characteristic textures [38–40]. Notable examples include the Schlieren texture, commonly seen in the nematic phase, and the focal conic texture, which is typical of the smectic A phase. The analysis of textures is a powerful diagnostic tool for identifying phase transitions, defects, and orientational ordering in liquid crystal samples.

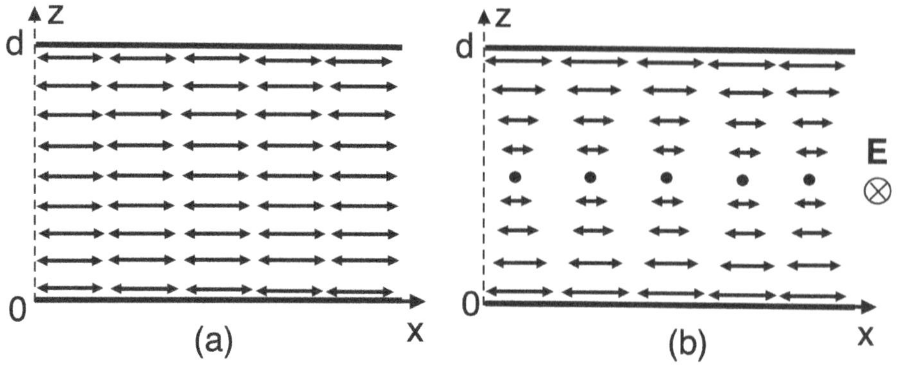

FIGURE 3.4
Schematic representation of the Freedericksz transition. The arrows indicate the nematic director between two glass plates. (a) $E < E_c$ and (b) $E > E_c$.

3.2.2 Freedericksz transition

When a liquid crystal cell is created by sandwiching liquid crystalline molecules between two glass plates, the properties of the glass surfaces induce either planar or homeotropic alignments. Planar alignment refers to the director being parallel to the glass surface, whereas homeotropic alignment denotes the director being perpendicular to the surface. Liquid crystals, in which the dielectric constant parallel to the long axis of the molecules ($\epsilon_{||}$) is greater than that perpendicular to the axis ($\epsilon_{\perp}$), exhibit positive dielectric anisotropy, characterized by $\Delta\epsilon = \epsilon_{||} - \epsilon_{\perp} > 0$. Conversely, when the perpendicular component dominates ($\epsilon_{||} < \epsilon_{\perp}$), the system has negative dielectric anisotropy, corresponding to $\Delta\epsilon < 0$.

When liquid crystalline molecules are confined between two glass plates and subjected to an external electric field, they tend to align in the direction that minimizes the total free energy (Fig. 3.4). The free energy density associated with the interaction between the liquid crystal and the electric field $\mathbf{E}$ is given by

$$f_{elec} = -\frac{1}{2}\Delta\epsilon(\mathbf{n}\cdot\mathbf{E})^2. \tag{3.30}$$

The free energy under an external field is discussed in Chapter 6 (see Eq. (6.2)). The sign of $\Delta\epsilon$ depends on the molecular structure of the liquid crystal: If $\Delta\epsilon > 0$, the free energy is minimized when the director $\mathbf{n}$ aligns parallel to the electric field $\mathbf{E}$. If $\Delta\epsilon < 0$, the free energy is minimized when $\mathbf{n}$ aligns perpendicular to $\mathbf{E}$. In the following discussion, we focus on the case $\Delta\epsilon > 0$, which is typical for many nematic liquid crystal materials used in display applications.

Consider a nematic liquid crystal confined between two parallel glass plates, as illustrated in Fig. 3.4(a). The plates are positioned at $z = 0$ and $z = d$, and the liquid crystal occupies the space between them. Strong planar anchoring conditions are imposed at the boundaries, meaning the liquid crystalline molecules are aligned parallel to the substrate surfaces at both walls. This boundary condition enforces a uniform alignment of the director $\mathbf{n}$ in the plane of the walls near $z = 0$ and $z = d$, even in the absence of an external field. In the absence of an electric field, the director is uniformly aligned along the x-axis, i.e., $\mathbf{n} = (1, 0, 0)$, throughout the system. This configuration minimizes the elastic free energy under the planar anchoring condition at both walls. When a uniform electric field $\mathbf{E} = (0, E, 0)$ is applied in the y-direction, the liquid crystalline molecules experience a torque due to dielectric anisotropy. For a liquid crystal with positive dielectric anisotropy ($\Delta\epsilon > 0$), this torque tends to reorient the director toward the direction of the electric field. As a result, the uniform planar alignment becomes unstable beyond a certain threshold field, and the director begins to tilt away from the x-axis in order to minimize the total free energy, which now includes both elastic and electric contributions.

In the following, we examine how the director configuration $\mathbf{n}(z)$ varies with z in thermal equilibrium under the influence of the electric field, considering both the elastic distortion energy and the coupling to the external field. As shown in Fig. 2.5, the director can be expressed as

$$\mathbf{n} = (\cos\Theta(z), \sin\Theta(z), 0), \tag{3.31}$$

where $\Theta(z)$ is the angle between the director and the x-axis as a function of z. Under the applied electric field, the total free energy density can be obtained by substituting Eq. (3.31) into Eqs. (3.12) and (3.30):

$$f_F + f_{elec} = \frac{1}{2}K_2\Big(\big(\frac{\partial\Theta}{\partial z}\big)^2 - \frac{1}{\xi^2}\sin^2\Theta(z)\Big), \tag{3.32}$$

where only the term related to the elastic constant K_2 for twist deformation remains. To simplify this expression, we define the characteristic length scale ξ as

$$\xi = \frac{1}{E}\sqrt{\frac{K_2}{\Delta\epsilon}}, \tag{3.33}$$

which is called the correlation length. This length characterizes the spatial extent over which the director responds to the external electric field, and essentially sets the scale of deformation away from the walls, where planar anchoring is imposed.

The total free energy (per unit area) of the system in the presence of the external field, assuming uniformity along the x- and y-directions, is given by

$$F = \int_0^d (f_F(\Theta(z)) + f_{elec}(\Theta(z)))dz, \tag{3.34}$$

where the boundary conditions assume that the director is in the ground state at the walls: $\Theta(0) = \Theta(d) = 0$. For small Θ, approximating $\sin^2\Theta \simeq \Theta^2 - \Theta^4/3$, we have

$$F = \frac{1}{2}K_2\int_0^d \Big(\big(\frac{\partial\Theta}{\partial z}\big)^2 - \frac{1}{\xi^2}(\Theta^2 - \Theta^4/3)\Big)dz. \tag{3.35}$$

Due to the symmetry of the liquid crystal cell, the distortion is largest at $z = d/2$. Therefore, we denote $\Theta(d/2) = \Theta_m$. As a solution satisfying the boundary conditions, we assume

$$\Theta(z) = \Theta_m \sin(qz), \tag{3.36}$$

$$q = \pi/d. \tag{3.37}$$

Substituting Eq. (3.36) into Eq. (3.35), the free energy is given by

$$F = \frac{d}{4}K_2q^2\Big(\big(1 - \frac{1}{(q\xi)^2}\big)\Theta_m^2 + \frac{1}{4(q\xi)^2}\Theta_m^4\Big), \tag{3.38}$$

which takes the form of a Landau-type expansion in terms of the order parameter Θ_m, representing the maximum deviation of the director from the initial planar alignment. This expression clearly illustrates a second-order phase transition: the coefficient of the quadratic term changes sign at the critical field, while the quartic term remains positive to ensure the

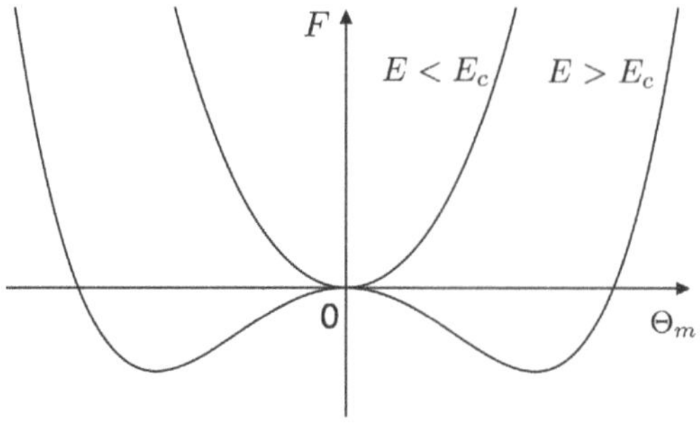

FIGURE 3.5
Free energy plotted as a function of Θ_m.

stability of the free energy. The phase transition point is determined when the coefficient of Θ_m^2 becomes zero: $q\xi = 1$, and the critical electric field is given by

$$E_c = \frac{\pi}{d}\sqrt{\frac{K_2}{\Delta\epsilon}}. \tag{3.39}$$

Using the critical electric field E_c, the critical condition can be rewritten as: $q\xi = E_c/E$. Figure 3.5 shows the dependence of the free energy (Eq. (3.38)) on the twist angle Θ_m. For $E < E_c$, $\Theta_m = 0$ is stable. For $E > E_c$, the free energy is minimized at a finite twist angle, and the solution $\Theta_m = 0$ becomes unstable.

The condition for minimizing the free energy (Eq. (3.38)) is given by $dF/d\Theta_m = 0$ and leads to

$$\frac{d}{4}K_2q^2\left(2(1-\frac{E^2}{E_c^2}) + \frac{E^2}{E_c^2}\Theta_m^2\right)\Theta_m = 0. \tag{3.40}$$

When $E < E_c$, the equilibrium state is $\Theta_m = 0$. On the other hand, for $E > E_c$, the solution for Θ_m is given by[3]

$$\begin{aligned}\Theta_m &= \sqrt{2}\sqrt{\frac{(E+E_c)(E-E_c)}{E^2}} \\ &\simeq \sqrt{2}\sqrt{\frac{2E_c(E-E_c)}{E_c^2}} = 2\sqrt{\frac{E-E_c}{E_c}}.\end{aligned} \tag{3.41}$$

Figure 3.6 schematically shows the angle Θ_m plotted against E. For $E < E_c$, the director remains in the ground state even in the presence of the external field. However, for $E > E_c$, the director twists in the direction parallel to the electric field. The order parameter Θ_m increases continuously from zero, indicating a second-order phase transition. Figure 3.4(b) shows a schematically representation of the liquid crystal cell where $E > E_c$ and $\Theta_m \simeq \pi/2$. This phenomenon is known as the Freedericksz transition and is the basis for the most fundamental type of liquid crystal display: the twisted nematic cell.

Multiplying both sides of Eq. (3.39) by the cell thickness d, the critical voltage V^* at which the Freedericksz transition occurs is obtained as

$$V^* = \pi\sqrt{\frac{K_2}{\Delta\epsilon}}. \tag{3.42}$$

[3]Near the critical electric field, we approximate $E \simeq E_c$ except for the term $E - E_c$.

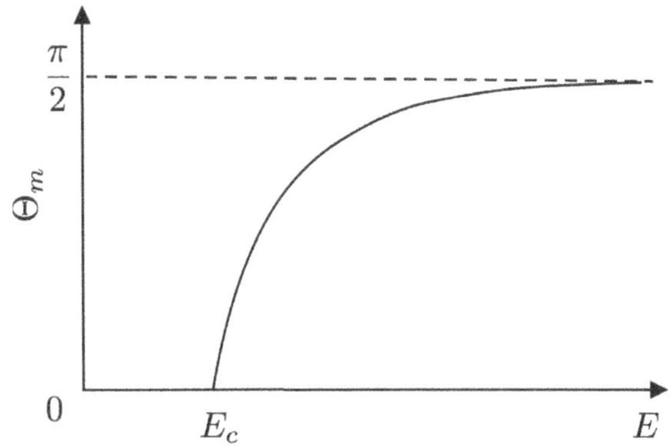

FIGURE 3.6
Schematic representation of the angle Θ_m plotted against the external field E (Fredericksz transition).

This critical voltage is independent of the cell thickness d and is determined solely by the intrinsic material parameters of the liquid crystal. For typical nematic materials, V^* is on the order of 1 volt.

A Freedericksz transition can also occur under the influence of a magnetic field H, analogous to the case with an electric field. When a magnetic field $\mathbf{H} = (0, H, 0)$ is applied to a nematic liquid crystal, the free energy density due to the magnetic field is given by

$$f_{mag} = -\frac{1}{2}\mu_0 \Delta\chi (\mathbf{n} \cdot \mathbf{H})^2, \tag{3.43}$$

where μ_0 is the permeability of free space and $\Delta\chi = \chi_{||} - \chi_{\perp}$ represents the magnetic anisotropy of liquid crystalline molecules. This expression has the same form as the electric field interaction term (Eq. (3.30)), and the analysis of the Freedericksz transition proceeds similarly. The critical magnetic field H_c is then given by:

$$H_c = \frac{\pi}{d}\sqrt{\frac{K_2}{\mu_0 \Delta\chi}}, \tag{3.44}$$

This critical field marks the threshold above which the director begins to reorient in response to the applied magnetic field.

We have examined the planar alignment as the ground state. A similar analysis can also be performed on the homeotropic alignment.

3.2.3 Director fluctuations

In the nematic phase, the director undergoes spatial and temporal fluctuations due to thermal motion in a system of volume V. In this subsection, we analyse the magnitude of orientational fluctuation. Furthermore, we examine how the application of a static electric field along the z-axis affects the stability of the system.

Consider the small changes in the director $\mathbf{n}$ induced by thermal fluctuations with the average orientation of the director along the z-axis in the ground state. As shown in Fig. 3.7(a), the director at a position $\mathbf{r}$, initially aligned along the z-axis, undergoes small fluctuations $\delta\mathbf{n}(\mathbf{r}) = (n_x, n_y, 0)$ due to thermal motion. Using $\{\mathbf{e}_x, \mathbf{e}_y, \mathbf{e}_z\}$ as the basis vectors, the

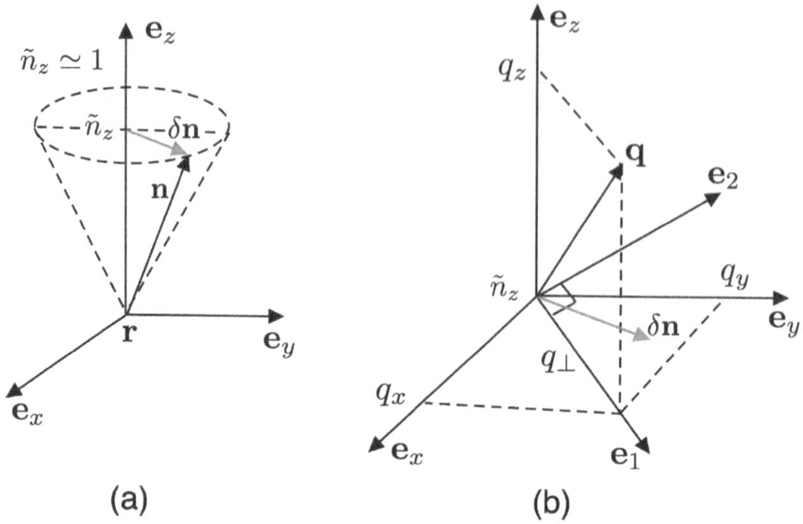

FIGURE 3.7
(a) Fluctuation of director. (b) Coordinate system $\{\mathbf{e}_1, \mathbf{e}_2, \mathbf{e}_z\}$.

director at a position $\mathbf{r}$ is expressed as

$$\mathbf{n}(\mathbf{r}) = n_x\mathbf{e}_x + n_y\mathbf{e}_y + \tilde{n}_z\mathbf{e}_z, \tag{3.45}$$

where we assume small changes: $\tilde{n}_z \simeq 1$, $n_x, n_y \ll 1$.

As discussed in Section 3.2.2, we apply an electric field $\mathbf{E} = (0, 0, E)$ along the z-axis to analyse the effect of an external field, which may also represent a magnetic field. The total free energy of the nematic phase under an external field, as derived in Eqs. (3.12) and (3.30), is given by

$$F = \int d\mathbf{r}(f_F + f_{elec}). \tag{3.46}$$

Substituting Eq. (3.45) into Eq. (3.30), we get

$$f_{elec} = -\frac{1}{2}\Delta\epsilon E^2(\tilde{n}_z)^2 = -\frac{1}{2}\Delta\epsilon E^2 + \frac{1}{2}\Delta\epsilon E^2(n_x^2 + n_y^2), \tag{3.47}$$

where the first term is constant and does not depend on the director. Since the director's z-component $\tilde{n}_z$ is fixed, the spatial fluctuations must only be considered for the (n_x, n_y) components. Therefore, using Eq. (3.45), the Frank elastic energy (Eq. (3.12)) is given by

$$\begin{aligned} f_F &= \frac{1}{2}K_1\left(\frac{\partial n_x}{\partial x} + \frac{\partial n_y}{\partial y}\right)^2 + \frac{1}{2}K_2\left(\frac{\partial n_x}{\partial y} - \frac{\partial n_y}{\partial x}\right)^2 \\ &\quad + \frac{1}{2}K_3\left[\left(\frac{\partial n_x}{\partial z}\right)^2 + \left(\frac{\partial n_y}{\partial z}\right)^2\right]. \end{aligned} \tag{3.48}$$

The fluctuation components $n_k(\mathbf{r})$, where $k = x, y$, can be expanded in a Fourier series as

$$n_k(\mathbf{r}) = \sum_{\mathbf{q}} n_k(\mathbf{q})e^{i\mathbf{q}\cdot\mathbf{r}}, \tag{3.49}$$

where $\mathbf{q}\cdot\mathbf{r} = q_x x + q_y y + q_z z$. Since $n_k(\mathbf{r})$ is real-valued function, the Fourier coefficients satisfy the condition $n_k(\mathbf{q}) = n_k^*(-\mathbf{q})$, where the asterisk (*) represents the complex conjugate. Substituting Eq. (3.49) into Eq. (3.47), the second term of Eq. (3.47) is given by

$$\begin{aligned}\int d\mathbf{r} n_x^2 &= \int d\mathbf{r} e^{i(\mathbf{q}+\mathbf{q}')\cdot\mathbf{r}} \sum_{\mathbf{q},\mathbf{q}'} n_x(\mathbf{q}) n_x^*(\mathbf{q}') \\ &= V\delta(\mathbf{q}+\mathbf{q}') \sum_{\mathbf{q},\mathbf{q}'} n_x(\mathbf{q}) n_x^*(\mathbf{q}') \\ &= V \sum_{\mathbf{q}} |n_x(\mathbf{q})|^2, \end{aligned} \tag{3.50}$$

and the first term of Eq. (3.48):

$$\begin{aligned}\int d\mathbf{r} \left(\frac{\partial n_x}{\partial x} + \frac{\partial n_y}{\partial y}\right)^2 &= V\delta(\mathbf{q}+\mathbf{q}') \\ &\times \sum_{\mathbf{q},\mathbf{q}'} i^2 \big(q_x n_x(\mathbf{q}) + q_y n_y(\mathbf{q})\big) \big(q'_x n_x(\mathbf{q}') + q'_y n_y(\mathbf{q}')\big)^* \\ &= V \sum_{\mathbf{q}} (-1) \big(q_x n_x(\mathbf{q}) + q_y n_y(\mathbf{q})\big) \big(-q_x n_x(-\mathbf{q}) - q_y n_y(-\mathbf{q})\big)^* \\ &= V \sum_{\mathbf{q}} |q_x n_x(\mathbf{q}) + q_y n_y(\mathbf{q})|^2, \end{aligned} \tag{3.51}$$

where we have used

$$\int d\mathbf{r} e^{i(\mathbf{q}+\mathbf{q}')} = V\delta(\mathbf{q}+\mathbf{q}'), \tag{3.52}$$

$$\frac{\partial n_x(\mathbf{r})}{\partial x} = \sum_{\mathbf{q}} n_x(\mathbf{q})(iq_x) e^{i\mathbf{q}\cdot\mathbf{r}}. \tag{3.53}$$

We then obtain the free energy in the wavevector space:

$$\begin{aligned} F &= V \sum_{\mathbf{q}} \frac{1}{2} \Big(K_1 |q_x n_x(\mathbf{q}) + q_y n_y(\mathbf{q})|^2 + K_2 |q_y n_x(\mathbf{q}) - q_x n_y(\mathbf{q})|^2 \\ &\quad + K_3 q_z^2 \big(|n_x(\mathbf{q})|^2 + |n_y(\mathbf{q})|^2\big) + \Delta\epsilon E^2 \big(|n_x(\mathbf{q})|^2 + |n_y(\mathbf{q})|^2\big) \Big), \end{aligned} \tag{3.54}$$

where, for a given wavevector $\mathbf{q}$, we consider a new coordinate system $\{\mathbf{e}_1, \mathbf{e}_2, \mathbf{e}_z\}$, obtained by rotating the axes $\{\mathbf{e}_x, \mathbf{e}_y\}$ within the x-y plane such that the $\mathbf{e}_2$ component vanishes (See Fig. 3.7(b)). In the new coordinate system, $\mathbf{q}$ is given by

$$\mathbf{q} = q_\perp \mathbf{e}_1 + q_z \mathbf{e}_z, \tag{3.55}$$

where $q_\perp^2 = q_x^2 + q_y^2$. If the components (n_x, n_y) of the director fluctuation $\delta\mathbf{n}$ in the new coordinate system $\{\mathbf{e}_1, \mathbf{e}_2\}$ are denoted as (n_1, n_2), then

$$\delta\mathbf{n} = n_1 \mathbf{e}_1 + n_2 \mathbf{e}_2. \tag{3.56}$$

In this new coordinate system, the K_1 term in Eq. (3.54) corresponds to the inner product $\mathbf{q}\cdot\delta\mathbf{n}$. Thus, we have

$$|q_x n_x(\mathbf{q}) + q_y n_y(\mathbf{q})|^2 = |\mathbf{q}\cdot\delta\mathbf{n}|^2 = q_\perp^2 |n_1(\mathbf{q})|^2. \tag{3.57}$$

For the K_2 term in Eq. (3.54), it corresponds to the z-component of the cross product $\mathbf{q} \times \delta\mathbf{n}$: $(\mathbf{q} \times \delta\mathbf{n})_z$, we then obtain

$$|q_y n_x(\mathbf{q}) - q_x n_y(\mathbf{q})|^2 = |(\mathbf{q} \times \delta\mathbf{n})_z|^2 = q_\perp^2 |n_2(\mathbf{q})|^2. \tag{3.58}$$

For the K_3 term in Eq. (3.54), we have

$$q_z^2\left(|n_x(\mathbf{q})|^2 + |n_y(\mathbf{q})|^2\right) = q_z^2\left(|n_1(\mathbf{q})|^2 + |n_2(\mathbf{q})|^2\right), \tag{3.59}$$

and all cross terms involving $q_x q_y n_x n_y$ in Eq. (3.54) vanish.

Substituting these Eqs.(3.57)–(3.59) into Eq. (3.54), the free energy is given by

$$\begin{aligned} F &= V\sum_{\mathbf{q}} \frac{1}{2}\big((K_1 q_\perp^2 + K_3 q_z^2 + \Delta\epsilon E^2)|n_1(\mathbf{q})|^2 \\ &\quad +(K_2 q_\perp^2 + K_3 q_z^2 + \Delta\epsilon E^2)|n_2(\mathbf{q})|^2\big). \end{aligned} \tag{3.60}$$

Using Eq. (3.60), the probability density of the $\alpha(=1,2)$ component $n_\alpha(\mathbf{q})$ of the director fluctuation at a certain wave vector $\mathbf{q}$ is given by the Boltzmann factor as

$$p(n_\alpha(\mathbf{q})) = \frac{1}{c_1}\exp\left(-\frac{V}{2k_BT}(K_\alpha q_\perp^2 + K_3 q_z^2 + \Delta\epsilon E^2)|n_\alpha(\mathbf{q})|^2\right), \tag{3.61}$$

where c_1 is a constant. The mean square of the α component of the director fluctuation can be obtained using Gaussian integration as

$$\begin{aligned} \langle|n_\alpha(\mathbf{q})|^2\rangle &= \int |n_\alpha(\mathbf{q})|^2 p(n_\alpha(\mathbf{q}))dn_\alpha(\mathbf{q}) \Big/ \int p(n_\alpha(\mathbf{q}))dn_\alpha \\ &= \frac{k_BT}{V(K_\alpha q_\perp^2 + K_3 q_z^2 + \Delta\epsilon E^2)}. \end{aligned} \tag{3.62}$$

In the absence of an electric field ($E = 0$), the fluctuation spectrum $\langle|n_\alpha(\mathbf{q})|^2\rangle$ diverges in the limit $|\mathbf{q}| \to 0$. This divergence is analogous to the critical behavior observed in concentration fluctuations near a critical point in binary mixtures. In the nematic phase, large-scale fluctuations of the director occur, leading to a visually observable phenomenon: the nematic phase often appears as a milky or cloudy solution under optical microscopy. This phenomenon is reminiscent of critical opalescence and results from the scattering of light due to the long-wavelength (small $|\mathbf{q}|$) fluctuations of the director field. Thus, even though the system is not at a thermodynamic critical point, the nematic phase exhibits fluctuation-driven turbidity due to the softness of the orientational order.

By using the one constant approximation $K_{\rm el} = K_1 = K_2 = K_3$, Eq. (3.62) is given by

$$\langle|n_x(\mathbf{q})|^2\rangle = \langle|n_y(\mathbf{q})|^2\rangle = \frac{k_BT}{VK_{el}(q^2 + \xi^{-2})}, \tag{3.63}$$

where ξ is the correlation length, given by

$$\xi = \frac{1}{E}\sqrt{\frac{K_{el}}{\Delta\epsilon}}. \tag{3.64}$$

As the electric field increases, the orientational fluctuations $\langle|n_k(\mathbf{q})|^2\rangle$ approach zero, which stabilises the ground state. This is how liquid crystal displays achieve clear images by applying an electric field to reduce the director fluctuations to zero.

The real-space correlation function of the orientational fluctuations is obtained by the Fourier inverse transform of Eq. (3.63) as

$$\begin{aligned}\langle|n_k(\mathbf{r})|^2\rangle &= \frac{V}{8\pi^3}\int\langle|n_x(\mathbf{q})|^2\rangle\exp(-i\mathbf{q}\cdot\mathbf{r})d\mathbf{q}\\ &= \frac{k_BT}{4\pi K_{el}}\left(\frac{e^{-r/\xi}}{r}\right).\end{aligned} \tag{3.65}$$

In the absence of an electric field: $\xi \to \infty$, the spatial fluctuations of the director exhibit long-range correlations of the form r^{-1}, similar to those at a critical point in binary mixtures, where the spatial fluctuations of the concentration exhibit long-range correlations [41].

4

Nemattic, Smectic A, and Columnar Liquid Crystals

In this chapter, we introduce two foundational theoretical approaches to understanding the stability of the nematic phase: the Onsager model [42], which is based on repulsive interactions (excluded volume effects), and the Maier-Saupe model [43], which incorporates attractive interactions between molecules. These models offer complementary perspectives on how orientational order emerges in systems of rod-like molecules. Furthermore, the mean field theories of biaxial nematic, smectic A, and columnar phases are introduced.

4.1 Onsager model for solutions of rod-like molecules

We begin by formulating the general free energy within the framework of the second virial approximation. Subsequently, we incorporate the orientational degrees of freedom of liquid crystalline molecules to derive a free energy expression that accounts for anisotropic interactions.

4.1.1 Free energy of solutions of rod-like molecules

Let us consider a solution containing N rod-like molecules dispersed in solvent molecules, confined within a system of volume V at temperature T. As illustrated in Fig. 4.1, the state of a rod-like molecule is characterized by its center-of-mass position vector $\mathbf{r}$ and its orientation vector $\boldsymbol{\Omega}$. We denote the number of molecules located at position $\mathbf{r}$ with orientation $\boldsymbol{\Omega}$ by $N(\mathbf{r}, \boldsymbol{\Omega})$. The number density of the molecules is then given by $\rho(\mathbf{r}, \boldsymbol{\Omega}) = N(\mathbf{r}, \boldsymbol{\Omega})/V$. The number density of molecules within the local region $d\mathbf{r}d\boldsymbol{\Omega}$, defined by position $\mathbf{r}$ to $\mathbf{r} + d\mathbf{r}$ and orientation $\boldsymbol{\Omega}$ to $\boldsymbol{\Omega} + d\boldsymbol{\Omega}$, is given by $\rho(\mathbf{r}, \boldsymbol{\Omega})d\mathbf{r}d\boldsymbol{\Omega}$. Therefore, integrating the entire system ensures the conservation of the total number of molecules, leading to the normalization condition:

$$\iint \rho(\mathbf{r}, \boldsymbol{\Omega}) d\mathbf{r} d\boldsymbol{\Omega} = N, \tag{4.1}$$

where $d\boldsymbol{\Omega}(= \sin\theta d\theta d\varphi)$ is the solid angle with $0 \le \theta \le \pi$ and $0 \le \varphi \le 2\pi$. The system's total volume is given by $V = \int d\mathbf{r}$.

The free energy F within the framework of the second virial approximation is given by

$$\begin{aligned} \beta F &= \iint \rho(\mathbf{r}, \boldsymbol{\Omega}) \left(\beta\mu_r^\circ - 1 + \ln \rho(\mathbf{r}, \boldsymbol{\Omega})\right) d\mathbf{r} d\boldsymbol{\Omega} \\ &\quad -\frac{1}{2} \iiiint \rho(\mathbf{r}, \boldsymbol{\Omega}) \rho(\mathbf{r}', \boldsymbol{\Omega}') \Phi_{12}(r_{12}; \boldsymbol{\Omega}, \boldsymbol{\Omega}') d\mathbf{r} d\boldsymbol{\Omega} d\mathbf{r}' d\boldsymbol{\Omega}', \end{aligned} \tag{4.2}$$

DOI: 10.1201/9781003737711-4

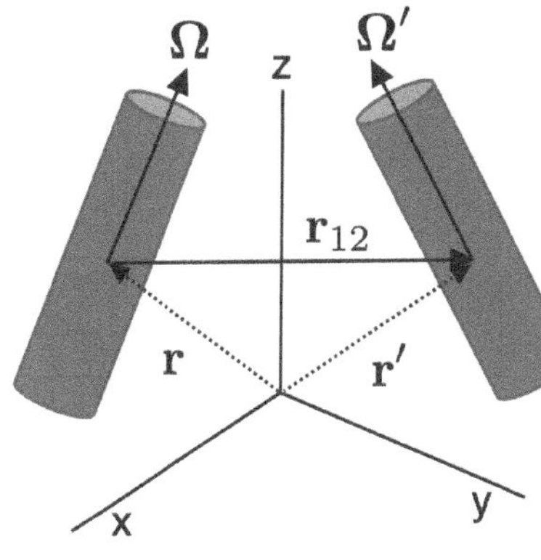

FIGURE 4.1
Orientation vector $\mathbf{\Omega}$ and the center-of-mass position vector $\mathbf{r}$ of a rod-like molecule with $\mathbf{r}_{12} = \mathbf{r}' - \mathbf{r}$.

where $\beta = (k_B T)^{-1}$, k_B is the Boltzmann constant, μ_r° is the standard chemical potential of a rod-like molecule, and

$$\Phi_{12}(r_{12}, \mathbf{\Omega}, \mathbf{\Omega}') = \exp\left(- \beta U_{12}(r_{12}; \mathbf{\Omega}, \mathbf{\Omega}')\right) - 1, \tag{4.3}$$

is the Mayer function, and U_{12} is the intermolecular interactions between two rod-like molecules in the configurations $(\mathbf{r}, \mathbf{\Omega})$ and $(\mathbf{r}', \mathbf{\Omega}')$. Using the distribution function $f_d(\mathbf{r}, \mathbf{\Omega})$ of the molecules, the number density can be expressed as

$$\rho(\mathbf{r}, \mathbf{\Omega}) = \rho f_d(\mathbf{r}, \mathbf{\Omega}), \tag{4.4}$$

where $\rho = N/V$ is the average number density of rods. The total number N of the rod-like molecules must be conserved, and then we have the normalization condition:

$$\frac{1}{V} \iint f_d(\mathbf{r}, \mathbf{\Omega}) d\mathbf{r} d\mathbf{\Omega} = 1. \tag{4.5}$$

Substituting Eq. (4.4) into Eq. (4.2), we obtain

$$\begin{aligned} \beta F &= N\left(\beta \mu_r^\circ + \ln \rho - 1\right) \\ &+ \rho \iint f_d(\mathbf{r}, \mathbf{\Omega}) \ln f_d(\mathbf{r}, \mathbf{\Omega}) d\mathbf{r} d\mathbf{\Omega} \\ &- \frac{1}{2}\rho^2 \iiiint f_d(\mathbf{r}, \mathbf{\Omega}) f_d(\mathbf{r}', \mathbf{\Omega}') \Phi_{12}(r_{12}; \mathbf{\Omega}, \mathbf{\Omega}') d\mathbf{r} d\mathbf{\Omega} d\mathbf{r}' d\mathbf{\Omega}'. \end{aligned} \tag{4.6}$$

The first term represents the translational entropy of the rod-like molecules and corresponds to the free energy of an ideal gas (isotropic phase). The first integral term shows the entropy change due to ordering. In the isotropic phase, the distribution function is constant $1/(4\pi)$ (Eq. (2.7)), so this term becomes a constant (or zero). In an ordered phase, the distribution function $f_d(\mathbf{r}, \mathbf{\Omega})$ exhibits spatial and orientational dependence on $\mathbf{r}$ and $\mathbf{\Omega}$, and then, this term does not vanish. The last term shows the interaction between rod-like molecules. Equation (4.6) represents the expression for the free energy under the second virial approximation, applicable to any ordered phase (e.g., liquid crystal phases or crystal phases).

4.1.2 Definition of the distribution function

Let us consider the following two distribution functions:

(1) The distribution function is uniform (random) with respect to the center-of-mass position of the molecules but non-uniform with respect to their orientation.

(2) The distribution function is non-uniform with respect to the center-of-mass position of the molecules but uniform with respect to their orientation.

4.1.2.1 Case (1)

Since the molecular distribution function is only a function of orientation, $f_d(\mathbf{r}, \mathbf{\Omega}) = f_d(\mathbf{\Omega})$, we can substitute this into the second term of Eq. (4.6). The second term now becomes

$$\begin{aligned}\text{second term} &= \rho \int d\mathbf{r} \int f_d(\mathbf{\Omega}) \ln \left[4\pi f_d(\mathbf{\Omega})\right] d\mathbf{\Omega} \\ &= N \int f_d(\mathbf{\Omega}) \ln \left[4\pi f_d(\mathbf{\Omega})\right] d\mathbf{\Omega}. \end{aligned} \tag{4.7}$$

In the isotropic phase, $f_d(\mathbf{\Omega}) = 1/(4\pi)$, so the second term in the free energy becomes zero. By changing the integration variable from $d\mathbf{r}'$ to $dr_{12} = d|\mathbf{r}' - \mathbf{r}|$, the third term of Eq. (4.6) is given by

$$\begin{aligned}\text{third term} &= -\frac{1}{2}\rho^2 \iiint f_d(\mathbf{\Omega}) f_d(\mathbf{\Omega}')\Phi_{12}(r_{12}; \mathbf{\Omega}, \mathbf{\Omega}') dr_{12} d\mathbf{\Omega} d\mathbf{\Omega}' \int d\mathbf{r}, \\ &= \frac{N^2}{2V} \iint f_d(\mathbf{\Omega}) f_d(\mathbf{\Omega}') A_2(\mathbf{\Omega}, \mathbf{\Omega}') d\mathbf{\Omega} d\mathbf{\Omega}', \end{aligned} \tag{4.8}$$

where

$$A_2(\mathbf{\Omega}, \mathbf{\Omega}') = -\int \Phi_{12}(r_{12}; \mathbf{\Omega}, \mathbf{\Omega}') dr_{12}, \tag{4.9}$$

is the intermolecular interaction, which depends on the orientations of rod-like molecules. Then the free energy is given as a function of $f_d(\mathbf{\Omega})$:

$$\begin{aligned}\beta F/V &= \rho(\beta\mu_r^\circ + \ln\rho - 1) \\ &\quad + \rho \int f_d(\mathbf{\Omega}) \ln \left[4\pi f_d(\mathbf{\Omega})\right] d\mathbf{\Omega} \\ &\quad + \frac{1}{2}\rho^2 \iint f_d(\mathbf{\Omega}) f_d(\mathbf{\Omega}') A_2(\mathbf{\Omega}, \mathbf{\Omega}') d\mathbf{\Omega} d\mathbf{\Omega}', \end{aligned} \tag{4.10}$$

and the normalization condition (Eq. (4.5)) is given by

$$\int f_d(\mathbf{\Omega}) d\mathbf{\Omega} = 1. \tag{4.11}$$

Equation (4.10) serves as the starting point for the Onsager model of the nematic phase [44].

4.1.2.2 Case (2)

In cases such as crystal phases, where the molecular distribution function depends only on position (e.g., due to periodicity), the distribution function $f_d(\mathbf{r}, \mathbf{\Omega})$ can be written as $f_d(\mathbf{r}, \mathbf{\Omega}) = f_d(\mathbf{r})$. Substituting this into the second term of Eq. (4.6), we obtain

$$\begin{aligned}\text{second term} &= \rho \int d\mathbf{\Omega} \int f_d(\mathbf{r}) \ln f_d(\mathbf{r}) d\mathbf{r} \\ &= N \int f_d(\tilde{\mathbf{r}}) \ln f_d(\tilde{\mathbf{r}}) d\tilde{\mathbf{r}}, \end{aligned} \tag{4.12}$$

where $\tilde{\mathbf{r}} = (4\pi/V)d\mathbf{r}$ and $\int d\boldsymbol{\Omega} = 4\pi$. The normalization condition (Eq. (4.5)) is given by

$$\int f_d(\tilde{\mathbf{r}})d\tilde{\mathbf{r}} = 1. \tag{4.13}$$

Using the relative orientation vector $\boldsymbol{\Omega}_{12} = \boldsymbol{\Omega}' - \boldsymbol{\Omega}$, the third term of Eq. (4.6) is given by

$$\begin{aligned} \text{third term} &= -\frac{1}{2}\rho^2 \iiint f_d(\mathbf{r})f_d(\mathbf{r}')\Phi_{12}(\mathbf{r},\mathbf{r}';\boldsymbol{\Omega}_{12})d\mathbf{r}d\mathbf{r}'d\boldsymbol{\Omega}_{12}\int d\boldsymbol{\Omega}, \\ &= \frac{N^2}{2V}\iint f_d(\tilde{\mathbf{r}})f_d(\tilde{\mathbf{r}}')A_2(\tilde{\mathbf{r}},\tilde{\mathbf{r}}')d\tilde{\mathbf{r}}d\tilde{\mathbf{r}}', \end{aligned} \tag{4.14}$$

where

$$A_2(\tilde{\mathbf{r}},\tilde{\mathbf{r}}') = -\left(\frac{V}{4\pi}\right)\int \Phi_{12}(\tilde{\mathbf{r}},\tilde{\mathbf{r}}';\boldsymbol{\Omega}_{12})d\boldsymbol{\Omega}_{12}. \tag{4.15}$$

We then obtain the free energy:

$$\begin{aligned} \beta F/V &= \rho(\beta\mu_r^\circ + \ln\rho - 1) \\ &\quad +\rho\int f_d(\tilde{\mathbf{r}})\ln f_d(\tilde{\mathbf{r}})d\tilde{\mathbf{r}} \\ &\quad +\frac{1}{2}\rho^2\iint f_d(\tilde{\mathbf{r}})f_d(\tilde{\mathbf{r}}')A_2(\tilde{\mathbf{r}},\tilde{\mathbf{r}}')d\tilde{\mathbf{r}}d\tilde{\mathbf{r}}'. \end{aligned} \tag{4.16}$$

This represents the starting point for discussing the free energy of crystal phases [45].

4.1.3 Excluded volume between rod-like molecules

In the following, we introduce the Onsager model to study the nematic phase. Under the normalization condition of Eq. (4.11), the distribution function that minimizes the free energy (Eq. (4.10)) can be found by using the method of Lagrange multipliers. The Lagrange multiplier λ_L is introduced to enforce the normalization condition on the distribution function:

$$\frac{\delta}{\delta f_d(\boldsymbol{\Omega})}\left[F\big(f_d(\boldsymbol{\Omega})\big) - \lambda_L\int f_d(\boldsymbol{\Omega})d\boldsymbol{\Omega}\right] = 0. \tag{4.17}$$

We then obtain

$$\rho(\ln f_d(\boldsymbol{\Omega}) + 1) + \rho^2\int f_d(\boldsymbol{\Omega}')A_2(\boldsymbol{\Omega},\boldsymbol{\Omega}')d\boldsymbol{\Omega}' - \lambda_L = 0. \tag{4.18}$$

After moving all constant terms to the right-hand side, the condition for the free energy to be minimized is given by

$$\ln f_d(\boldsymbol{\Omega}) + \rho\int f_d(\boldsymbol{\Omega}')A_2(\boldsymbol{\Omega},\boldsymbol{\Omega}')d\boldsymbol{\Omega}' = \lambda_L. \tag{4.19}$$

The distribution function in the thermal equilibrium state, in the presence of an arbitrary $A_2(\boldsymbol{\Omega},\boldsymbol{\Omega}\prime)$, satisfies the following condition:

$$f_d(\boldsymbol{\Omega}) = \frac{1}{Z_0}\exp\left[-\rho\int f_d(\boldsymbol{\Omega}')A_2(\boldsymbol{\Omega},\boldsymbol{\Omega}')d\boldsymbol{\Omega}'\right], \tag{4.20}$$

where Z_0 is determined from the normalization condition (Eq. (4.11)). As discussed in Eq. (2.10), the orientational order parameter for the nematic phase is given by

$$S = \int P_2(\cos\theta)f_d(\boldsymbol{\Omega})d\boldsymbol{\Omega}. \tag{4.21}$$

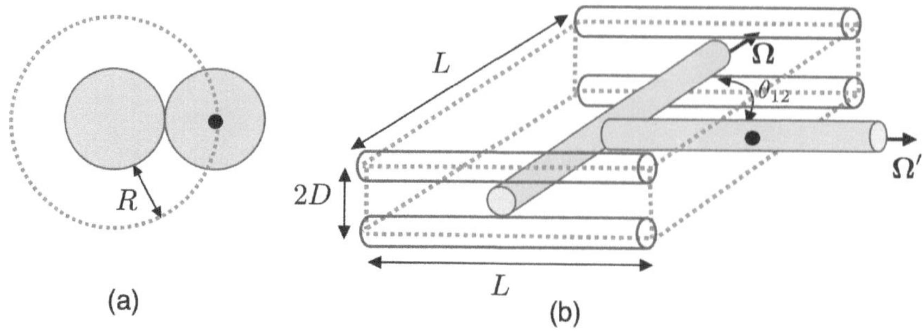

FIGURE 4.2
Excluded volume of spherical molecules (a) and rod-like molecules (b).

In the Onsager model, the intermolecular interaction term $A_2(\mathbf{\Omega}, \mathbf{\Omega'})$ considers the excluded volume between two rod-like molecules. This model is often referred to as a repulsive interaction model.

Consider the excluded volume between two spherical molecules as shown in Fig. 4.2(a). When one spherical molecule is fixed, and the other is rotated around it while keeping in contact, a region inside the dashed line emerges where the center of mass of the other molecules cannot enter. This region is referred to as the excluded volume. For a spherical particle with radius R, the excluded volume is given by $v_{exc} = 4\pi(2R)^3/3$. For a rod-like molecule with length L and diameter D, a region inside the dashed line shows the excluded volume as shown in Fig 4.2(b). In the context of Eq. (4.3), we consider the excluded volume for the intermolecular interaction $U_{12}(r)$. Considering the two contact rod-like molecules with the angle $\theta_{12}(\mathbf{\Omega}, \mathbf{\Omega}')$ between their orientations fixed, it can be expressed as $U_{12}(r) = \infty$ when the molecules are in contact, and $U_{12}(r) = 0$ when the molecules are in non-contact. The Mayer function $\Phi_{12}(\mathbf{r}, \mathbf{\Omega}, \mathbf{\Omega}')$ can be written as

$$\Phi_{12}(r_{12}, \mathbf{\Omega}, \mathbf{\Omega}') = \begin{cases} -1 & \text{(contact)} \\ 0 & \text{(non-contact)} \end{cases} \tag{4.22}$$

and then the excluded volume interaction is given by

$$\begin{aligned} A_2(\mathbf{\Omega}, \mathbf{\Omega}') &= -\int \Phi_{12}(r_{12}; \mathbf{\Omega}, \mathbf{\Omega}')dr_{12} \\ &\simeq 2L^2D|\mathbf{\Omega} \times \mathbf{\Omega}'| = 2L^2D|\sin\theta_{12}|. \end{aligned} \tag{4.23}$$

We here neglect the effects at the ends of the molecules for sufficiently long rods ($L \gg D$). The excluded volume decreases as the two molecules become more parallel, and the contribution from the A_2 term in Eq. (4.10) becomes smaller, reducing the free energy.

Furthermore, $|\sin\theta_{12}|$ can be expanded using the Legendre polynomials $P_{2n}(\cos\theta)$ as follows:

$$|\sin\theta_{12}| = \frac{\pi}{4} - \sum_{n=1}^{\infty} B_{2n} P_{2n}(\cos\theta) P_{2n}(\cos\theta'), \tag{4.24}$$

where

$$B_{2n} = \frac{\pi(4n+1)(2n-3)!!(2n-1)!!}{2^{2n+2}n!(n+1)!}, \tag{4.25}$$

and the angle θ (θ') represents the angle between the director $\mathbf{n}$ and the molecular orientation vector $\mathbf{\Omega}$ ($\mathbf{\Omega}'$). Expanding and approximating Eq. (4.24) at $n = 1$, we obtain

$$|\sin\theta_{12}| = \frac{\pi}{4} - \frac{5\pi}{32}P_2(\cos\theta)P_2(\cos\theta'), \tag{4.26}$$

where the orientational order parameter $P_2(\cos\theta)$ appears. Substituting this into Eq. (4.23), we obtain

$$A_2(\mathbf{\Omega}, \mathbf{\Omega}') = L^2D\left(\frac{\pi}{2} - \frac{5\pi}{16}P_2(\cos\theta)P_2(\cos\theta')\right). \tag{4.27}$$

Substituting Eq. (4.27) into Eq. (4.20) and incorporating all terms independent of the angle θ into the constant Z_0, the orientational distribution function is given by:

$$f_d(\cos\theta) = \frac{1}{Z_0}\exp\left[\Gamma_0 S P_2(\cos\theta)\right], \tag{4.28}$$

where

$$\Gamma_0 = \frac{5L}{4D}\phi, \tag{4.29}$$

is referred to as the "nematic factor." A larger value of the parameter indicates a stronger tendency for the system to form a nematic phase. The quantity $\phi = v_r\rho$ represents the volume fraction (i.e., concentration) of the rod-like molecules, where $v_r = (\pi/4)D^2L$ is the volume of a single rod-like molecule. In the original Onsager theory, the orientational distribution function was assumed to follow a Gaussian distribution with respect to the orientational angle, and its form was determined by minimising the free energy [42]. In this text, we instead expand Eq. (4.26) in terms of Legendre polynomials, which makes the solution analytically tractable. Using Eq. (4.11), the normalization constant is given by

$$\begin{aligned} Z_0 &= 2\pi\int_0^\pi \exp\left[\Gamma_0 S P_2(\cos\theta)\right]\sin\theta d\theta \\ &= 4\pi\int_0^1 \exp\left[\Gamma_0 S P_2(x)\right]dx, \end{aligned} \tag{4.30}$$

where $x = \cos\theta$. Substituting Eq. (4.28) into Eq. (4.21), the order parameter is determined by the self-consistent equation:

$$S = \frac{I_1(\Gamma_0 S)}{I_0(\Gamma_0 S)}, \tag{4.31}$$

where we define

$$I_m(\Gamma_0 S) = \int_0^1 [P_2(x)]^m \exp\left[\Gamma_0 S\left(\frac{3}{2}x^2 - \frac{1}{2}\right)\right]dx, \tag{4.32}$$

$m = 0, 1, 2, \cdots$ and $Z_0 = 4\pi I_0(\Gamma_0 S)$. Since Γ_0 is a function of the concentration ϕ, the order parameter S also becomes a function of ϕ.

Figure 4.3 shows the dependence of the orientational distribution function on $x (= \cos\theta)$, as described by Eq. (4.28). The parameter Γ_0 is fixed at 4.6, while the orientational order parameter is varied as S = 0, 0.3, and 0.6. The distribution function is symmetric with respect to $x = 0$. When $S = 0$, the system is in the isotropic phase, and $f_d(x)$ becomes constant: $f_d(x) = 1/(4\pi)$. As S increases, $f_d(x)$ becomes sharply peaked near $x = \pm 1$, indicating that a larger fraction of the liquid crystalline molecules align parallel to the director.

In the Onsager theory, only the excluded volume is considered. Thus, the free energy shown in Eq. (4.10) can be expressed as a contribution purely from entropy S_e: $F = -TS_e$. Furthermore, the entropy contribution can be divided into two components:

$$S_e = S_{or} + S_{pc}, \tag{4.33}$$

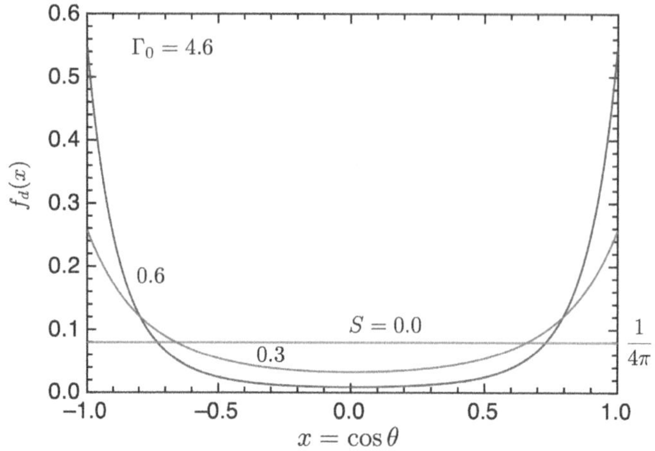

FIGURE 4.3
Orientational distribution function of rod-like molecules plotted aginst $x(=\cos\theta)$ for various values of the order parameter S.

where the first component, S_{or}, represents the entropy change associated with orientational ordering and corresponds to the second term in Eq. (4.10). By substituting the orientational distribution function given in Eq. (4.28) into this entropy term, we obtain

$$\begin{aligned} S_{or}/(k_B\rho) &= -\int f_d(\mathbf{\Omega}) \ln \left[4\pi f_d(\mathbf{\Omega})\right] d\mathbf{\Omega} \\ &= -\Gamma_0 S^2 + \ln I_0(\Gamma_0 S). \end{aligned} \tag{4.34}$$

The second component, S_{pc}, corresponds to the entropy change associated with the excluded volume effect, often referred to as the packing entropy. This contribution is represented by the third term in Eq. (4.10). By substituting Eq. (4.27) into this term, we obtain

$$\begin{aligned} S_{pc}/(k_B\rho) &= -\frac{1}{2}\rho \iint f_d(\mathbf{\Omega}) f_d(\mathbf{\Omega}') A_2(\mathbf{\Omega}, \mathbf{\Omega}') d\mathbf{\Omega} d\mathbf{\Omega}'. \\ &= -\frac{4}{5}\Gamma_0\big(1 - \frac{5}{8}S^2\big). \end{aligned} \tag{4.35}$$

Thus, the free energy $F(= F_{iso} + F_{nem})$ given in Eq. (4.10) can be expressed as the sum of the isotropic free energy $F_{\rm iso}$ and the nematic free energy $F_{\rm nem}$ as follows:

$$f_{LC} = v_r \beta F/V \quad = \quad f_{iso} + f_{nem}, \tag{4.36}$$

$$f_{iso} = v_r \beta F_{iso}/V = \phi\big(\beta\mu_r^\circ + \ln\phi - 1 + \frac{L}{D}\phi\big), \tag{4.37}$$

$$f_{nem} = v_r \beta F_{nem}/V = \phi\big(\frac{1}{2}\Gamma_0 S^2 - \ln I_0(\Gamma_0 S)\big), \tag{4.38}$$

where the term involving $\ln v_r$ is constant and has been absorbed into the reference chemical potential μ_r°. When $S = 0$, corresponding to the isotropic phase, we recover $f_{\rm nem} = 0$.

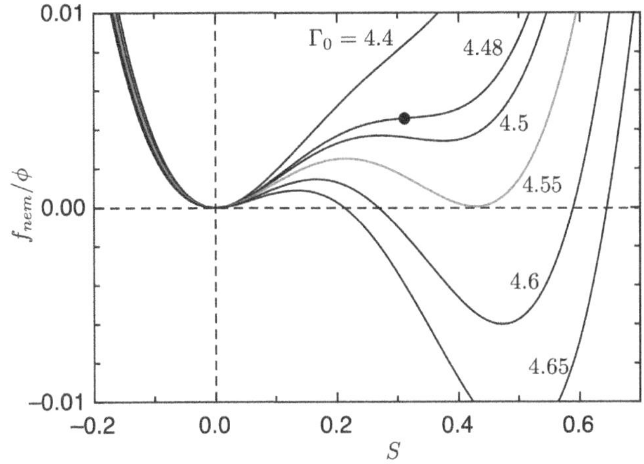

FIGURE 4.4
Nematic free energy (Eq. (4.38)) plotted against the order parameter S for various values of Γ_0. The first-order isotropic–nematic phase transition occurs at $\Gamma_0 = 4.55$.

4.1.4 Entropy-driven nematic ordering

Figure 4.4 shows the dependence of the nematic free energy per unit volume fraction, f_{nem}/ϕ, on the orientational order parameter S, for different values of the nematic factor Γ_0. As previously discussed, Γ_0 is a function of the concentration ϕ; thus, smaller values of Γ_0 correspond to lower concentrations. From this figure, one can observe how variations in Γ_0 affect the shape of the free energy landscape. As Γ_0 increases, the free energy develops a local minimum at finite S, indicating the emergence of a stable nematic phase. This behavior is consistent with the framework of Landau theory for first-order phase transitions.

(1) For $\Gamma_0 < \Gamma_0^c$ ($= 4.48$: critical point), the free energy f_{nem} has a single minimum at $S = 0$. Therefore, the isotropic phase (I) is stable in thermal equilibrium. At $\Gamma_0^c = 4.48$, the critical point (indicated by a black circle) emerges.

(2) For $\Gamma_0^c < \Gamma_0 < \Gamma_0^{NI}(= 4.55)$, f_{nem} has two minima: one at $S = 0$ and another at $S \simeq 0.4$. However, since $f_{nem}(S = 0) < f_{nem}(S \simeq 0.4)$, the isotropic phase ($S = 0$) remains thermodynamically stable, while the minimum at $S \simeq 0.4$ corresponds to a metastable nematic phase.

(3) When $\Gamma_0 = \Gamma_0^{NI}$, $f_{nem}(S = 0) = f_{nem}(S_{NI} = 0.43) = 0$. At this point, the order parameter exhibits a discontinuous jump from 0 to 0.43, indicating a first-order nematic-isotropic phase transition (NIT).

(4) For $\Gamma_0 > \Gamma_0^{NI}$, the minimum of the free energy always satisfies $f_{nem}(S = 0) > f_{nem}(S \neq 0)$. Thus, the nematic phase (N) becomes thermodynamically stable.

Figure 4.5 shows the value of the orientational order parameter (S) as a function of $\Gamma_0(= 5L\phi/(4D))$. At $\Gamma_0 = \Gamma_0^{NI}$, the NIT occurs, where the value of S jumps from 0 to 0.43. Between Γ_0^c and Γ_0^{NI}, a metastable nematic phase appears. As the value of Γ_0 increases, the value of the orientational order parameter also increases. Using $\Gamma_0^{NI} = 4.55$ and Eq. (4.29), the NIT concentration is given by

$$\phi_{NI} = 3.63D/L. \tag{4.39}$$

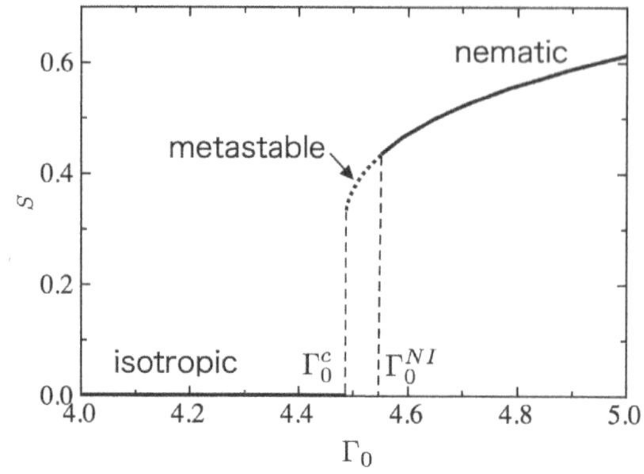

FIGURE 4.5
Orientational order parameter S plotted against Γ_0. When $\Gamma_0 < \Gamma_0^{NI}$, the isotropic phase appears, while when $\Gamma_0 > \Gamma_0^{NI}$, the nematic phase appears. The region where $\Gamma_0^c < \Gamma_0 < \Gamma_0^{NI}$ indicates the metastable region.

At concentrations above ϕ_{NI}, the excluded volume interactions between rod-like molecules promote parallel alignment, resulting in the emergence of the nematic phase.

The Onsager theory plays a central role in explaining why rod-like molecules spontaneously align at high concentrations. As shown in Eq. (4.33), the free energy of the nematic phase arises purely from entropic contributions, and is expressed as $F = -TS_e$. Thus, the thermal equilibrium state corresponds to the configuration that minimizes the free energy, or equivalently, maximizes the entropy. In this framework, the orientational ordering of molecules is not driven by energetic interactions, but rather by the system's tendency to increase translational entropy by reducing excluded volume through alignment.

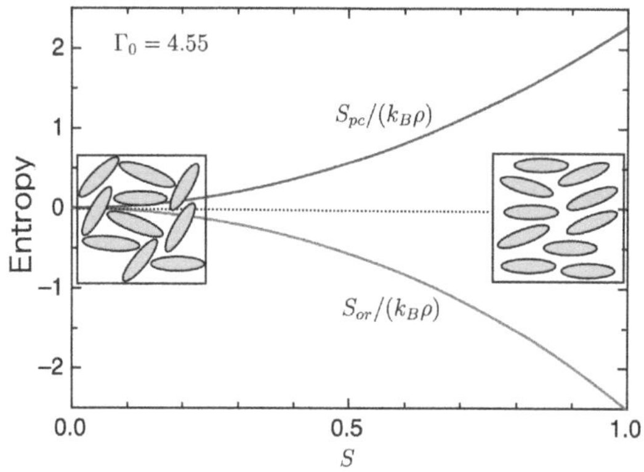

FIGURE 4.6
Entropy terms S_{or}(Eq. (4.34)) and S_{pc} (Eq. (4.35)) plotted against the order parameter S. The nematic free energy is given by $F_{nem} = -T(S_{or} + S_{pc})$.

Figure 4.6 shows the two entropy components, S_{or} (Eq. (4.34)) and S_{pc} (Eq. (4.35)), plotted as functions of the orientational order parameter S for $\Gamma_0 = 4.55$, which is just above the NIT point. Since the first term in Eq. (4.35) is constant, it has been subtracted so that the entropy is zero at $S = 0$, corresponding to the isotropic phase. In this phase, both S_{or}

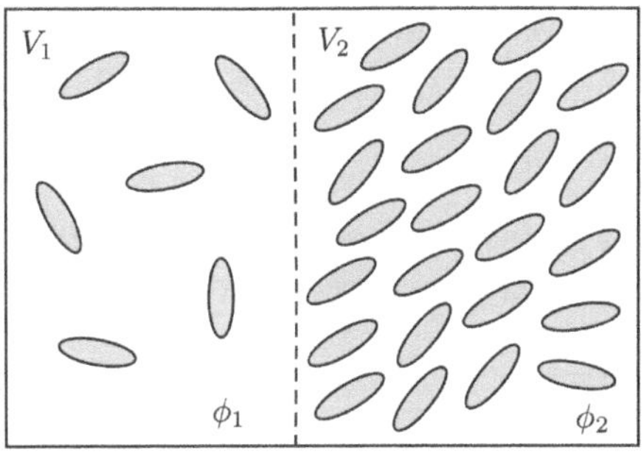

FIGURE 4.7
Phase separation between isotropic and nematic phases.

and S_{pc} are zero. As the value of S increases, the entropy due to orientational ordering, S_{or}, decreases (i.e., becomes more negative), reflecting the loss of rotational degrees of freedom as the molecules align. In contrast, the packing entropy S_{pc}, which accounts for the excluded volume effect, increases with S, as molecular alignment reduces mutual obstruction and enhances translational freedom. This gain in translational entropy lowers the total free energy of the system, making the nematic phase thermodynamically favorable above ϕ_{NI}. As illustrated in the inset of the figure, at the same concentration, molecules in the isotropic phase are randomly oriented but closely packed, limiting their translational motion. In the nematic phase, alignment reduces orientational entropy ($S_{or} < 0$) but simultaneously increases translational entropy ($S_{pc} > 0$) by alleviating crowding through orientational order. This entropic trade-off underlies the emergence of the nematic phase in rod-like molecules. Such entropy-driven phase transitions are also fundamental in other systems, such as the crystallization of colloidal particles [46].

The essence of the Onsager theory reveals itself in everyday scenarios, for example, how cars arrange themselves in a crowded parking lot or how people position themselves in a packed train. In both cases, some alignment or ordering is accepted ($S_{or} < 0$) in exchange for greater mobility and space efficiency ($S_{pc} > 0$). The system sacrifices orientational freedom to maximize translational freedom, as rod-like molecules do in forming the nematic phase.

4.1.5 Nematic-isotropic phase separation

Here, we consider phase separation in a solution of rod-like molecules based on the free energy (Eq. (4.36)). Suppose the system separates into two distinct phases, as illustrated in Fig. 4.7, where phase 1 occupies a volume V_1 with concentration ϕ_1, and phase 2 occupies a volume V_2 with concentration ϕ_2. Since the total volume of the system is conserved, the following condition holds:

$$V = V_1 + V_2. \tag{4.40}$$

Additionally, conservation of the total number of particles gives

$$\phi V = \phi_1 V_1 + \phi_2 V_2, \tag{4.41}$$

where $\phi V = v_r \rho V = v_r N$, with N being the total number of rod-like molecules in the system. The right-hand side of Eq. (4.41) represents the number of molecules in each phase, N_1 and N_2, respectively. Thus, we have $N = N_1 + N_2$. The total free energy of the two-phase system, $f_p(\phi)$, is given by the sum of the free energies of each phase:

$$f_p(\phi) = \big(V_1 f_{LC}(\phi_1) + V_2 f_{LC}(\phi_2)\big)/V. \tag{4.42}$$

Using Eqs. (4.40) and (4.41), we can rewrite this expression as:

$$f_p(\phi) = \frac{\phi_2 - \phi}{\phi_2 - \phi_1} f_{LC}(\phi_1) + \frac{\phi - \phi_1}{\phi_2 - \phi_1} f_{LC}(\phi_2), \tag{4.43}$$

which represents the free energy of the phase-separated system in terms of the free energies of the two phases and their respective concentrations.

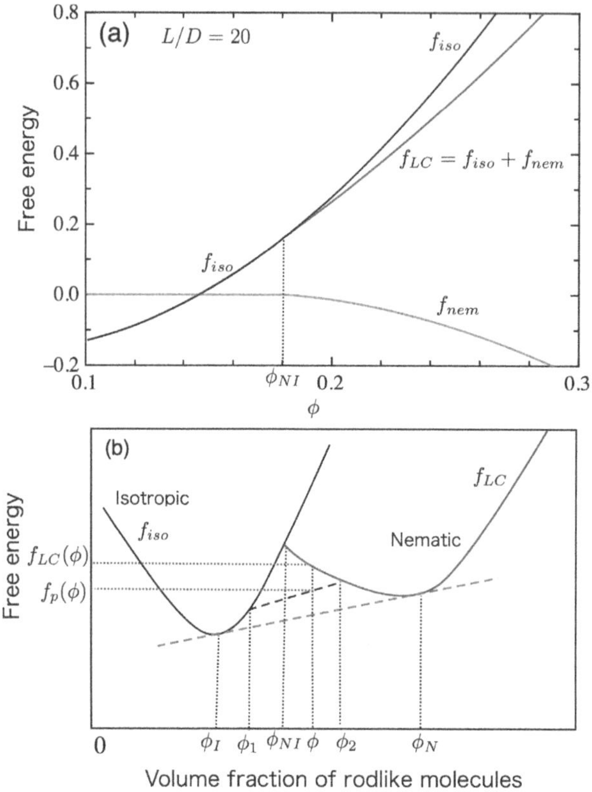

FIGURE 4.8
(a) The total free energy $f_{LC} = f_{iso} + f_{nem}$ is plotted as a function of the concentration ϕ. For $\phi > \phi_{NI}$, the contribution from the nematic free energy f_{nem} lowers the total free energy of the system, resulting in $f_{LC} < f_{iso}$. At $\phi = \phi_{NI}$, the free energy exhibits a discontinuous kink, indicating a first-order NIT. (b) The behavior of free energy near the NIT Concentration ϕ_{NI}. When the solution with concentration ϕ undergoes phase separation into two phases with concentrations ϕ_1 and ϕ_2, the free energy decreases from $f_{LC}(\phi)$ to $f_p(\phi)$. The free energy decreases as ϕ_1 and ϕ_2 approach ϕ_I and ϕ_N, respectively. At the concentrations ϕ_I and ϕ_N, where the common tangent can be drawn to the free energy curve, the free energy reaches its minimum, and phase separation occurs.

Figure 4.8(a) plots the free energy $f_{LC} = f_{iso} + f_{nem}$ (Eq. (4.36)) as a function of concentration for $L/D = 20$. Figure 4.8(b) shows an enlarged view of the free energy near the NIT concentration ϕ_{NI}. For $\phi > \phi_{NI}$, the free energy of the nematic phase f_{nem} becomes negative, and the total free energy of the system f_{LC} is lower than the free energy of the isotropic phase f_{iso}. The free energy curve discontinuously changes from f_{iso} to f_{LC} at ϕ_{NI}, indicating a first-order NIT. The free energy decreases as the molecules align. Figure 4.8(b)

shows the behavior of the free energy near the NIT concentration ϕ_{NI}. When a homogeneous solution of concentration ϕ undergoes phase separation into two coexisting phases with concentrations ϕ_1 and ϕ_2, the system's free energy decreases from $f_{LC}(\phi)$ to $f_p(\phi)$, the latter representing the free energy of the phase-separated state. The total free energy continues to decrease as ϕ_1 and ϕ_2 approach the coexistence concentrations ϕ_I (for the isotropic phase) and ϕ_N (for the nematic phase), respectively. At these concentrations, a common tangent can be drawn to the free energy curve, indicating the equilibrium coexistence of the two phases. This construction minimizes the system's free energy and defines the binodal line of the phase diagram. The phase separation concentrations ϕ_I and ϕ_N are determined by the condition where the slopes of the tangent lines to the free energy curves are equal (a common tangent method):

$$\left(\frac{df_{LC}}{d\phi}\right)_{\phi=\phi_I} = \left(\frac{df_{LC}}{d\phi}\right)_{\phi=\phi_N}. \tag{4.44}$$

The concentrations of the two coexisting phases in thermal equilibrium can be determined by equating the chemical potential and osmotic pressure in each phase. The osmotic pressure of the solution is given by

$$\Pi = -\left(\frac{\partial F}{\partial V}\right)_N, \tag{4.45}$$

and the chemical potential of the rod-like molecules:

$$\mu_r = \left(\frac{\partial F}{\partial N}\right)_V. \tag{4.46}$$

Using Eq. (4.36) and the volume fraction $\phi = v_r N/V$, we can rewrite the expressions for osmotic pressure and chemical potential as follows:

$$\begin{aligned} -\Pi = \left(\frac{\partial F}{\partial V}\right)_N &= \frac{1}{v_r\beta}\left(\frac{\partial V f_{LC}}{\partial V}\right)_N \\ &= \frac{1}{v_r\beta}\left(f_{LC} - \phi\left(\frac{\partial f_{LC}}{\partial \phi}\right)_T\right), \end{aligned} \tag{4.47}$$

$$\begin{aligned} \mu_r = \left(\frac{\partial F}{\partial N}\right)_V &= \frac{V}{v_r\beta}\left(\frac{\partial f_{LC}}{\partial N}\right)_V \\ &= \frac{1}{\beta}\left(\frac{\partial f_{LC}}{\partial \phi}\right)_T, \end{aligned} \tag{4.48}$$

where the free energy f_{LC} is a function of S and ϕ, then we have

$$\left(\frac{df_{LC}}{d\phi}\right)_T = \left(\frac{\partial f_{LC}}{\partial \phi}\right)_S + \left(\frac{\partial f_{LC}}{\partial S}\right)_\phi \left(\frac{dS}{d\phi}\right). \tag{4.49}$$

The second term corresponds to the condition that determines the orientational order parameter that minimizes the free energy: $(\partial f_{LC}/\partial S)_\phi = 0$, which implies that it vanishes at equilibrium. Therefore, in Eqs. (4.47) and (4.48), we treat S as a constant and differentiate f_{LC} with respect to ϕ.[1] We then obtain

$$v_r\beta\Pi = \phi + \frac{L}{D}\phi^2 - \frac{5L}{8D}\phi^2 S^2, \tag{4.50}$$

[1]Note that although the function $I_0(\Gamma_0 S)$ depends on ϕ, the partial derivative with respect to ϕ at constant S can be evaluated using the chain rule: $(\partial I_0/\partial\phi)_S = (\partial I_0/\partial\Gamma_0)_S(\partial\Gamma_0/\partial\phi)_S$.

$$\beta\mu_r = \beta\mu_r^\circ + \ln\phi + 2\frac{L}{D}\phi - \ln I_0(\Gamma_0 S). \tag{4.51}$$

The concentrations ϕ_I and ϕ_N of the two coexisting phases are determined by the conditions that the chemical potentials and osmotic pressures are equal in both phases:

$$\Pi(\phi_I, S=0) = \Pi(\phi_N, S), \tag{4.52}$$

$$\mu_r(\phi_I, S=0) = \mu_r(\phi_N, S). \tag{4.53}$$

The solution of the simultaneous equations can be easily shown to be the same as the concentration obtained from the common tangent method (Eq. (4.44)). The orientational order parameter S is derived as a function of concentration by numerically solving Eq. (4.31). By substituting Eq. (4.50) into Eq. (4.52), the concentration ϕ_I is given as a function of ϕ_N:

$$x_I = \frac{1}{2}\left[-1 + \sqrt{1 + 4\left(x_N + x_N^2 - \frac{5}{8}x_N^2 S^2\right)}\right], \tag{4.54}$$

where we define $x_N = (L/D)\phi_N$ and $x_I = (L/D)\phi_I$. By substituting Eq. (4.54) into Eq. (4.53), we can numerically determine the coexisting concentrations as

$$\phi_I = 3.504\frac{D}{L}, \tag{4.55}$$

$$\phi_N = 3.872\frac{D}{L}. \tag{4.56}$$

The NIT concentration (Eq. (4.39)) lies between these two concentrations: ϕ_I and ϕ_N.
According to the Onsager trial function, it gives the following results [42]:

$$\phi_I = 3.34\frac{D}{L}, \qquad \phi_N = 4.486\frac{D}{L}, \qquad S_{NI} = 0.848. \tag{4.57}$$

The bilinear expansion in Eq. (4.24) with higher order terms yields [47]:

$$\phi_i = 3.290\frac{D}{L}, \qquad \phi_n = 4.191\frac{D}{L}, \qquad S_{NI} = 0.792. \tag{4.58}$$

Figure 4.9 shows the phase diagram of a rod-like molecular solution plotted on the L/D-ϕ plane, based on Eqs. (4.55) and (4.56). The solid lines indicate the coexisting concentrations of the isotropic and nematic phases, ϕ_I and ϕ_N, respectively. The region between these curves represents the two-phase coexistence region: I+N phase separation. According to Onsager theory, this coexistence region is entirely metastable; the free energy curve, as illustrated in Fig. 4.8, remains convex downward with respect to ϕ, and no unstable region satisfying $(\partial^2 f_{LC}/\partial\phi^2)_T < 0$ appears. As the axial ratio L/D of the rod-like molecules increases, the NIT occurs at lower concentrations. The phase separation observed in Fig. 1.6 appears to correspond to the behavior of rod-like molecules with $L/D \simeq 30$, as depicted in Fig. 4.9.

Although the Onsager theory is strictly valid in the limit $L/D \gg 1$, it nevertheless provides a robust qualitative framework for understanding the emergence of nematic order. The fundamental mechanism underlying nematic ordering, namely the excluded volume interactions between rod-like molecules, remains applicable even outside the strict $L/D \gg 1$ limit. In practical systems, several additional factors influence the NIT, including intermolecular attractive interactions, molecular flexibility, and polydispersity in molecular weight. The effective diameter of the rod-like molecules can vary with ionic strength, further influencing the phase behavior. These effects can shift the transition concentration, alter the sharpness of the phase boundary, or even modify the nature of the phase transition itself.

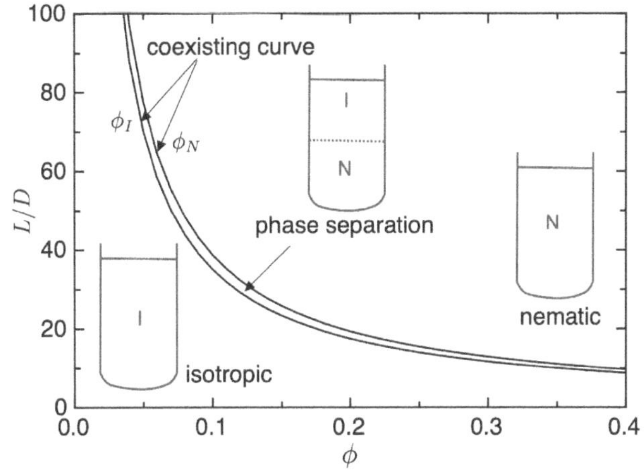

FIGURE 4.9
Phase diagram on the axial ratio $L/D - \phi$ plane. The solid curve shows the coexistence line. Inside the coexistence curves is the two-phase separation region (I + N).

4.2 Maier-Saupe model for nematic phases

This section introduces the Maier-Saupe model of nematic ordering, which accounts for intermolecular attractive interactions [43]. The theory is particularly suitable for describing the nematic ordering of low-molecular-weight liquid crystals with small axial ratios L/D.

In the mean field approximation, the intermolecular interaction term $A_2(\mathbf{\Omega}, \mathbf{\Omega}')$ in Eq. (4.20) can be replaced using the orientational order parameter S as the following one-body potential:

$$A_2(\mathbf{\Omega}) = -v_{rr}\nu_M S P_2(\cos\theta), \tag{4.59}$$

or equivalently,

$$A_2(\mathbf{\Omega}, \mathbf{\Omega}') = -v_{rr}\nu_M P_2(\cos\theta) P_2(\cos\theta'), \tag{4.60}$$

where the negative sign indicates an attractive interaction, $v_{rr} = (\pi/4)L^2D$ is the average excluded volume of two rod-like molecules,[2] and $\nu_M = U_M/(k_BT)(> 0)$ shows the attractive interaction between rod-like molecules that depends on their orientation (the Maier-Saupe interaction parameter for the nematic phase). The interaction parameter ν_M is dimensionless, and higher values of ν_M imply that two rod-like molecules tend to align in the same direction. Using the volume fraction $\phi = v_r\rho$ and $v_r = (\pi/4)D^2L$, substituting Eq. (4.59) into Eq. (4.10) yields:

$$\text{2nd term of Eq. (4.10)} = \frac{\phi}{v_r}\int f_d(\mathbf{\Omega}) \ln\left[4\pi f_d(\mathbf{\Omega})\right] d\mathbf{\Omega}, \tag{4.61}$$

$$\begin{aligned}\text{3rd term of Eq. (4.10)} &= -\frac{1}{2}\rho^2 v_{rr}\nu_M S \int P_2(\cos\theta) f_d(\mathbf{\Omega}) d\Omega \int f_d(\mathbf{\Omega}') d\mathbf{\Omega}' \\ &= -\frac{1}{2v_r}\nu_M \phi^2 S^2.\end{aligned} \tag{4.62}$$

[2] v_{rr} is defined as the excluded volume when two rod-like molecules are perpendicular to each other.

We then obtain the dimensionless free energy density (Eq. (4.10)) of the Maier-Saupe model:

$$f_{LC} = v_r \beta F/V = f_{iso} + f_{nem}, \tag{4.63}$$

where the isotropic free energy f_{iso} is given by Eq. (4.37) and the nematic free energy is expressed as

$$f_{nem} = v_r \beta F_{nem}/V \quad = \quad \phi \int f_d(\mathbf{\Omega}) \ln \left[4\pi f_d(\mathbf{\Omega})\right] d\mathbf{\Omega} - \frac{1}{2}\left(\frac{L}{D}\right)\nu_M \phi^2 S^2. \tag{4.64}$$

Substituting Eq. (4.59) into Eq. (4.20), the distribution function is given by

$$f_d(\mathbf{\Omega}) = \frac{1}{Z_0} \exp\left[\Gamma_0 S P_2(\cos\theta)\right], \tag{4.65}$$

where

$$\Gamma_0 = \frac{L}{D}\nu_M \phi. \tag{4.66}$$

Substituting Eq. (4.65) into Eq. (4.64), the nematic free energy becomes

$$f_{nem} = \phi\left(\frac{1}{2}\Gamma_0 S^2 - \ln I_0(\Gamma_0 S)\right), \tag{4.67}$$

which is identical in form to the nematic free energy obtained from the Onsager model (Eq. (4.38)). The free energy curve derived here remains consistent with the results shown in Fig. 4.4. Likewise, the behavior of the order parameter across the phase transition, as illustrated in Fig. 4.5, can be directly applied. The only difference between the repulsive and attractive interaction models lies in the definition of the parameter Γ_0. In the attractive model, the NIT can be analysed not only in terms of concentration ϕ but also in terms of temperature, via the interaction parameter $\nu_M = U_M/(k_B T)$. As a result, the transition temperature T_{NI} corresponds to the condition $\Gamma_0 = 4.55$. Using Eq. (4.66), the NIT temperature can be expressed as

$$T_{NI} = \frac{L}{4.55D}\left(\frac{U_M}{k_B}\right)\phi. \tag{4.68}$$

For $T < T_{NI}$, the nematic phase emerges due to attractive interactions between liquid crystalline molecules. The NIT temperature increases with increasing concentration ϕ, reflecting the dilution effect introduced by the presence of solvent molecules. When $\phi = 1$, the system corresponds to a pure liquid crystalline molecule, and the NIT temperature represents its intrinsic transition point. For instance, if the transition temperature is $T_{NI}(\phi = 1) = 300$ K and the axial ratio is $L/D = 2$, the interaction strength U_M can be estimated to be approximately $U_M \sim 10^{-20}$ J. This value is on the order of van der Waals interactions, perhaps a few times stronger, but still relatively weak. This suggests that the intermolecular potential responsible for aligning the molecules in a common direction is relatively weak.

Mean field theory has been extended to determine the temperature-composition phase diagram of binary mixtures of nematic liquid crystals [48, 49]. In Chapter 7, we introduce a hybrid model that combines the effects of excluded volume interactions, as described in the Onsager theory, with the attractive interactions considered in the Maier-Saupe theory. This unified approach provides a more comprehensive framework for describing the nematic phase in a broader class of liquid crystal systems.

4.3 Landau expansion of the nematic free energy

The Landau-de Gennes theory of phase transitions expresses the free energy as a power series expansion of the order parameter, typically up to the fourth-order term. Expanding the exponential function $\exp[y(x)]$ in the definition of $I_0(\Gamma_0 S)$, Eq. (4.32), where the argument $y(x) = \Gamma_0 S(3x^2/2 - 1/2)$ is assumed to be small, we obtain a power series up to the fourth order:

$$\begin{aligned} I_0(\Gamma_0 S) &= \int_0^1 \exp\left[y(x)\right] dx \\ &= \int_0^1 \left[1 + y(x) + \frac{1}{2!}y(x)^2 + \frac{1}{3!}y(x)^3 + \frac{1}{4!}y(x)^4 + \cdots\right] dx. \end{aligned} \tag{4.69}$$

After integrating term by term and taking the logarithm of $I_0(\Gamma_0 S)$, we obtain an approximate expression in powers of $\Gamma_0 S$:

$$\ln I_0(\Gamma_0 S) \simeq \frac{1}{10}(\Gamma_0 S)^2 + \frac{1}{105}(\Gamma_0 S)^3 - \frac{1}{700}(\Gamma_0 S)^4 \cdots . \tag{4.70}$$

Substituting Eq. (4.70) into Eq. (4.67), we obtain an approximate expression for the dimensionless nematic free energy density in terms of the orientational order parameter S:

$$f_{nem} = \Gamma_0 \phi \left[C_1 S^2 - C_2 S^3 + C_3 S^4\right], \tag{4.71}$$

where the coefficients are given by

$$C_1 = \frac{1}{2}(1 - \frac{\Gamma_0}{5}), \tag{4.72}$$

$$C_2 = \frac{\Gamma_0^2}{105}, \tag{4.73}$$

$$C_3 = \frac{\Gamma_0^3}{700}. \tag{4.74}$$

This form is directly analogous to the Landau expansion used to describe first-order phase transitions. The cubic term $-C_2 S^3$ is what allows for a discontinuous jump in S at the transition point, one of the key features distinguishing first-order from second-order phase transition. Note that this approximate expression for the free energy is valid only near the NIT ($\Gamma_0 < 5$).

The value of S in thermal equilibrium can be determined by the condition that minimizes the free energy, $(\partial f_{\text{nem}}/\partial S)_T = 0$, and is given by

$$S = \frac{3C_2}{8C_3}\left(1 + \sqrt{1 - \frac{32 C_1 C_3}{9 C_2^2}}\right). \tag{4.75}$$

At the NIT point, as shown in Fig. 4.4, the condition $f_{\text{nem}}(S = 0) = f_{\text{nem}}(S) = 0$ holds and we obtain the following relation:

$$4 C_1 C_3 = C_2^2, \tag{4.76}$$

for the NIT temperature. By substituting Eqs. (4.72)–(4.74), we obtain $\Gamma_0^{NI} = 4.32$ at the NIT. This value is close to the numerically calculated value of 4.55, shown in Fig. 4.4. The value of the orientational order parameter at the NIT is $S_{NI} = C_2/2C_3 = 0.77$. The value of the orientational order parameter jumps from 0 to 0.77 at the NIT and increases toward one as Γ_0 further increases. Equation (4.71) demonstrates the first-order phase transition discussed in Fig. 4.4. It is also necessary for the coefficients C_2 and C_3 to be positive. If $C_2 = 0$, the system undergoes a second-order phase transition, where the order parameter increases continuously from zero.

The free energy can also be constructed using a power series of the tensor order parameter ($Q_{\alpha\beta}$), introduced in Section 2.2. Therefore, with positive coefficients $a_1 > 0$, $a_2 > 0$, and $a_3 > 0$, it can be expanded as follows:

$$f_{nem} = a_1 Q_{\alpha\beta}Q_{\beta\alpha} - a_2 Q_{\alpha\beta}Q_{\beta\gamma}Q_{\gamma\alpha} + a_3\left(Q_{\alpha\beta}Q_{\beta\alpha}\right)^2, \tag{4.77}$$

where the linear term Q_{ij} cannot appear in the free energy expansion. This is a key aspect of nematic order: the order parameter tensor is traceless and symmetric, allowing it to describe uniaxial or biaxial orientational order while containing no polar vector direction. Using Eq. (2.37), we have[3]

$$\begin{aligned} Q_{\alpha\beta}Q_{\alpha\beta} &= S^2\left(\frac{3}{2}n_\alpha n_\beta - \frac{1}{2}\delta_{\alpha\beta}\right)\left(\frac{3}{2}n_\alpha n_\beta - \frac{1}{2}\delta_{\alpha\beta}\right) \\ &= S^2\left(\frac{9}{4}n_\alpha^2 n_\beta^2 - \frac{3}{2}n_\alpha n_\beta\delta_{\alpha\beta} + \frac{1}{4}\delta_{\alpha\beta}\delta_{\alpha\beta}\right) \\ &= S^2\left(\frac{9}{4}n_\alpha^2 n_\beta^2 - \frac{3}{2}n_\alpha n_\alpha + \frac{1}{4}\delta_{\alpha\alpha}\delta_{\alpha\alpha}\right) \\ &= S^2\left(\frac{9}{4} - \frac{3}{2} + \frac{3}{4}\right) = \frac{3}{2}S^2, \end{aligned} \tag{4.78}$$

and

$$Q_{\alpha\beta}Q_{\beta\gamma}Q_{\gamma\alpha} = \frac{3}{4}S^3. \tag{4.79}$$

Therefore, Eq. (4.77) can be rewritten as a power series expansion in terms of the scalar order parameter S:

$$f_{nem} = C_1 S^2 - C_2 S^3 + C_3 S^4, \tag{4.80}$$

where the coefficients are defined as

$$C_1 = \frac{3}{2}a_1 = \frac{1}{2}\left(1 - \frac{\Gamma_0}{5}\right), \tag{4.81}$$

$$C_2 = \frac{3}{4}a_2 = \frac{\Gamma_0}{9}, \tag{4.82}$$

$$C_3 = \frac{9}{4}a_3 = \frac{\Gamma_0}{6}. \tag{4.83}$$

The last three expressions define the coefficients, which are obtained from the dynamical mean-field theory [50]. This formulation allows the description of phase transitions across the full range of Γ_0. With this expression for the free energy, one can readily explore the dependence of f_{nem} on S for different values of Γ_0, as shown in Fig. 4.4.

4.4 Simulations of nematic liquid crystals

In this section, we introduce a simple simulation model that helps illustrate how nematic ordering emerges. As shown in Fig. 4.10, rod-like molecules are placed on lattice points

[3] $\delta_{\alpha\beta}\delta_{\alpha\beta} = \delta_{\alpha\alpha}\delta_{\alpha\alpha} = \delta_{xx}\delta_{xx} + \delta_{yy}\delta_{yy} + \delta_{zz}\delta_{zz} = 3$

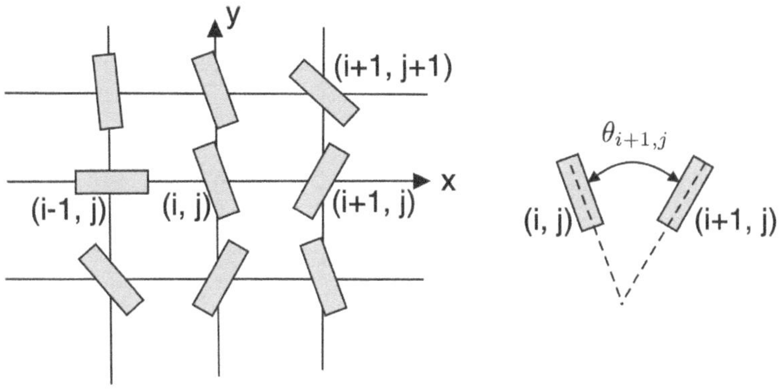

FIGURE 4.10
Lebwohl-Lasher (LL) model for a nematic phase. Rod-like molecules are placed on lattice points (i, j), with each molecule's center of mass fixed at the lattice point, allowing only rotational degrees of freedom.

(i, j), with each molecule's center of mass fixed at the lattice point, allowing only rotational degrees of freedom. The attractive interaction used in the Maier-Saupe theory, as discussed in Eq. (4.59), is adopted here to define the interaction energy between neighboring molecules. The dimensionless interaction energy of the Lebwohl–Lasher (LL) model [51] is given by

$$\begin{aligned} U_{LL}/(k_BT) &= -\nu_M \sum_{(i,j)} P_2(\cos\theta_{i,j}) \\ &= -\nu_M \sum_{i,j=1}^{N} \big(P_2(\cos\theta_{i-1,j}) + P_2(\cos\theta_{i+1,j}) \\ &\quad +P_2(\cos\theta_{i,j+1}) + P_2(\cos\theta_{i,j-1})\big), \end{aligned} \tag{4.84}$$

where N is the total number of rod-like molecules, and (i, j) denotes the lattice coordinates. Only interactions between nearest-neighbor lattice sites are considered; that is, the four adjacent sites $(i \pm 1, j)$ and $(i, j \pm 1)$. Diagonal interactions are neglected. For instance, the orientational interaction term $P_2(\cos\theta_{i+1,j})$ involves the angle between the molecular orientations at sites (i, j) and $(i+1, j)$. This angle is computed as $\cos\theta_{i+1,j} = \mathbf{\Omega}(i, j)\cdot\mathbf{\Omega}(i+1, j)$, where $\mathbf{\Omega}(i, j)$ and $\mathbf{\Omega}(i + 1, j)$ are unit vectors indicating the orientation of the molecules at the respective lattice points.

Figure 4.11 shows a snapshot from a Metropolis Monte Carlo simulation under periodic boundary conditions, using the attractive interaction parameter $\nu_M = 3$. Short line segments represent the orientation of rod-like molecules fixed at each lattice point. The global orientational order parameter is $S = 0.73$, and the central arrow indicates the average orientation direction of the system. Two characteristic liquid crystal defects, known as disclinations, appear in the figure. Point A shows an $s = +1/2$ disclination, while point B corresponds to an $s = -1/2$ disclination (see Section 3.2.1 for details). These disclinations arise due to thermal fluctuations in the system. As the attractive interaction parameter ν_M increases, the molecules align more uniformly, and the disclinations tend to disappear.

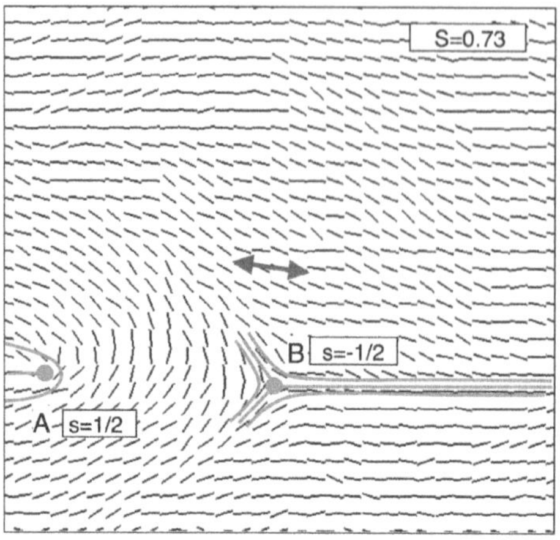

FIGURE 4.11
Snapshot of Monte Carlo simulation of the LL model.

4.5 Biaxial nematic phases of bent-shaped liquid crystals

In this section, we introduce the uniaxial and biaxial nematic phases that can arise in bent-shaped molecules. Before proceeding, the author recommends that readers review Section 2.2.2. By employing the tensor order parameters introduced in Section 2.2.2, we construct the free energy and discuss the resulting phase transitions.

4.5.1 Free energy of bent-shaped liquid crystals

Consider a thermotropic bent-shaped molecule, or banana-shaped molecule, consisting of two mesogens with the axial ratio $n_r = L/D$, where L is the length and D is the diameter of the mesogen (see Fig. 2.11). Let $f_d(\mathbf{\Omega})$ be the orientational distribution function with the solid angle $d\mathbf{\Omega} = \sin\theta d\theta d\phi d\psi$. Using the tensor order parameter $Q_{\alpha\beta}$ of Eq. (2.38) and the Maier-Saupe model of Eq. (4.64), the dimensionless nematic free energy density for the thermotropic bent-shaped molecules is given by [33]

$$\begin{aligned} f_{nem} &= \int f_d(\mathbf{\Omega}) \ln(8\pi^2 f_d(\mathbf{\Omega}))d\mathbf{\Omega} - \frac{1}{2}n_r\nu_M\frac{2}{3}\mathcal{Q}_{\alpha\beta}\mathcal{Q}_{\alpha\beta} \\ &= \int f_d(\mathbf{\Omega}) \ln(8\pi^2 f_d(\mathbf{\Omega}))d\mathbf{\Omega} - \frac{1}{2}n_r\nu_M(S_n^2 + \frac{1}{3}B^2), \end{aligned} \tag{4.85}$$

where the first term shows the entropy changes due to orientational ordering, $\nu_M = U_M/(k_BT)$ is the Maier-Saupe interaction parameter between molecules. The order parameters S_n and B are defined in Eqs. (2.61) and (2.64), respectively.

Using the distribution function $f_d(\mathbf{\Omega})$, the four order parameters, introduced in Section 2.2.2, are given by

$$S = \int P_2(\cos\theta) f_d(\mathbf{\Omega}) d\mathbf{\Omega}, \tag{4.86}$$

$$U = \int W_0(\theta, \psi) f_d(\mathbf{\Omega}) d\mathbf{\Omega}, \tag{4.87}$$

$$B_1 = \int W_1(\theta,\phi) f_d(\mathbf{\Omega}) d\mathbf{\Omega}, \tag{4.88}$$

and

$$B_2 = \int W_2(\theta,\phi,\psi) f_d(\mathbf{\Omega}) d\mathbf{\Omega}, \tag{4.89}$$

where we define

$$W_0(\theta,\psi) = \sin^2\theta \cos 2\psi, \tag{4.90}$$

$$W_1(\theta,\phi) = \sin^2\theta \cos 2\phi, \tag{4.91}$$

and

$$W_2(\theta,\phi,\psi) = \frac{1}{2}(1+\cos^2\theta)\cos 2\phi \cos 2\psi - \cos\theta \sin 2\phi \sin 2\psi. \tag{4.92}$$

In a thermal equilibrium state, the orientational distribution function $f_d(\mathbf{\Omega})$ is determined by the derivation of the free energy with respect to this function:

$$\left(\frac{\delta f_{nem}}{\delta f_d(\mathbf{\Omega})}\right)_T = 0, \tag{4.93}$$

under the normalization condition:

$$\int_0^{\pi}\int_0^{2\pi}\int_0^{2\pi} f_d(\mathbf{\Omega}) \sin\theta d\theta d\phi d\psi = 1. \tag{4.94}$$

We then obtain

$$f_d(\theta,\phi,\psi) = \frac{1}{Z_0} \exp\left[n_r \nu_M \left(S_n \Lambda_u(\theta,\psi) + \frac{1}{3} B \Lambda_b(\theta,\phi,\psi)\right)\right], \tag{4.95}$$

where the coefficients are given by

$$\Lambda_u(\theta,\psi) = P_2(\cos\chi_B) P_2(\cos\theta) + \frac{3}{4}\sin^2\chi_B W_0(\theta,\psi), \tag{4.96}$$

and

$$\Lambda_b(\theta,\phi,\psi) = \frac{3}{2}\left(P_2(\cos\chi_B) W_1(\theta,\phi) + \sin^2\chi_B W_2(\theta,\phi,\psi)\right). \tag{4.97}$$

The constant Z_0 in Eq. (4.95) is determined by the normalization condition (Eq. (4.94)): $Z_0 = I_{[0000]}$, where the function $I_{[hijk]}$ is defined as

$$\begin{aligned} I_{[hijk]} &= \int_{-1}^{1}\int_0^{2\pi}\int_0^{2\pi} [P_2(\cos\theta)]^h [W_0(\theta,\psi)]^i [W_1(\theta,\phi)]^j [W_2(\theta,\phi,\psi)]^k \\ &\times \exp\left[n_r \nu_M \left(S_n \Lambda_u(\theta,\psi) + \frac{1}{3} B \Lambda_b(\theta,\phi,\psi)\right)\right] d(\cos\theta) d\phi d\psi, \end{aligned} \tag{4.98}$$

with $h,i,j,k = 0,1$. Note that the function $I_{[hijk]}$ contains the four order parameters, S, U, B_1, and B_2 through the order parameters S_n and B. Then the four orientational order parameters are determined by the four coupled self-consistent equations:

$$S = I_{[1000]}/I_{[0000]}, \tag{4.99}$$

$$U = I_{[0100]}/I_{[0000]}, \tag{4.100}$$

$$B_1 = I_{[0010]}/I_{[0000]}, \tag{4.101}$$

$$B_2 = I_{[0001]}/I_{[0000]}. \tag{4.102}$$

Substituting the distribution function (Eq. (4.95)) into Eq. (4.85), the free energy is given by

$$f_{nem} = \frac{1}{2}n_r\nu_M\left(S_n^2 + \frac{1}{3}B^2\right) - \ln(I_{[0000]}/(8\pi^2)). \tag{4.103}$$

To analyze the uniaxial and biaxial nematic phases, it is sufficient to consider two order parameters, S_n and B, rather than the full set of four: S, U, B_1, and B_2. The uniaxial order parameter is given by

$$S_n = \int_{-1}^{1}\int_{0}^{2\pi}\int_{0}^{2\pi} \Lambda_u(\theta,\psi)f_d(\theta,\phi,\psi)d(\cos\theta)d\phi d\psi, \tag{4.104}$$

and a biaxial order parameter

$$B = \int_{-1}^{1}\int_{0}^{2\pi}\int_{0}^{2\pi} \Lambda_b(\theta,\phi,\psi)f_d(\theta,\phi,\psi)d(\cos\theta)d\phi d\psi. \tag{4.105}$$

Note that the values of S_n and B are included in the distribution function (4.95).[4]

4.5.2 Biaxial-uniaxial phase transitions

Using Eqs. (2.54) and (2.60), the scalar orientational order parameters are given by

$$S_l = l_\alpha Q_{\alpha\beta} l_\beta = -\frac{1}{2}(S_n + B), \tag{4.106}$$

and

$$S_m = m_\alpha Q_{\alpha\beta} m_\beta = -\frac{1}{2}(S_n - B), \tag{4.107}$$

where $Q_{xx} = S_l$, $Q_{yy} = S_m$, and $Q_{zz} = S_n$.

Three uniaxial nematic phases, as shown in Fig. 2.12, can be characterized by using two order parameters S_n and S. In the N_U^a phase, the molecular axis **a** rotates around the director **n**, leading to $S_n < 0$ and $S < 0$. The N_U^b phase appears when $S_n > 0$ and $S < 0$, whereas the N_U^c phase occurs for $S_n > 0$ and $S > 0$. The order parameter U has the same sign as S_n. In all uniaxial nematic phases, $S_n \neq 0$ and $S_l = S_m > 0$ hold. The biaxial nematic phase $N_B^i (i = a, b, c)$ is defined by the condition $B \neq 0$.

In the numerical calculations, we define the reduced temperature $\tau = 1/\nu_M (= k_B T/U_M)$ and we take $n_r = 4$. Figure 4.12 shows a typical phase diagram in the temperature (T/T_{LP}) versus bend angle (θ_B) plane. The temperature is normalized by the Landau point temperature, T_{LP}. The solid curve represents the first-order nematic-isotropic phase transition (NIT), while the dotted curve corresponds to the second-order biaxial-uniaxial transition (BUT). The two closed circles indicate the Landau points, where the first- and the second-order phase transitions meet. Depending on the bend angle, three different uniaxial nematic

[4]Triple integrals require substantial computation time, so using parallel processing is recommended.

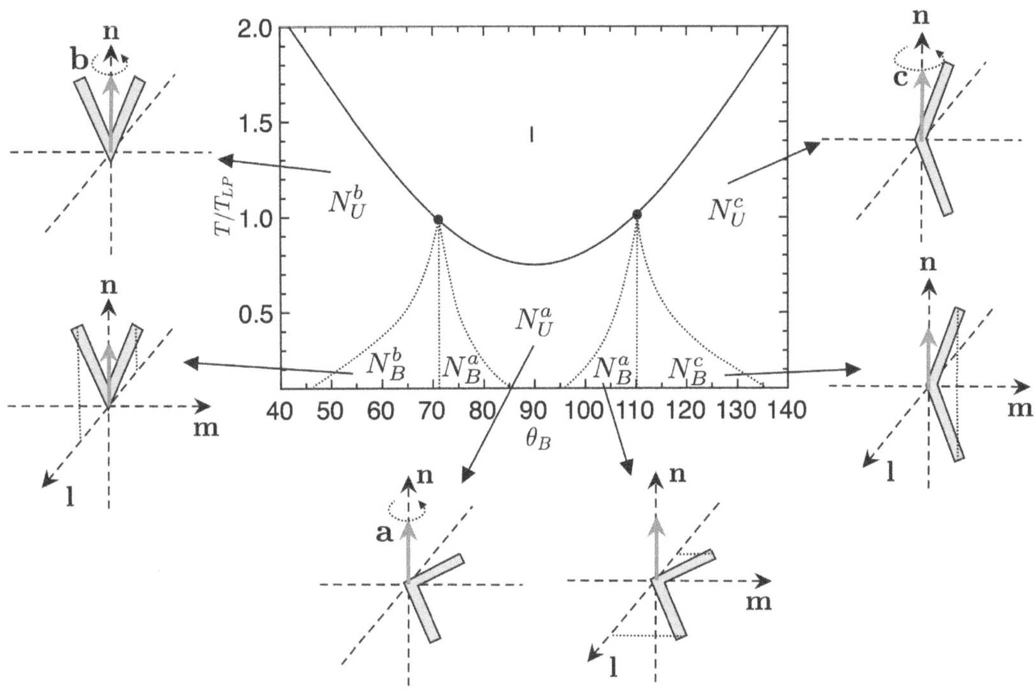

FIGURE 4.12
Phase diagram on the temperature-bend angle plane. The temperature is normalized by the Landau point temperature T_{LP}. The solid curve shows the first-order NIT, and the dotted curves correspond to the second-order BUT. The two closed circles show the Landau points. Three uniaxial, N_U^i, and three biaxial N_B^i phases are shown in the phase diagram ($i = a, b, c$).

phases appear: N_U^a, N_U^b, and N_U^c, where $N_U^i (i = a, b, c)$ denotes a uniaxial nematic phase characterized by the predominant alignment of the molecular axis $\mathbf{i} = \mathbf{a}, \mathbf{b}, \mathbf{c}$ along the director $\mathbf{n}$. Additionally, three biaxial nematic phases, N_B^a, N_B^b, and N_B^c, are found, each derived from the corresponding uniaxial phase N_U^a, N_U^b, and N_U^c, respectively, by imposing a two-fold rotational symmetry in the plane perpendicular to $\mathbf{n}$ (see Fig. 2.12). The uniaxial nematic phases N_U^a and N_U^c for bend angles $\theta_B > 90°$ have also been referred to as N_U^- and N_U^+, respectively, in previous studies [34, 52].

Figure 4.13(a) shows the order parameters S_n, S_m, S_l, and B plotted against the temperature for the bend angle $\theta_B = 115°$. With decreasing temperature, a first-order NIT occurs from the isotropic phase to the uniaxial nematic phase N_U^c, characterized by $S_n > 0$, $S_m = S_l < 0$, and $S > 0$, where the molecular axis $\mathbf{c}$ tends to align with the director $\mathbf{n}$, as shown in Fig. 4.12. Upon further cooling, the biaxial order parameter B increases continuously from zero, indicating a second-order BUT from the uniaxial nematic N_U^c phase to the biaxial nematic phase N_B^c. In the N_B^c phase, we have $S_n > 0$ and $S_m \neq S_l < 0$. In the limit of high orientational order, perfect alignment of the molecules is achieved, with the triad $\{\mathbf{l}, \mathbf{m}, \mathbf{n}\} = \{\mathbf{a}, \mathbf{b}, \mathbf{c}\}$, as illustrated in Fig. 4.12. In this case, the orientational correlations become $\langle P_2(\mathbf{a}\cdot\mathbf{l})\rangle = \langle P_2(\mathbf{b}\cdot\mathbf{m})\rangle = 1$, $\langle P_2(\mathbf{c}\cdot\mathbf{m})\rangle = \langle P_2(\mathbf{c}\cdot\mathbf{l})\rangle = \langle P_2(\mathbf{b}\cdot\mathbf{n})\rangle = \langle P_2(\mathbf{a}\cdot\mathbf{n})\rangle = -1/2$. As the temperature decreases, the order parameters in the biaxial N_B^c phase approach the limiting values: $S \to 1$, $U \to 0$, $B_1 \to 0$, and $B_2 \to 1$.

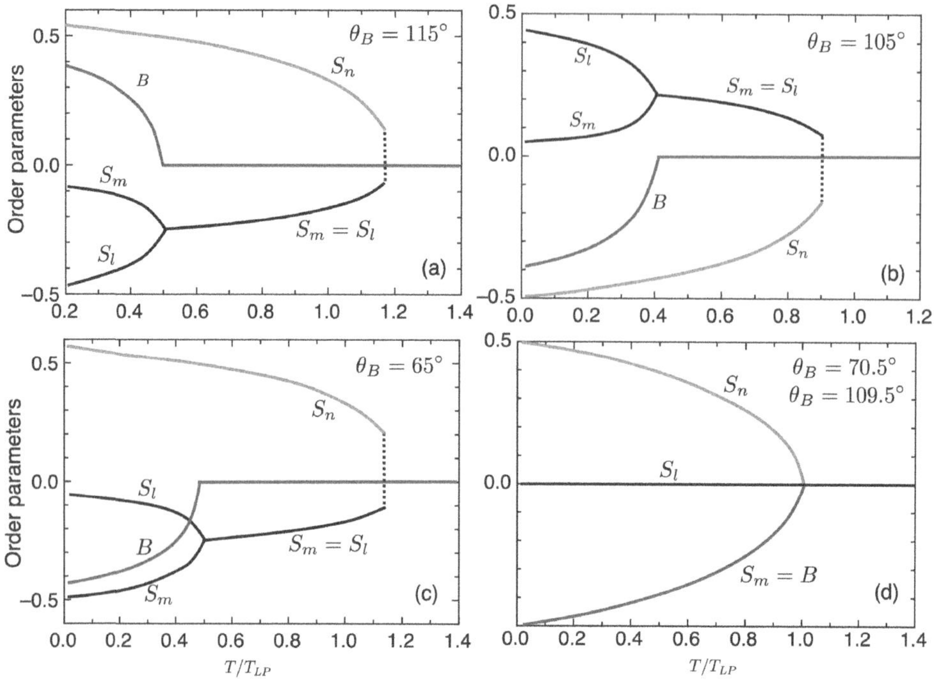

FIGURE 4.13
Order parameters S_n, S_m, S_l, and B plotted against the temperature for the bent-shaped molecule of the bend angle $\theta_B = 115°$ (a), $\theta_B = 105°$ (b), $\theta_B = 65°$ (c), and $\theta_B = 70.5°$, $109.5°$ (d).

Figure 4.13(b) shows the order parameters S_n, S_m, S_l, and B plotted against the temperature for the bend angle $\theta_B = 105°$, or disk-like molecules. With decreasing temperature, a first-order NIT occurs from the isotropic phase to the uniaxial nematic phase N_U^a, characterized by $S_n < 0$, $S_m = S_l > 0$, and $S < 0$, where the molecular axis $\mathbf{a}$ tends to align with the director $\mathbf{n}$, as illustrated in Fig. 4.12. The order parameter $\langle P_2(\mathbf{a} \cdot \mathbf{n}) \rangle$ is positive, indicating the tendency of the molecular frame $\mathbf{a}$ to align parallel to the director $\mathbf{n}$. Upon further cooling, a second-order BUT takes place from the uniaxial nematic phase N_U^a to the biaxial nematic phase N_B^a, where $S_n < 0$ and $S_m \neq S_l$. At lower temperatures, perfect alignment of the molecules is achieved with the triad $\{\mathbf{l}, \mathbf{m}, \mathbf{n}\} = \{\mathbf{c}, \mathbf{b}, \mathbf{a}\}$, as shown in Fig. 4.12. In this limit, the order parameters satisfy $\langle P_2(\mathbf{c} \cdot \mathbf{l}) \rangle = \langle P_2(\mathbf{b} \cdot \mathbf{m}) \rangle = \langle P_2(\mathbf{a} \cdot \mathbf{n}) \rangle = 1$, and $\langle P_2(\mathbf{b} \cdot \mathbf{n}) \rangle = \langle P_2(\mathbf{a} \cdot \mathbf{l}) \rangle = \langle P_2(\mathbf{a} \cdot \mathbf{m}) \rangle = -1/2$. Therefore, as the temperature decreases, the order parameters in the biaxial phase N_B^a approach the limiting values: $S \to -1/2$, $U \to -1$, $B_1 \to -1$, $B_2 \to 1/2$.

Figure 4.13(c) shows the order parameters S_n, S_m, S_l, and B plotted against the temperature for the bend angle $\theta_B = 65°$, or V-shaped molecules. With decreasing temperature, a first-order NIT occurs from the isotropic phase to the uniaxial nematic phase N_U^b, characterized by $S_n > 0$, $S_m = S_l < 0$, and $S < 0$, where the molecular axis $\mathbf{b}$ tends to align with the director $\mathbf{n}$, as illustrated in Fig. 4.12. Upon further cooling, a second-order BUT takes place from the uniaxial nematic N_U^b phase to the biaxial nematic phase N_B^b. In the N_B^b phase, we have $S_n > 0$, $S_m \neq S_l$, and $B < 0$. At lower temperatures, the molecules become perfectly aligned with the triad $\{\mathbf{l}, \mathbf{m}, \mathbf{n}\} = \{\mathbf{c}, \mathbf{a}, \mathbf{b}\}$, as illustrated in Fig. 4.12. In this perfectly ordered state, the orientational order parameters satisfy $\langle P_2(\mathbf{c} \cdot \mathbf{l}) \rangle = \langle P_2(\mathbf{a} \cdot \mathbf{m}) \rangle = 1$, and $\langle P_2(\mathbf{c} \cdot \mathbf{m}) \rangle = \langle P_2(\mathbf{a} \cdot \mathbf{l}) \rangle = \langle P_2(\mathbf{b} \cdot \mathbf{l}) \rangle = \langle P_2(\mathbf{b} \cdot \mathbf{m}) \rangle = -1/2$. As the temperature continues to decrease, the order parameters in the biaxial N_B^b phase asymptotically approach the values $S = -1/2$, $U = 1$, $B_1 = -1$, and $B_2 = -1/2$, respectively.

Figure 4.13(d) shows the order parameters plotted as functions of temperature for molecules with a bend angle of $\theta_B = 70.5°$, and for the tetrahedral angle $\theta_B = 109.5°$, corresponding to the Landau point. As the temperature decreases, a second-order phase transition occurs from the isotropic phase to the biaxial nematic phase, characterized by $S_n \neq 0$ and $B \neq 0$. One of the scalar order parameters, S_l, remains zero, while the other two, S_n and S_m, change continuously. We may consider a parallelogram-shaped molecule composed of two internal angles, $\theta_B = 70.5°$ and $\theta_B = 109.5°$. In this configuration, the molecular axis **b** associated with the angle $\theta_B = 109.5°$ aligns parallel to the director **l**, while the molecular axis **c** associated with $\theta_B = 70.5°$ also becomes parallel to **l**. This implies that the angle between the director **l** and each of the arm vectors $\mathbf{u}_A$ and $\mathbf{u}_B$ is fixed at 54.75°, yielding $P_2(\mathbf{u}_A \cdot \mathbf{l}) = 0$ and $P_2(\mathbf{u}_B \cdot \mathbf{l}) = 0$ in the biaxial nematic phase. Using Eq. (4.107), this geometric constraint leads to a symmetry in the scalar order parameters, where $S_n = -S_m = -B$ holds throughout the entire temperature range of the biaxial nematic phase. In the limit of high orientational order, the molecular axes align as $\{\mathbf{l}, \mathbf{m}, \mathbf{n}\} = \{\mathbf{c}, \mathbf{a}, \mathbf{b}\}$ for $\theta_B = 70.5°$, where the order parameters approach the values $S \to -1/2$, $U \to 1$, $B_1 \to -1$, and $B_2 \to -1/2$. In contrast, for $\theta_B = 109.5°$, the alignment becomes $\{\mathbf{l}, \mathbf{m}, \mathbf{n}\} = \{\mathbf{b}, \mathbf{a}, \mathbf{c}\}$, where the order parameter S tends to 1 and B_2 tends to -1. In this case, both U and B_1 tend to zero as the temperature decreases.

4.5.3 Landau expansion and phase transitions

We derive the Landau expansion (see Section 4.3) of the free energy (Eq. (4.103)) and discuss the corresponding phase transitions in Fig. 4.12. This approach provides a useful qualitative understanding of the phase behavior.

Expanding the exponential term $\exp[A(\theta, \phi, \psi)]$ of the function $I_{[0000]}$ in Eq. (4.103) up to the fourth-order term in $A(\theta, \phi, \psi)$, assuming

$$A(\theta, \phi, \psi) = n_r \nu_M \left(S_n \Lambda_u(\theta, \psi) + \frac{1}{3} B \Lambda_b(\theta, \phi, \psi) \right), \tag{4.108}$$

is sufficiently small, we obtain

$$\begin{aligned} \frac{I_{[0000]}}{8\pi^2} &= \frac{1}{8\pi^2} \int_{-1}^{1} \int_0^{2\pi} \int_0^{2\pi} \exp\left[A(\theta, \phi, \psi)\right] d(\cos\theta) d\phi d\psi \\ &= \frac{1}{8\pi^2} \int_{-1}^{1} \int_0^{2\pi} \int_0^{2\pi} d(\cos\theta) d\phi d\psi \left[1 + \frac{1}{2} A(\theta, \phi, \psi)^2 \right. \\ &\quad \left. + \frac{1}{3!} A(\theta, \phi, \psi)^3 + \frac{1}{4!} A(\theta, \phi, \psi)^4 + \cdots \right] \\ &\simeq 1 + \frac{1}{10} (n_r \nu_M)^2 (S_n^2 + \frac{1}{3} B^2) g_1 + \frac{1}{210} (n_r \nu_M S_n)^3 g_2 \\ &\quad + \frac{1}{280} (n_r \nu_M)^4 (S_n^2 + \frac{1}{3} B^2)^2 {g_1}^2, \end{aligned} \tag{4.109}$$

where

$$g_1 = 3\cos^4 \chi_B - 3\cos^2 \chi_B + 1, \tag{4.110}$$

and

$$g_2 = \left(3\cos^2 \chi_B - 1\right)\left(3\cos^2 \chi_B - 2\right). \tag{4.111}$$

Substituting Eq. (4.109) into (4.103), the free energy can be expanded in powers of the order parameters S_n and B:

$$\begin{aligned}\beta f_{nem} &= \nu_M\Big(C_1(S_n^2+\frac{1}{3}B^2)-C_2S_n^3+C_3(S_n^2+\frac{1}{3}B^2)^2\Big),\\ &= \nu_M\Big(C_1S_n^2-C_2S_n^3+C_3S_n^4+\frac{1}{3}(C_1+2C_3S_n^2)B^2+\frac{1}{9}C_3B^4\Big),\end{aligned} \tag{4.112}$$

where the coefficients are given by

$$C_1=\frac{1}{2}(1-\frac{n_r\nu_M}{5}g_1), \tag{4.113}$$

$$C_2=\frac{(n_r\nu_M)^2}{210}g_2, \tag{4.114}$$

and

$$C_3=\frac{(n_r\nu_M)^3}{700}g_1^2. \tag{4.115}$$

Note that the free energy expression in Eq. (4.112) is valid only in the vicinity of the NIT. When $\chi_B=0$: $g_1=1$ and $g_2=2$, Eq. (4.112) reduces to the uniaxial nematic free energy given in Eq. (4.71). The coefficient C_3 is always positive, whereas the coefficient C_2 can take either positive or negative values depending on the bend angle θ_B, or g_2. When $C_2=0$ or $g_2=0$, we have $\cos^2\chi_B=1/3$ or $2/3$, and the NIT becomes a second-order phase transition, occurring at bend angles $\theta_B=70.5°$ or $109.5°$. Using the conditions of the Landau point: $C_1=0$ and $C_2=0$, the Landau point temperature is given by $T_{LP}=(n_r/5)g_1(U_M/k_B)=(n_r/15)(U_M/k_B)$.

The equilibrium value of S_n is obtained by minimizing the free energy, i.e., by solving

$$\left(\frac{\partial f_{nem}}{\partial S_n}\right)_{T,B}=0. \tag{4.116}$$

The result is given by

$$S_n=\frac{1}{8C_3}\left[3C_2+\sqrt{9C_2^2-32C_1C_3}\right]. \tag{4.117}$$

When $C_2>0$ or $g_2>0$, i.e., for $\theta_B<70.5°$ or $\theta_B>109.5°$, the equilibrium order parameter S_n takes positive values (Fig. 4.13(a), (c), (d)). Conversely, when $C_2<0$ or $g_2<0$, i.e., for $70.5°<\theta_B<109.5°$, S_n becomes negative (Fig. 4.13(b)). This implies that the cubic term $-C_2S_n^3$ in Eq. (4.112) can always be negative to minimize the free energy, thereby favoring a first-order NIT, except at the Landau points where $C_2=0$. The first-order NIT occurs at $C_1=C_2^2/4C_3$. Using this relation with Eq. (4.117), the order parameter S_n at the NIT is given by

$$S_n^{NI}=\frac{C_2}{2C_3}\left(=\frac{2C_1}{C_2}\right). \tag{4.118}$$

We also derive an approximate expression for the NIT temperature T_{NI} as a function of the bend angle:

$$\frac{T_{NI}}{T_{LP}}=3\left[g_1+\frac{5}{126}\left(\frac{g_2}{g_1}\right)^2\right], \tag{4.119}$$

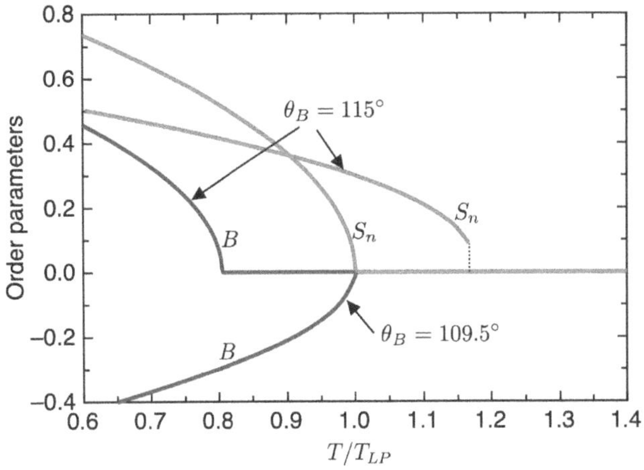

FIGURE 4.14
The approximate order parameters S_n (Eq. (4.117)) and B (Eq. (4.123)) are plotted as functions of temperature for $\theta_B = 115°$ and $109.5°$. The calculated values qualitatively reproduce the trends that appeared in the numerical results shown in Fig. 4.13(a) and (d).

where the temperature is normalized by the Landau point temperature T_{LP}. Since the second term is relatively small, it can be neglected in most practical cases. Using Eq. (4.119), we obtain $T_{NI}/T_{LP} \simeq 0.75$ for a bend angle of $\theta_B = 90°$, which is consistent with the result shown in Fig. 4.12.

The biaxial nematic phase appears when

$$C_1 + 2C_3 S_n^2 = 0, \tag{4.120}$$

in Eq. (4.112). We assume $S_n \simeq cS_n^{NI}$ with a constant $c > 1$, where S_n^{NI} is given by Eq. (4.118). Since the value of S_n at the BUT (S_n^{BU}) is larger than that at the NIT, as shown in Fig. 4.13(a), this assumption is reasonable. Substituting Eq. (4.118) into Eq. (4.120), we obtain the expression for the BUT temperature T_{BU} as

$$\frac{T_{BU}}{T_{LP}} = 3\left[g_1 - \frac{5c^2}{63}\left(\frac{g_2}{g_1}\right)^2\right]. \tag{4.121}$$

The equilibrium value of the biaxial order parameter B is given by

$$\left(\frac{\partial f_{nem}}{\partial B}\right)_{T,S_n} = 0, \tag{4.122}$$

and we obtain

$$B = \pm\sqrt{-\frac{3}{2}\left(\frac{C_1}{C_3} + 2S_n^2\right)}. \tag{4.123}$$

Figure 4.14 shows the approximate order parameters S_n (Eq. (4.117)) and B (Eq. (4.123)) plotted as functions of temperature for $\theta_B = 115°$ and $109.5°$ (the Landau point), with $c = 4$. The calculated order parameters qualitatively reproduce the behavior, shown in Fig. 4.13(a) and (d).

4.6 Smectic A liquid crystals

In this Section, we introduce McMillan theory [53] for the smectic A phase, which extends the Maier-Saupe theory of the nematic phase.

4.6.1 Free energy of the smectic A phase

In the smectic A phase, molecules align along the normal direction of the smectic layers (taken as the z-axis), and the positions of their centers of mass vary periodically along this direction (see Section 2.1.4). Let the periodicity of the center-of-mass positions be L_s. Using a Fourier series expansion, the intermolecular potential in Eq. (4.6) can be expressed as a one-body potential, as follows (see also Eq. (4.59)):

$$\Phi_{12}(\mathbf{r},\mathbf{\Omega}) = v_{rr}\nu_M S P_2(\cos\theta) \sum_{q_x,q_y,q_z} \sigma_{q_x q_y q_z} e^{i\mathbf{q}\cdot\mathbf{r}}, \tag{4.124}$$

$$\mathbf{q} = \frac{2\pi}{L_s}(q_x\mathbf{i} + q_y\mathbf{j} + q_z\mathbf{k}), \tag{4.125}$$

where $\mathbf{r} = (x, y, z)$ and $\mathbf{q} = (q_x, q_y, q_z)$ are the position and wavevectors, respectively, and q_x, q_y, and q_z are integers. The vectors $\mathbf{i}$, $\mathbf{j}$, and $\mathbf{k}$ are the unit vectors in the x-, y-, and z-directions, respectively.

For simplicity, we assume that the periodicity of the ordered phase is identical in all directions, i.e., $L_s = L_x = L_y = L_z$. In the smectic A phase with a one-dimensional periodicity along the z-axis, we expand using $\mathbf{r} = (0, 0, z)$ and $\mathbf{q} = (0, 0, q_z)$, where $-1 \leq q_z \leq 1$. The expansion then takes the following form [54]:[5]

$$\begin{aligned} \sum_{q_x,q_y,q_z} \sigma_{q_x q_y q_z} e^{i\mathbf{q}\cdot\mathbf{r}} &= \sigma_{000} + \sigma_{001} e^{i\frac{2\pi z}{L_s}} + \sigma_{00-1} e^{-i\frac{2\pi z}{L_s}} \\ &= 1 + \alpha_M \sigma \cos(2\pi z/L_s), \end{aligned} \tag{4.126}$$

where we take $\sigma_{000} = 1$, $\sigma_{001} = \sigma_{00-1} = \sigma$. The quantity $\sigma\,(= \langle\cos(2\pi\tilde{z})\rangle)$ represents the average molecular density modulation along the z-direction, where $\tilde{z}(= z/L_s)$ denotes the normalized position along the z-axis. The interaction that characterizes the smectic A phase is governed by the constant α_M. When $\alpha_M = 0$, the interaction potential reduces to the nematic interaction term ν_M, corresponding to the absence of positional order. The parameter α_M thus serves as a control parameter for the transition between the nematic and smectic A phases, influencing the degree of molecular layering along the z-axis. Substituting Eq. (4.126) into Eq. (4.124), we obtain

$$\Phi_{12}(\tilde{z}, \cos\theta) = v_{rr}\nu_M \left[S + \alpha_M \sigma_A \cos(2\pi\tilde{z})\right] P_2(\cos\theta), \tag{4.127}$$

where the order parameter of the smectic A phase is given by (see Eq. (2.15))

$$\sigma_A = S\sigma = \langle P_2(\cos\theta)\cos(2\pi\tilde{z})\rangle. \tag{4.128}$$

According to McMillan theory, the constant α_M (McMillan interaction parameter) is given by the following expression:

$$\alpha_M = 2\exp\left[-\left(\frac{\pi r}{L_s}\right)^2\right], \tag{4.129}$$

[5]Note: Euler's formula is $\exp(i\psi) = \cos\psi + i\sin\psi$.

where r is a characteristic length, this expression indicates how the interaction strength α_M decays with the periodicity of the smectic phase. The constant α_M ranges from 0 to 2 and characterizes the strength of the intermolecular interactions responsible for smectic layering. The periodicity of the smectic A phase is typically on the order of the molecular length L. As the number of alkyl chains at the ends of the liquid crystalline molecules increases, the value of α_M tends to increase. This increase reflects the enhanced intermolecular interactions arising from longer or more flexible terminal chains, which promote stronger positional ordering and stabilize the smectic phase.

The molecular distribution function $f_d(\mathbf{r}, \mathbf{\Omega})$ depends on the position along the z-axis and the orientational angle. Therefore, it can be expressed as $f_d(\tilde{z}, \cos\theta)$. Using this distribution function, the two order parameters are given by

$$S = \int_0^1 \int_0^1 P_2(\cos\theta) f_d(\tilde{z}, \cos\theta) d\tilde{z} d(\cos\theta), \tag{4.130}$$

$$\sigma_A = \int_0^1 \int_0^1 P_2(\cos\theta) \cos(2\pi\tilde{z}) f_d(\tilde{z}, \cos\theta) d\tilde{z} d(\cos\theta), \tag{4.131}$$

with the normalization condition:

$$\int_0^1 \int_0^1 f_d(\tilde{z}, \cos\theta) d\tilde{z} d(\cos\theta) = 1. \tag{4.132}$$

By substituting Eq. (4.127) into Eq. (4.6) and following the same procedure as in Eqs. (4.7) and (4.8), the dimensionless free energy density can be written as

$$f_{LC} = v_r \beta F/V = f_{iso} + f_A, \tag{4.133}$$

where the free energy f_{iso} of the isotropic phase is given by Eq. (4.37) and the free energy f_A of the smectic A phase is given by

$$\begin{aligned} f_A &= \phi \int_0^1 \int_0^1 f_d(\tilde{z}, \cos\theta) \ln f_d(\tilde{z}, \cos\theta) d\tilde{z} d(\cos\theta) \\ &\quad -\frac{1}{2} n_r \nu_M \phi^2 (S^2 + \alpha_M \sigma_A^2). \end{aligned} \tag{4.134}$$

When $S = \sigma_A = 0$, the system is in the isotropic phase; when $S \neq 0$ and $\sigma_A = 0$, it corresponds to the nematic phase; and when both $S \neq 0$ and $\sigma_A \neq 0$, the system is in the smectic A phase. The distribution function in thermal equilibrium is determined by the condition

$$\left(\frac{\delta f_A}{\delta f_d(\tilde{z}, \cos\theta)} \right)_T = 0, \tag{4.135}$$

from which we obtain

$$f_d(\tilde{z}, x) = \frac{1}{Z_0} \exp[n_r \nu_M \phi A(\tilde{z}, x)], \tag{4.136}$$

$$A(\tilde{z}, x) = S P_2(x) + \alpha_M \sigma_A P_2(x) \cos(2\pi\tilde{z}), \tag{4.137}$$

where $x = \cos\theta$. The constant Z_0 is determined from the normalization condition in Eq. (4.132) as

$$Z_0 = \int_0^1 \int_0^1 \exp\left[n_r \nu_M \phi A(\tilde{z}, x)\right] d\tilde{z} dx. \tag{4.138}$$

Using the distribution function, Eq. (4.136), the two order parameters given by Eqs. (4.130) and (4.131) are expressed as the self-consistent equations for S and σ_A:

$$S = \frac{1}{Z_0} \int_0^1 \int_0^1 P_2(x) \exp[A(\tilde{z}, x)] d\tilde{z} dx, \tag{4.139}$$

$$\sigma_A = \frac{1}{Z_0} \int_0^1 \int_0^1 P_2(x) \cos(2\pi \tilde{z}) \exp[A(\tilde{z}, x)] d\tilde{z} dx. \tag{4.140}$$

Note that the function $A(\tilde{z}, x)$ depends on the order parameters S and σ_A.

To simplify the calculations, we employ the decoupling approximation $f_d(\tilde{z}, \cos\theta) = f_d(\tilde{z}) f_d(\cos\theta)$, whereby the double integrals in equations such as Eq. (4.130) can be reduced to single integrals. Even with this approximation, the numerical results remain nearly identical to those obtained from exact calculations [55]. Substituting Eq. (4.136) into Eq. (4.134), the free energy of the smectic A phase is given by

$$f_A = \frac{1}{2} n_r \nu_M \phi^2 (S^2 + \alpha_M \sigma_A^2) - \phi \ln Z_0. \tag{4.141}$$

To discuss the thermal equilibrium state, it is necessary to numerically determine the values of S and σ_A that minimize this free energy. The value of order parameters S and σ_A can be determined by Eqs. (4.139) and (4.140).

4.6.2 Nematic-smectic A phase transitions

We present the calculation results for pure liquid crystalline molecules with an axial ratio of $n_r = L/D = 2$ and a volume fraction $\phi = 1$. The dimensionless temperature parameter τ (defined as $\tau = 1/\nu_M = k_B T/U_M$) is used, where larger values of τ correspond to higher temperatures. Two key parameters characterize the material: ν_M, which determines the NIT temperature, and the dimensionless smectic interaction parameter α_M (see Eq. (4.129)), which represents the strength of the interaction for smectic ordering.

Figure 4.15 (a), (b), and (c) show the temperature dependence of the order parameters S and σ_A for various values of α_M. (a) For $\alpha_M = 1.1$, both order parameters exhibit a discontinuous jump, indicating a direct first-order isotropic-smectic A phase transition (IS$_A$T). (b) At $\alpha_M = 0.8$, a first-order NIT occurs at higher temperatures, followed by a first-order nematic-smectic-A transition (NS$_A$T) at lower temperatures. (c) For $\alpha_M = 0.65$, the smectic order parameter σ_A increases continuously from zero at $\tau = 0.37$, following the NIT. This behavior indicates that the NS$_A$T becomes second-order. Figure 4.15(d) shows the free energy of the smectic-A phase, f_A, plotted as a function of the temperature parameter τ. The three curves correspond to the values of α_M used in Fig. 4.15(a), (b), and (c). The free energy of the isotropic phase, f_{iso}, is independent of S and is set to zero as a reference. For $\alpha_M = 1.1$, the free energy exhibits a kink at the first-order IS$_A$T transition temperature. For $\alpha_M = 0.8$, the free energy shows kinks at the first-order NIT and NS$_A$T transition temperatures. For $\alpha_M = 0.65$, the free energy displays a kink at the first-order NIT and a change in slope at the second-order NS$_A$T transition temperature ($\tau \simeq 0.37$). This corresponds to the continuous onset of the smectic order parameter σ_A from zero, as shown in Fig. 4.15(c).

Figure 4.16 shows the free energy difference Δf_A plotted against the order parameter σ_A near the second-order NS$_A$T temperature for $\alpha_M = 0.65$ (Fig. 4.15(c)).The free energy f_A is a function of the two order parameters, S and σ_A, and the free energy difference from the nematic phase is calculated as $\Delta f_A = f_A(\sigma_A, S = 0.7) - f_A(\sigma_A = 0, S = 0.7)$, where S is fixed at 0.7. The curve of $\tau = 0.372$ represents the point where the continuous

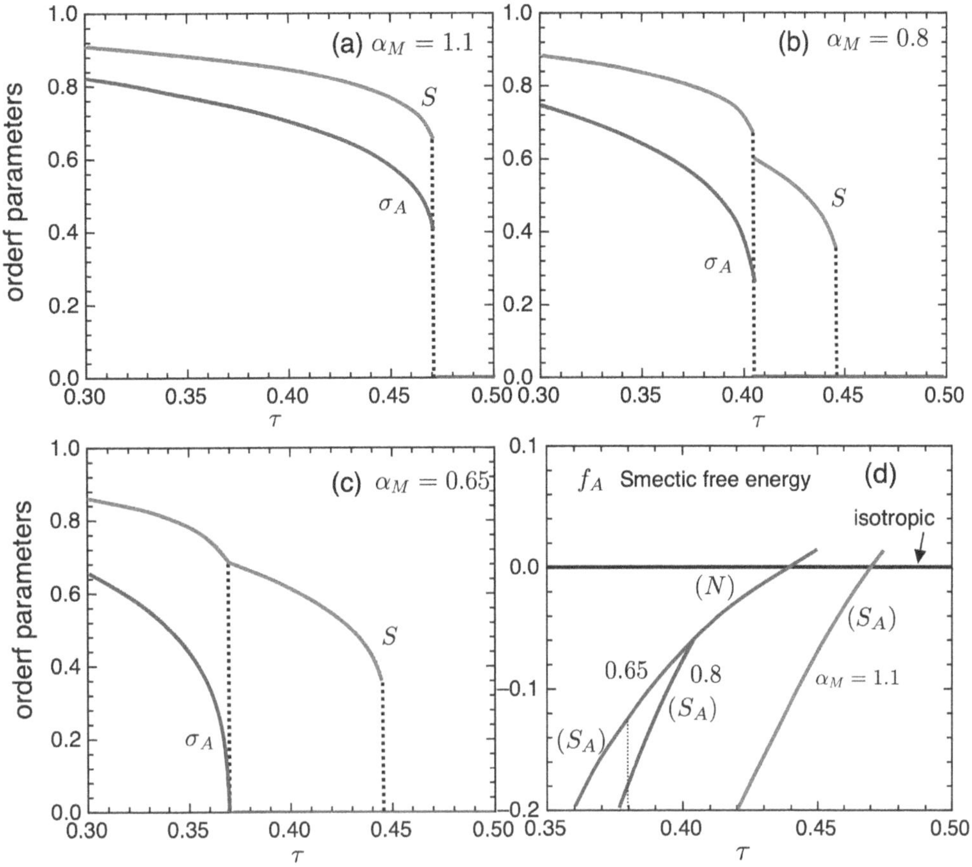

FIGURE 4.15
Order parameters S and σ_A plotted against temperature τ for various values of α_M: (a) $\alpha_M = 1.1$, (b) $\alpha_M = 0.8$, (c) $\alpha_M = 0.65$. (d) Smectic free energy f_A plotted against the temperature τ for various values of $\alpha_M = 1.1, 0.8, 0.65$.

NS$_A$T occurs. For $\tau < 0.372$ on the lower-temperature side, the free energy curve attains its minimum at $\sigma_A > 0$, indicating the stability of the smectic A phase.

Near the second-order NS$_A$T, we assume that S remains constant, $S = S_0$. Under the decoupling approximation, Eq. (4.137) can be rewritten as

$$A(\tilde{z}, x) = A(\tilde{z}) = S_0 \alpha_M \sigma_A \cos(2\pi\tilde{z}). \tag{4.142}$$

Substituting Eq. (4.142) into Eq. (4.136), we obtain the distribution function of the smectic A phase:

$$f_d(\tilde{z}) = \frac{1}{Z_0} \exp[\Gamma_A \sigma_A \cos(2\pi\tilde{z})], \tag{4.143}$$

where

$$\Gamma_A = n_r \nu_M \alpha_M \phi S_0. \tag{4.144}$$

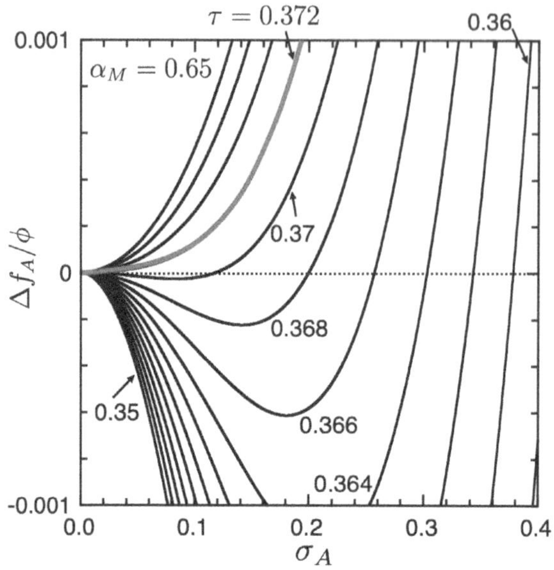

FIGURE 4.16
Free energy difference Δf_A plotted against the order parameter σ_A near the second-order NS_AT temperature for $\alpha_M = 0.65$.

The normalization constant of the distribution function, $\int_0^1 f_d(\tilde{z})d\tilde{z} = 1$, can be expanded in a power series of σ_A as

$$\begin{aligned} Z_0 &= \int_0^1 \exp[\Gamma_A \sigma_A \cos(2\pi\tilde{z})]d\tilde{z} \\ &\simeq 1 + \frac{1}{4}(\Gamma_A\sigma_A)^2 + \frac{1}{64}(\Gamma_A\sigma_A)^4 + \cdots . \end{aligned} \tag{4.145}$$

Taking the logarithm of Z_0 in Eq. (4.141), the Landau expansion[6] of the free energy (Eq. (4.141)) is expressed as (see Section 4.3)

$$f_A \simeq C_1\sigma_A^2 + C_3\sigma_A^4, \tag{4.146}$$

where the coefficients are given by

$$C_1 = 1 - \frac{1}{2}S_0\Gamma_A, \tag{4.147}$$

and

$$C_3 = \frac{1}{32}S_0\Gamma_A^3. \tag{4.148}$$

The key difference from the free energy of a first-order phase transition (Eq. (4.80)) is the absence of the cubic term σ_A^3, which characterizes the Landau theory for second-order phase transitions. The sign of C_1 changes at the NS_AT transition temperature T_{NS_A}. Using Eq. (4.147), the NS_AT transition temperature is estimated as $\tau \simeq 0.32$. In Fig. 4.16, for $\tau > 0.372$, we have $C_1 > 0$, whereas for $\tau < 0.372$, $C_1 < 0$, corresponding to the stability conditions of the S_A phase.

[6] Using $\log(1 + x) \simeq x - x^2/2$.

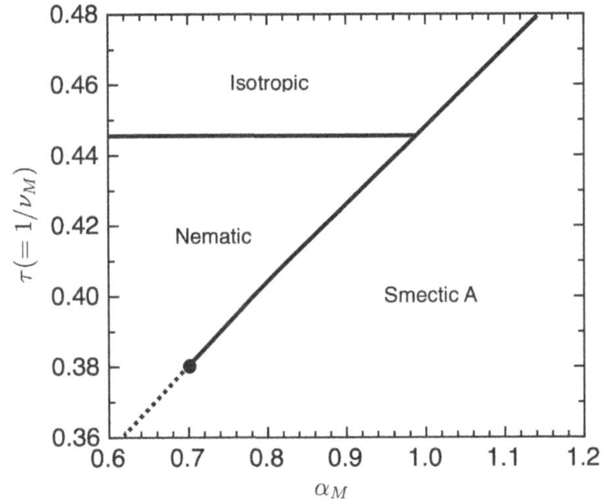

FIGURE 4.17
Phase diagram on the temperature-intermolecular interaction parameter α_M plane. Solid lines represent the first-order phase transitions, while a dashed line indicates the second-order phase transition. The intersection of the first-order and second-order transition lines corresponds to the TCP.

Figure 4.17 shows the phase diagram in the temperature-α_M plane. Solid lines represent first-order phase transitions, while dashed lines indicate second-order phase transitions. The intersection of the first-order and second-order transition lines at $\alpha_M \simeq 0.7$ is known as the tricritical point (TCP). For large values of α_M (corresponding to liquid crystalline molecules with long alkyl chains), specifically $\alpha_M > 0.98$, only the isotropic and smectic-A phases appear with temperature variation. In contrast, for $\alpha_M < 0.98$, three phases, isotropic, nematic, and smectic-A, are present. This type of phase diagram qualitatively accounts for a wide range of experimental observations (see Fig. 1.3).

4.7 Hexagonal columnar phase

At the end of this chapter, we briefly introduce the hexagonal columnar phase (see Section 2.1.5). Using the order parameter (σ_{hcl}: Eq. (2.20)) of the hexagonal columnar phase, the nematic-columnar phase transition has been studied by extending McMillan's mean field model [30]. In this case, the distribution function (Eq. (4.136)) can be rewritten as

$$f_d(\tilde{x}, \tilde{y}, \cos\theta) = \frac{1}{Z_0} \exp[n_r \nu_M \phi A(\tilde{x}, \tilde{y}, \cos\theta)], \tag{4.149}$$

where

$$\begin{aligned} A(\tilde{x}, \tilde{y}, \cos\theta) &= SP_2(\cos\theta) \\ &\quad + \alpha_M \sigma_{hcl} P_2(\cos\theta) \big(\cos(\mathbf{q}_1 \cdot \mathbf{r}) + \cos(\mathbf{q}_2 \cdot \mathbf{r}) + \cos(\mathbf{q}_3 \cdot \mathbf{r}) \big). \end{aligned} \tag{4.150}$$

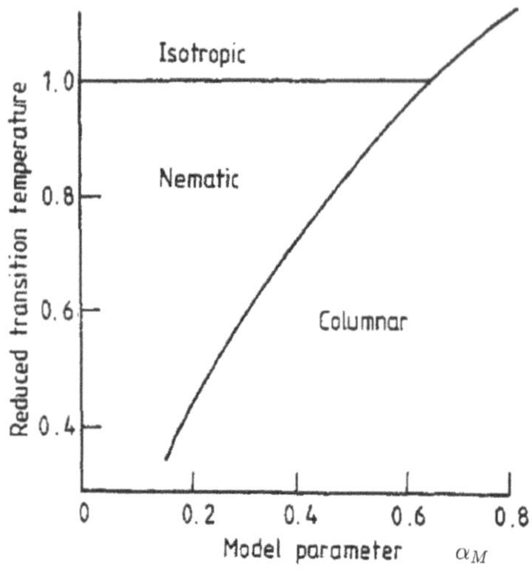

FIGURE 4.18
Phase diagram on the temperature-interaction parameter α_M plane. Solid lines indicate the phase boundaries of the hexagonal, nematic, and isotropic phases. All the phase transitions are of first order. Reprinted with permission from Ref. [30].

Similar to Eq. (4.141), the free energy of the hexagonal columnar phase is given by

$$f_{hcl} = \frac{1}{2} n_r \nu_M \phi^2 (S^2 + \alpha_M \sigma_{hcl}^2) - \phi \ln Z_0. \tag{4.151}$$

Minimizing the free energy with respect to the order parameters S and σ_{hcl}, we can calculate the phase diagram.

Figure 4.18 shows the phase diagram on the temperature-intermolecular interaction parameter α_M plane. The temperature range of the nematic phase decreases with increasing α_M, and for $\alpha_M > 0.64$ the columnar phase transforms directly to the isotropic phase. The columnar-nematic transition is always of first order, because a Landau expansion of the free energy contains a non-vanishing cubic term in the order parameter σ_{hcl}.

Various extensions and refinements of McMillan theory have been proposed to improve its accuracy [29, 56–58]. Furthermore, mean field theoretical frameworks have also been developed for other liquid crystal phases, including the smectic C and chiral smectic C phases [59,60], as well as the ferroelectric nematic (N_F) phase [61], where rod-like molecules exhibit not only long-range orientational order, as in a conventional nematic, but also a spontaneous macroscopic polarization [62, 63].

5

Cholesteric Liquid Crystals

This chapter presents the statistical mechanical theory for describing the cholesteric phase (Ch) in systems of liquid crystalline molecules and rod-like polymer solutions. The theoretical treatments of cholesteric phases were developed in the 1970s by Straley [36], Goossen [64], and Lin-Liu et al. [65,66]. Broadly speaking, Lin-Liu's theory is based on attractive interactions, whereas the models proposed by Straley and Goossen are based on excluded volume interactions. On the other hand, a theory of cholesteric polymer solutions based on the scaled particle theory (SPT) has also been developed [67], in which the chain stiffness and flexibility of polymers play crucial roles. In this chapter, we introduce a mean field theory of the cholesteric phase developed by Lin-Liu and coworkers. Building upon the framework of intermolecular interactions introduced in Chapter 4, we extend the description by incorporating a chiral contribution to the intermolecular potential.

5.1 Cholesteric free energy

The director $\mathbf{n}$ in the nematic phase is uniform over macroscopic regions. However, in the cholesteric phase, as shown in Fig. 5.1(a), it rotates in a helical manner. This helical structure is formed by chiral molecules, which do not possess mirror symmetry and are referred to as the chiral nematic phase (N*). The pitch length (p) is defined as the distance over which the director twists by 2π along the axis of the helix. Typically, the pitch length is on the order of the wavelength of visible light. By varying the pitch length through temperature, electric fields, or impurity concentrations, the color of the cholesteric phase can be changed to correspond to visible light wavelengths. This property is used in various liquid crystal sensors. The helical structure can be unwound by applying electric or magnetic fields, leading to a phase transition to the nematic phase. This is the fundamental principle behind liquid crystal displays (LCDs). Cholesteric phases also play an important role in structures such as viruses and biomacromolecules [68].

5.1.1 Scalar and vector

The director of the cholesteric phase, $\mathbf{n}$, exhibits a spatial twist along the helical axis. Let $\mathbf{p}$ denote the unit vector in the direction of the helical axis, which is assumed to be parallel to the z-axis. As illustrated in Fig. 5.1(b), the director rotates in the x-y plane with an angle $\omega(z) = (2\pi/p)z$ relative to the x-axis. Accordingly, the director at position z is given by

$$\mathbf{n}(z) = (\cos qz,\ \sin qz,\ 0), \tag{5.1}$$

where q is the helical wavenumber, and the pitch length p is related to q by $p = 2\pi/|q|$. A positive helical wave number, $q > 0$, corresponds to a right-handed helix, while a negative value, $q < 0$, corresponds to a left-handed helix. The unit vector $\mathbf{\Omega}$ representing the

DOI: 10.1201/9781003737711-5

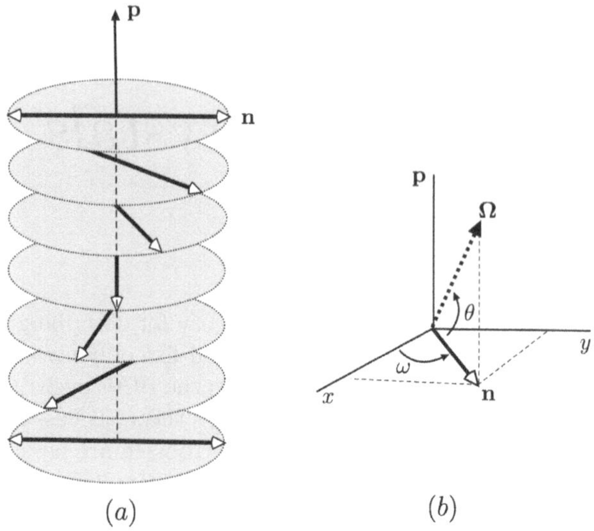

FIGURE 5.1
(a) Helical structure of the director **n** in the cholesteric liquid crystal phase. (b) The relationship between the director **n** and the long-axis vector **Ω** of a rod-like molecule.

direction of the long axis of a rod-like molecule possesses uniaxial symmetry, meaning it rotates around the director **n** while maintaining a fixed tilt angle θ with respect to it. Consequently, the molecular orientation distribution function does not depend on the azimuthal angle. However, since the director **n** varies along the z-axis, the orientational distribution function of rod-like molecules depends only on the inner product between the local director at position z, $\mathbf{n}(z)$, and the molecular orientation $\boldsymbol{\Omega}$; that is, $\mathbf{n}(z) \cdot \boldsymbol{\Omega} = \cos\theta$. Accordingly, the distribution function can be expressed as

$$f(\mathbf{r}, \boldsymbol{\Omega}) = f\big(\mathbf{n}(z) \cdot \boldsymbol{\Omega}\big), \tag{5.2}$$

which implies that the spatial dependence on $\mathbf{r} = (0, 0, z)$ is incorporated through the z-dependent director $\mathbf{n}(z)$.

Let us define the director field more generally. As illustrated in Fig. 5.2, we introduce a local orthonormal coordinate system based on the helical axis **p** and a perpendicular unit vector $\mathbf{n}_0$. The third axis of this system is given by the vector product $\mathbf{p} \times \mathbf{n}_0$. Let **r** denote the position vector, and let $\omega(\mathbf{r})$ represent the azimuthal angle of the director $\mathbf{n}(\mathbf{r})$ measured from $\mathbf{n}_0$. The director field is then expressed as

$$\mathbf{n}(\mathbf{r}) = \mathbf{n}_0 \cos\omega(\mathbf{r}) + (\mathbf{p} \times \mathbf{n}_0) \sin\omega(\mathbf{r}), \tag{5.3}$$

where the azimuthal angle is defined by

$$\omega(\mathbf{r}) = q\, \mathbf{p} \cdot \mathbf{r}. \tag{5.4}$$

In the special case, where $\mathbf{p} = (0, 0, 1)$ and $\mathbf{r} = (0, 0, z)$, we obtain $\mathbf{p} \cdot \mathbf{r} = z$, resulting in $\omega(z) = qz$. This formulation highlights that a complete description of the cholesteric helical structure involves both the dot product and the cross product, capturing the local rotation of the director around the helical axis.

Here, we review vector analysis, focusing on scalar quantities (dot products) and vector quantities (cross products), which will be necessary for the discussions in the following sections.

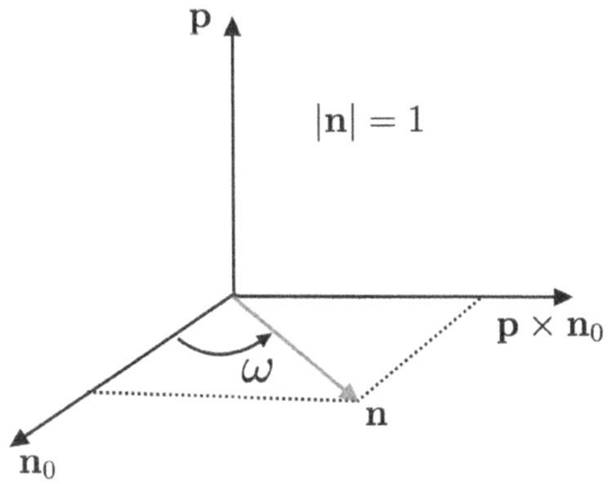

FIGURE 5.2
Director $\mathbf{n}$ in a local coordinate system defined by the pitch axis $\mathbf{p}$ and a perpendicular vector $\mathbf{n_0}$.

1. In general, the dot product of two arbitrary polar vectors $\mathbf{\Omega}$ and $\mathbf{\Omega}'$ (i.e., $\mathbf{\Omega} \cdot \mathbf{\Omega}'$) yields a scalar quantity that possesses only magnitude.[1] If the components of the vector $\mathbf{\Omega}$ are denoted as Ω_α with $\alpha = x, y, z$, the dot product can be expressed using Einstein summation notation as:

$$\mathbf{\Omega} \cdot \mathbf{\Omega}' = \Omega_x \Omega'_x + \Omega_y \Omega'_y + \Omega_z \Omega'_z = \Omega_\alpha \Omega'_\alpha, \tag{5.5}$$

where repeated indices imply summation. Importantly, the dot product is symmetric with respect to the order of the vectors: $\mathbf{\Omega} \cdot \mathbf{\Omega}' = \mathbf{\Omega}' \cdot \mathbf{\Omega}$.

2. The cross product of two vectors $\mathbf{\Omega}$ and $\mathbf{\Omega}'$: $(\mathbf{\Omega} \times \mathbf{\Omega}')$, results in a vector quantity that possesses both magnitude and direction. If $\mathbf{e}_\alpha$ denotes the unit basis vector along the $\alpha(= x, y, z)$ axis, the cross product can be expressed as

$$\mathbf{\Omega} \times \mathbf{\Omega}' = \epsilon_{\alpha\beta\gamma} \Omega_\beta \Omega'_\gamma \, \mathbf{e}_\alpha, \tag{5.6}$$

where $\epsilon_{\alpha\beta\gamma}$ is the Levi-Civita symbol (Eq. (2.57)). The indices (α, β, γ) follow the order of the 3D coordinate axes (x, y, z), and the value is 1 when the order is (x, y, z) (i.e., $\epsilon_{xyz} = 1$), -1 when the order is swapped, such as (y, x, z) (i.e., $\epsilon_{yxz} = -1$), and one again for a further swap, such as (y, z, x) (i.e., $\epsilon_{yzx} = 1$). If the same letter appears in any two indices, for example, (y, y, z), the value is 0 (i.e., $\epsilon_{yyz} = 0$). The direction of the cross product $\mathbf{\Omega} \times \mathbf{\Omega}'$ is given by the right-hand rule: if you rotate a right-hand screw from the direction of $\mathbf{\Omega}$ to $\mathbf{\Omega}'$ by 180 degrees, the direction the screw moves is the direction of the cross product. For example, in Fig. 5.2, the direction of the cross product $\mathbf{n_0} \times \mathbf{n}$ is along the $\mathbf{p}$-axis.[2] It is also important to note the antisymmetric nature of the cross product: swapping the two vectors reverses the direction of the result: $\mathbf{\Omega} \times \mathbf{\Omega}' = -\mathbf{\Omega}' \times \mathbf{\Omega}$.

3. Considering the scalar triple product involving the cross product $\mathbf{\Omega} \times \mathbf{\Omega}'$ and the position vector $\mathbf{r}$, we have the identity: $\mathbf{\Omega} \times \mathbf{\Omega}' \cdot \mathbf{r} = -\mathbf{\Omega}' \times \mathbf{\Omega} \cdot \mathbf{r}$. This reflects

[1]There are two types of vectors: polar vectors (such as those representing position or orientation) and axial vectors (such as those representing angular velocity or magnetic fields).

[2]The magnitude of the cross product is the area of the parallelogram formed by $\mathbf{n_0}$ and $\mathbf{n}$, i.e., $|\mathbf{n_0} \times \mathbf{n}| = |\sin\omega|$.

the antisymmetric nature of the cross product. Using Eq. (5.6), the scalar triple product can be expressed in component form as:

$$\begin{aligned}\mathbf{\Omega} \times \mathbf{\Omega}' \cdot \mathbf{r} &= \epsilon_{\alpha\beta\gamma}\Omega_\beta\Omega'_\gamma \, \mathbf{e}_\alpha \cdot \mathbf{r} \\ &= \epsilon_{\alpha\beta\gamma}\Omega_\beta\Omega'_\gamma \, r_\alpha, \end{aligned} \tag{5.7}$$

where r_α denotes the α-th component of the vector $\mathbf{r}$. This scalar triple product is an example of a pseudoscalar and a quantity that changes sign under improper transformations such as spatial inversion (parity transformation). In contrast, the product of two pseudoscalars results in a true scalar, which remains invariant under such transformations. For reference, the scalar triple product of the form $\mathbf{\Omega} \times \mathbf{\Omega}' \cdot \mathbf{\Omega}$ is always zero due to the orthogonality of the cross product to both vectors involved: $\mathbf{\Omega} \times \mathbf{\Omega}' \cdot \mathbf{\Omega} = 0$.

The intermolecular interaction $U_{12}(\mathbf{r}, \mathbf{\Omega}; \mathbf{r}', \mathbf{\Omega}')$ is a scalar quantity. This is because interaction potentials are physical quantities that must remain invariant under coordinate transformations and do not depend on the choice of coordinate system. In other words, U_{12} must preserve rotational symmetry and is therefore a scalar field. The variables $\mathbf{r}$ and $\mathbf{\Omega}$ are vectors, representing the position and orientation of molecules, respectively. The interaction potential between two molecules located at $\mathbf{r}$ and $\mathbf{r}'$ is determined by their relative distance $r_{12} = |\mathbf{r}' - \mathbf{r}|$, so the interaction may be written as $U_{12}(r_{12})$ in the simplest cases. A classic example is the Coulomb interaction between two point charges q_1 and q_2, separated by a distance r_{12}: $U_{12}(r_{12}) = q_1 q_2/(4\pi\epsilon_0 r_{12})$, which is clearly a scalar quantity. Here, ϵ_0 is the vacuum permittivity. The corresponding Coulomb force is obtained by taking the gradient of this scalar potential:

$$\mathbf{F} = -\nabla U_{12}(r_{12}) = -\left(\frac{\partial U_{12}}{\partial r_{12}}\right) \mathbf{e}_r, \tag{5.8}$$

where $\mathbf{e}_r$ is the unit vector in the direction of $\mathbf{r}_{12} (= \mathbf{r}' - \mathbf{r})$. This result is a vector quantity. In the case of rod-like or liquid crystalline molecules, the intermolecular interaction may depend not only on the distance r_{12} but also on the orientations of the molecules. In such cases, the interaction is expressed as $U_{12} = U_{12}(r_{12}, \mathbf{\Omega}, \mathbf{\Omega}')$, where $\mathbf{\Omega}$ and $\mathbf{\Omega}'$ describe the orientations of the two molecules. The intermolecular potential U_{12}, which enters the expression for the free energy, is a scalar quantity with units of energy (e.g., joules [J] or calories [cal]). In contrast, the intermolecular force $\mathbf{F}$, derived from the potential via differentiation, is a vector with units of force (newtons [N = J/m]).

5.1.2 Twist interactions

The simplest model of a chiral rod-like molecule that exhibits broken mirror symmetry would be a rod-like molecule with a helical thread (see Fig. 5.3).[3] Although real molecules are more complex, we consider here a rod-like molecular model with length L and diameter D, featuring a helical thread (without considering its details). When the helical threads overlap, a twisted state emerges, giving rise to a right-handed (or left-handed) helical structure. The intermolecular interaction between two rod-like molecules, one located at position $\mathbf{r}$ with its long-axis direction $\mathbf{\Omega}$, and the other at position $\mathbf{r}'$ with its long-axis direction $\mathbf{\Omega}'$, can

[3] As shown in Fig. 5.3, when a molecule with an oblique helical thread is reflected in a mirror, it cannot be superimposed onto its original form, just like the right and left hands. Such molecules are called chiral molecules. On the other hand, the conventional rod-like molecules discussed in Chapter 4 can be superimposed onto their mirror images and are referred to as achiral molecules. The former exhibits broken mirror symmetry.

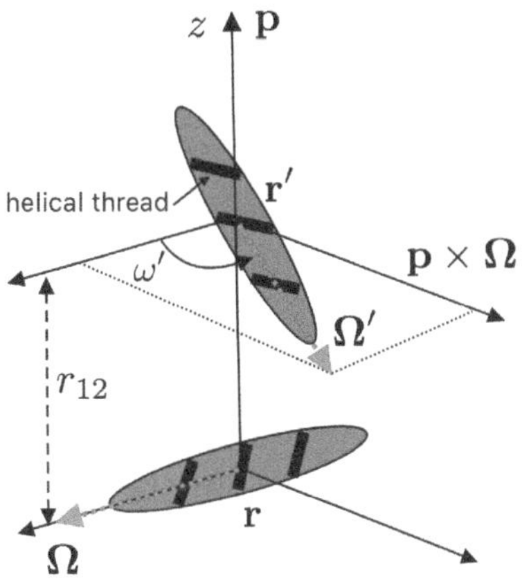

FIGURE 5.3
Two chiral rod-like molecules, one with orientation $\boldsymbol{\Omega}$ located at position $\mathbf{r}$, and the other with orientation $\boldsymbol{\Omega}'$ located at position $\mathbf{r}'$.

generally be expressed in terms of the following five scalar variables:

$$\begin{aligned} U_{12}(\mathbf{r}, \boldsymbol{\Omega}; \mathbf{r}', \boldsymbol{\Omega}') &= U_{12}(r_{12},\ \hat{\mathbf{r}}_{12} \cdot \boldsymbol{\Omega},\ \hat{\mathbf{r}}_{12} \cdot \boldsymbol{\Omega}', \\ &\quad \boldsymbol{\Omega} \cdot \boldsymbol{\Omega}',\ \boldsymbol{\Omega} \times \boldsymbol{\Omega}' \cdot \hat{\mathbf{r}}_{12}), \end{aligned} \tag{5.9}$$

where $\hat{\mathbf{r}}_{12}(= \mathbf{r}_{12}/r_{12})$ is the unit vector, and $\hat{\mathbf{r}}_{12} \cdot \boldsymbol{\Omega}$, $\hat{\mathbf{r}}_{12} \cdot \boldsymbol{\Omega}'$ represent the scalar products of the intermolecular distance vector with the molecular long-axis directions. Here, $\hat{\mathbf{r}}_{12} \cdot \boldsymbol{\Omega}$ and $\hat{\mathbf{r}}_{12} \cdot \boldsymbol{\Omega}'$ are not relevant to the cholesteric phase and are therefore ignored (these quantities become important in the smectic phase). The term $\boldsymbol{\Omega} \cdot \boldsymbol{\Omega}'$ denotes the scalar product of the orientation directions of the two rod-like molecules. These four scalar quantities remain unchanged in sign when the two vectors are interchanged ($\boldsymbol{\Omega} \cdot \boldsymbol{\Omega}' = \boldsymbol{\Omega}' \cdot \boldsymbol{\Omega}$). On the other hand, the last scalar triple product, $\boldsymbol{\Omega} \times \boldsymbol{\Omega}' \cdot \hat{\mathbf{r}}_{12}$, is a pseudoscalar that changes sign upon permutation of its elements. By separating these variables with different properties, the intermolecular interaction can be expressed as

$$\begin{aligned} U_{12}(\mathbf{r}, \boldsymbol{\Omega}; \mathbf{r}', \boldsymbol{\Omega}') &= U_N(r_{12},\ \boldsymbol{\Omega} \cdot \boldsymbol{\Omega}') \\ &\quad + (\boldsymbol{\Omega} \times \boldsymbol{\Omega}' \cdot \hat{\mathbf{r}}_{12}) U_X(r_{12},\ \boldsymbol{\Omega} \cdot \boldsymbol{\Omega}'). \end{aligned} \tag{5.10}$$

The first term represents the intermolecular interaction that describes the nematic phase, while the second term is specific to the cholesteric phase. The intermolecular interaction U_N is an even function of $\boldsymbol{\Omega} \cdot \boldsymbol{\Omega}'$: $U_N(-\boldsymbol{\Omega} \cdot \boldsymbol{\Omega}') = U_N(\boldsymbol{\Omega} \cdot \boldsymbol{\Omega}')$. In contrast, U_X is an odd function: $U_X(-\boldsymbol{\Omega} \cdot \boldsymbol{\Omega}') = -U_X(\boldsymbol{\Omega} \cdot \boldsymbol{\Omega}')$. Therefore, the second term in Eq. (5.10) is a product of two pseudoscalars, making it a true scalar quantity.

We here briefly explain how the pseudoscalar $(\boldsymbol{\Omega} \times \boldsymbol{\Omega}' \cdot \hat{\mathbf{r}}_{12})$ describes "twist." As shown in Fig. 5.3, if the orientation vectors of two rod-like molecules are given by $\boldsymbol{\Omega}$ and $\boldsymbol{\Omega}'$, then $\boldsymbol{\Omega}'$ can be expressed as[4]

$$\boldsymbol{\Omega}' = \boldsymbol{\Omega} \cos \omega' + (\mathbf{p} \times \boldsymbol{\Omega}) \sin \omega'. \tag{5.11}$$

[4]In Fig. 5.2, replacing the vectors $\mathbf{n}_0$ and $\mathbf{n}$ with $\boldsymbol{\Omega}$ and $\boldsymbol{\Omega}'$, respectively, is the same operation.

By setting the unit vector along the helical axis as $\mathbf{p} = (0, 0, 1)$ and using Eq. (5.11), we obtain

$$\begin{aligned}
\boldsymbol{\Omega} \times \boldsymbol{\Omega}' \cdot \hat{\mathbf{r}}_{12} &= (\boldsymbol{\Omega} \times (\mathbf{p} \times \boldsymbol{\Omega}) \sin\omega') \cdot \hat{\mathbf{r}}_{12} \\
&= \mathbf{p} \cdot \hat{\mathbf{r}}_{12} \sin\omega' \\
&= \hat{r}_{12,z} \sin\omega', \qquad (5.12)
\end{aligned}$$

where $\hat{r}_{12,z}$ denotes the z-component of the unit vector $\hat{\mathbf{r}}_{12}$, In the case of helical rotation along the z-axis, $\hat{\mathbf{r}}_{12} = (0, 0, 1)$ holds. Clearly, Eq. (5.12) reverses when ω' is replaced with $-\omega'$, it is classified as a pseudoscalar. Consequently, the chiral interaction term U_X in Eq. (5.10) must also be a pseudoscalar.

The intermolecular potential in Eq. (5.10) can be expanded in terms of Legendre polynomials $P_l(x)$, as follows:

$$U_N(r_{12},\ \boldsymbol{\Omega} \cdot \boldsymbol{\Omega}') = \sum_{l=0}^{\infty} U_{2l}(r_{12}) P_{2l}(\boldsymbol{\Omega} \cdot \boldsymbol{\Omega}'), \qquad (5.13)$$

$$U_X(r_{12},\ \boldsymbol{\Omega} \cdot \boldsymbol{\Omega}') = \sum_{l=0}^{\infty} U_{2l+1}(r_{12}) P_{2l+1}(\boldsymbol{\Omega} \cdot \boldsymbol{\Omega}'), \qquad (5.14)$$

where U_N denotes the non-chiral (even) part of the interaction potential, and U_X represents the chiral (odd) part. The use of even and odd-order Legendre polynomials reflects the symmetry and parity properties of each component with respect to the molecular orientations. In the following, we approximate by truncating the series at $l = 1$. This yields:

$$U_N(r_{12},\ \boldsymbol{\Omega} \cdot \boldsymbol{\Omega}') = U_0(r_{12}) + U_2(r_{12}) P_2(\boldsymbol{\Omega} \cdot \boldsymbol{\Omega}'), \qquad (5.15)$$

$$U_X(r_{12},\ \boldsymbol{\Omega} \cdot \boldsymbol{\Omega}') = U_1(r_{12}) P_1(\boldsymbol{\Omega} \cdot \boldsymbol{\Omega}'), \qquad (5.16)$$

where the Legendre polynomials are defined as $P_0(\xi) = 1$, $P_1(\xi) = \xi$, $P_2(\xi) = (3/2)(\xi^2 - 1/3)$. Higher-order terms beyond $P_2(\xi)$ are neglected in this approximation. Clearly, the $U_2(r_{12})$ term corresponds to the attractive intermolecular interaction that, in the nematic phase, is described by the Maier-Saupe theory introduced in Section 4.2. In the Onsager theory, it corresponds to the excluded volume term (Eq. (4.29)). The $U_1(r_{12})$ term corresponds to the intermolecular interaction that characterizes the twist interaction in the cholesteric phase (pseudoscalar potential). We will proceed with the approximation up to $P_2(\xi)$ in the following discussion. Additionally, the $U_0(r_{12})$ term can be neglected, as it does not depend on molecular orientation.

The intermolecular interaction $U_2(r_{12})$ is assumed to follow a square-well potential form (a similar square-well potential is also assumed for the chiral interaction U_1). As illustrated in Fig. 5.4(a), it is defined as:

$$U_2(r_{12}, \boldsymbol{\Omega}, \boldsymbol{\Omega}') = \begin{cases} \infty, & 0 < r_{12} < \sigma_{12} \quad \text{(in contact)} \\ -u_2(r_{12}), & \sigma_{12} < r_{12} < \sigma_{12} + d_0 \quad \text{(short range interaction)} \\ 0, & \sigma_{12} + d_0 < r_{12} \quad \text{(not in contact)} \end{cases} \qquad (5.17)$$

where the function σ_{12} denotes the minimum allowed distance between the centers of mass of two rod-like molecules in contact, defined for a fixed relative angle $\theta_{12}(\boldsymbol{\Omega}, \boldsymbol{\Omega}')$ between their orientations. The interaction $U_{12}(r_{12})$ is assumed to be a short-range interaction on the molecular scale d_0. Accordingly, for the nematic interaction U_N (Eq. (5.15)), which accounts for both the excluded volume effect (entropic contribution) and the attractive interaction between rod-like molecules, the expression can be written as:

$$\begin{aligned}
\beta U_N(r_{12};\ \boldsymbol{\Omega} \cdot \boldsymbol{\Omega}') &= 2DL^2 |\boldsymbol{\Omega} \times \boldsymbol{\Omega}'| \delta(R_{12}) \\
&\quad - \beta u_2(r_{12}) P_2(\boldsymbol{\Omega} \cdot \boldsymbol{\Omega}'), \qquad (5.18)
\end{aligned}$$

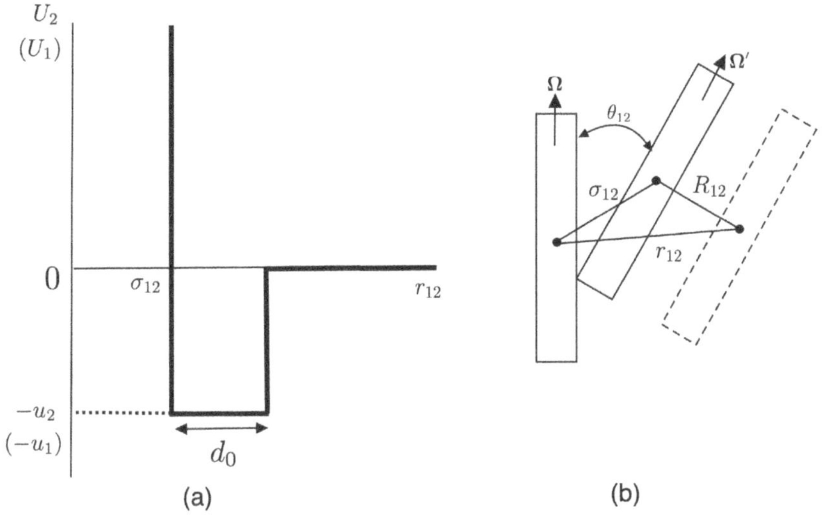

FIGURE 5.4
(a) Square-well potential and (b) the contact distance σ_{12} between two rod-like molecules.

where $R_{12} = \sigma_{12} - r_{12}$ represents the deviation of the actual intermolecular distance r_{12} from the contact distance σ_{12} (Fig. 5.4(b)). The first term corresponds to the excluded volume effect, represented by a delta function (representing contact interactions), while the second term describes the orientation-dependent attractive potential. The Kronecker delta function $\delta(R_{12})$ has the dimension of per unit volume and takes the value $\delta(R_{12}) = 1$ when $R_{12} = 0$ (i.e., when the molecules are in contact) and $\delta(R_{12}) = 0$ otherwise. In the Lin-Liu theory [65] of the cholesteric phase, the excluded volume in the first term is not taken into account.

Expanding the first term in terms of the Legendre polynomials (see Eq. (4.26)),

$$\begin{aligned} |\boldsymbol{\Omega} \times \boldsymbol{\Omega}'| &= |\sin\theta_{12}| \\ &\simeq \frac{\pi}{4} - \frac{5\pi}{32} P_2(\boldsymbol{\Omega} \cdot \boldsymbol{\Omega}'), \end{aligned} \tag{5.19}$$

and Eq. (5.18) is given by

$$\beta U_N(r_{12}; \boldsymbol{\Omega} \cdot \boldsymbol{\Omega}') = \frac{\pi}{2} D L^2 \delta(R_{12}) - \beta \tilde{u}_2(r_{12}) P_2(\boldsymbol{\Omega} \cdot \boldsymbol{\Omega}'), \tag{5.20}$$

where

$$\beta \tilde{u}_2(r_{12}) = \beta u_2(r_{12}) + \frac{5\pi}{16} D L^2 \delta(R_{12}). \tag{5.21}$$

We also have the twist interaction (Eq. (5.16))

$$U_X(r_{12},\ \boldsymbol{\Omega} \cdot \boldsymbol{\Omega}') = -\beta u_1(r_{12}) P_1(\boldsymbol{\Omega} \cdot \boldsymbol{\Omega}'). \tag{5.22}$$

The intermolecular interactions $u_1(r_{12})$ and $u_2(r_{12})$ are assumed to be short-range potentials acting over a distance of approximately d_0, as shown in Fig. 5.4.

Substituting Eqs. (5.20) and (5.22) into Eq. (5.10), we obtain

$$\begin{aligned} U_{12}(r_{12}; \boldsymbol{\Omega} \cdot \boldsymbol{\Omega}') &= \frac{\pi}{2} D L^2 \delta(R_{12}) - \beta \tilde{u}_2(r_{12}) P_2(\boldsymbol{\Omega} \cdot \boldsymbol{\Omega}') \\ &\quad - \beta u_1(r_{12})(\boldsymbol{\Omega} \times \boldsymbol{\Omega}' \cdot \hat{\mathbf{r}}_{12}) P_1(\boldsymbol{\Omega} \cdot \boldsymbol{\Omega}'). \end{aligned} \tag{5.23}$$

The first term represents the excluded volume interaction, the second term represents the nematic interaction, and the third term represents the twist interaction. By considering these intermolecular interactions, the free energy is ultimately expressed as a quadratic function of the pitch wavenumber. Below, we will derive this result in detail.

5.1.3 Free energy of the cholesteric phase

We consider a solution containing N rod-like molecules dispersed in a solvent, occupying a total volume V. The number density of rod-like molecules at position $\mathbf{r}$ with orientation $\boldsymbol{\Omega}$ is denoted as $\rho(\mathbf{r}, \boldsymbol{\Omega}) = \rho f_d(\mathbf{n}(z) \cdot \boldsymbol{\Omega})$, where $\rho = N/V$ and $f_d(\mathbf{n}(z) \cdot \boldsymbol{\Omega})$ is the orientational distribution function. Using the second virial approximation Eq. (4.10), the free energy is given by

$$\begin{aligned}\beta F &= N\big(\beta\mu_r^\circ + \ln\rho - 1\big) \\ &\quad + N\int f_d(\mathbf{n}(z)\cdot\boldsymbol{\Omega}) \ln\big[4\pi f_d(\mathbf{n}(z)\cdot\boldsymbol{\Omega})\big] d\boldsymbol{\Omega} \\ &\quad + \frac{1}{2}\rho^2 \iiiint f_d(\mathbf{n}(z)\cdot\boldsymbol{\Omega}) f_d(\mathbf{n}(z)\cdot\boldsymbol{\Omega}')\beta U_{12}(r_{12};\boldsymbol{\Omega}\cdot\boldsymbol{\Omega}')d\mathbf{R}, \qquad (5.24)\end{aligned}$$

where $\beta = 1/(k_B T)$, $d\boldsymbol{\Omega} = 2\pi\sin\theta d\theta$, and we define $d\mathbf{R} = d\mathbf{r}d\mathbf{r}'d\boldsymbol{\Omega}d\boldsymbol{\Omega}'$. The distribution function satisfies the normalization condition:

$$\int f_d(\mathbf{n}(z)\cdot\boldsymbol{\Omega})d\boldsymbol{\Omega} = 1. \qquad (5.25)$$

Substituting Eq. (5.23) into Eq. (5.24), we obtain

$$\begin{aligned}\beta F &= \beta F_{iso} + N\int f_d(\mathbf{n}(z)\cdot\boldsymbol{\Omega}) \ln\big[4\pi f_d(\mathbf{n}(z)\cdot\boldsymbol{\Omega})\big] d\boldsymbol{\Omega} \\ &\quad + \beta(F_1 + F_2). \qquad (5.26)\end{aligned}$$

The first term

$$\beta F_{iso} = N\big(\beta\mu_r^\circ + \ln\rho - 1 + (\pi/4)DL^2\rho\big), \qquad (5.27)$$

represents the free energy of the isotropic phase. The final term in Eq. (5.27) is derived as follows:

$$\begin{aligned}&\frac{1}{2}\rho^2 \iiiint f_d(\mathbf{n}(z)\cdot\boldsymbol{\Omega}) f_d(\mathbf{n}(z)\cdot\boldsymbol{\Omega}')\frac{\pi}{2}DL^2\delta(R_{12})d\mathbf{r}d\mathbf{r}'d\boldsymbol{\Omega}d\boldsymbol{\Omega}', \\ &= \frac{1}{2}\rho^2 \int f_d(\mathbf{n}(z)\cdot\boldsymbol{\Omega})d\Omega \int f_d(\mathbf{n}(z)\cdot\boldsymbol{\Omega}')d\boldsymbol{\Omega}' \iint \frac{\pi}{2}DL^2\delta(R_{12})d\mathbf{r}d\mathbf{r}' \\ &= \frac{1}{2}\rho^2 \int \frac{\pi}{2}DL^2\delta(R_{12})dr_{12} \int d\mathbf{r} \\ &= \frac{\pi}{4}DL^2\rho N, \qquad (5.28)\end{aligned}$$

where in the final expression, the integration variables are changed to $\mathbf{r}_{12}$ and $\mathbf{r}$. The term F_1 in Eq. (5.26) shows the chiral interaction term:

$$\begin{aligned}\beta F_1 &= -\frac{1}{2}\rho^2 \iiiint f_d(\mathbf{n}(\mathbf{r})\cdot\boldsymbol{\Omega}) f_d(\mathbf{n}(\mathbf{r}')\cdot\boldsymbol{\Omega}') \\ &\quad \beta u_1(r_{12})(\boldsymbol{\Omega}\times\boldsymbol{\Omega}'\cdot\hat{\mathbf{r}}_{12})P_1(\boldsymbol{\Omega}\cdot\boldsymbol{\Omega}')d\mathbf{R}, \qquad (5.29)\end{aligned}$$

and F_2 shows the free energy of the nematic phase:

$$\beta F_2 = -\frac{1}{2}\rho^2 \iiiint f_d(\mathbf{n}(\mathbf{r})\cdot\mathbf{\Omega}) f_d(\mathbf{n}(\mathbf{r}')\cdot\mathbf{\Omega}')\beta\tilde{u}_2(r_{12})P_2(\mathbf{\Omega}\cdot\mathbf{\Omega}')d\mathbf{R}. \quad (5.30)$$

To proceed further with the calculation, we introduce the tensor order parameter ($Q_{\alpha\beta}$) at position $\mathbf{r}$ (refer to the Section 2.2), defined as

$$\begin{aligned} Q_{\alpha\beta}(\mathbf{r}) &= \langle\frac{3}{2}\Omega_\alpha\Omega_\beta - \frac{1}{2}\delta_{\alpha\beta}\rangle \\ &= \int\left(\frac{3}{2}\Omega_\alpha\Omega_\beta - \frac{1}{2}\delta_{\alpha\beta}\right) f(\mathbf{n}(\mathbf{r})\cdot\mathbf{\Omega})d\mathbf{\Omega}, \end{aligned} \quad (5.31)$$

where Ω_α denotes the α-component ($\alpha = x, y, z$) of the unit vector $\mathbf{\Omega}$ along the long axis of a rod-like molecule. Using Eq. (5.31), when averaging with respect to the orientation $d\mathbf{\Omega}$, we obtain

$$\begin{aligned} \langle P_2(\mathbf{\Omega}\cdot\mathbf{\Omega}')\rangle &= \langle\frac{3}{2}(\mathbf{\Omega}\cdot\mathbf{\Omega}')^2 - \frac{1}{2}\delta_{\alpha\beta}\rangle \\ &= \langle\frac{3}{2}\Omega_\alpha\Omega_\beta\Omega'_\alpha\Omega'_\beta - \frac{1}{2}\delta_{\alpha\beta}\rangle \\ &= \frac{2}{3}\big(Q_{\alpha\beta}(\mathbf{r}) + \frac{1}{2}\delta_{\alpha\beta}\big)\big(Q_{\alpha\beta}(\mathbf{r}') + \frac{1}{2}\delta_{\alpha\beta}\big) - \frac{1}{2}\delta_{\alpha\beta} \\ &= \frac{2}{3}Q_{\alpha\beta}(\mathbf{r})Q_{\alpha\beta}(\mathbf{r}'). \end{aligned} \quad (5.32)$$

Using Eq. (5.7), we have

$$\begin{aligned} \langle(\mathbf{\Omega}\times\mathbf{\Omega}'\cdot\hat{\mathbf{r}}_{12})P_1(\mathbf{\Omega}\cdot\mathbf{\Omega}')\rangle &= \langle\epsilon_{\alpha\beta\gamma}\ \Omega_\beta\Omega'_\gamma\ \mathbf{e}_\alpha\cdot\hat{\mathbf{r}}_{12}\ \Omega_\mu\Omega'_\mu\rangle \\ &= \langle\epsilon_{\alpha\beta\gamma}\ \hat{r}_{12,\alpha}\Omega_\beta\Omega_\mu\Omega'_\gamma\Omega'_\mu\rangle \\ &= \frac{4}{9}\epsilon_{\alpha\beta\gamma}\ \hat{r}_{12,\alpha}\ Q_{\beta\mu}(\mathbf{r})Q_{\gamma\mu}(\mathbf{r}'), \end{aligned} \quad (5.33)$$

where $\hat{r}_{12,\alpha} = r_{12,\alpha}/r_{12}$ represents the α-component of the unit vector $\hat{\mathbf{r}}_{12}$.[5] Substituting Eq. (5.33) into Eq. (5.29), we obtain

$$\beta F_1 = -\frac{1}{2}\rho^2 \iint \beta u_1(r_{12})\frac{4}{9}\epsilon_{\alpha\beta\gamma}\ \hat{r}_{12,\alpha}\ Q_{\beta\mu}(\mathbf{r})Q_{\gamma\mu}(\mathbf{r}')d\mathbf{r}d\mathbf{r}', \quad (5.34)$$

and substituting Eq. (5.32) into Eq. (5.30), we obtain

$$\beta F_2 = -\frac{1}{2}\rho^2 \iint \beta\tilde{u}_2(r_{12})\frac{2}{3}Q_{\alpha\beta}(\mathbf{r})Q_{\alpha\beta}(\mathbf{r}')d\mathbf{r}d\mathbf{r}'. \quad (5.35)$$

In Eqs. (5.34) and (5.35), the tensor order parameter is defined at different positions $\mathbf{r}$ and $\mathbf{r}'$. Assuming that $Q_{\alpha\beta}(\mathbf{r}')$ varies smoothly in space around $\mathbf{r}$ (referred to as the long-wavelength approximation), we perform a Taylor expansion around $\mathbf{r}_{12}$:

$$\begin{aligned} Q_{\alpha\beta}(\mathbf{r}') &= Q_{\alpha\beta}(\mathbf{r}+\mathbf{r}_{12}) \\ &= Q_{\alpha\beta}(\mathbf{r}) + r_{12,\kappa}\ \partial_\kappa Q_{\alpha\beta}(\mathbf{r}) + \frac{1}{2!}r_{12,\kappa}^2\ \partial_\kappa\partial_\kappa Q_{\alpha\beta}(\mathbf{r})\cdots, \end{aligned} \quad (5.36)$$

where $\partial_\kappa = \partial/\partial\kappa$ and $r_{12,\kappa}$ ($\kappa = x, y, z$) are the components of the vector $\mathbf{r}_{12}$.

---

[5] We use Tr $Q_{\alpha\alpha} = 0$ and $\epsilon_{\alpha\beta\gamma}\delta_{\alpha\beta}$=0.

Substituting Eq. (5.36) into Eq. (5.34), we obtain

$$\begin{aligned}\beta F_1 &= -\frac{2}{9}\rho^2 \iint \beta u_1(r_{12})\epsilon_{\alpha\beta\gamma}\, \hat{r}_{12,\alpha}\Big[Q_{\beta\mu}(\mathbf{r})Q_{\gamma\mu}(\mathbf{r}) \\ &\quad + r_{12,\kappa}\, Q_{\beta\mu}(\mathbf{r})\partial_\kappa Q_{\gamma\mu}(\mathbf{r}) \\ &\quad + \frac{1}{2} r_{12,\kappa}^2\, Q_{\beta\mu}(\mathbf{r})\partial_\kappa\partial_\kappa Q_{\gamma\mu}(\mathbf{r})\Big] d\mathbf{r}d\mathbf{r}_{12}. \end{aligned} \tag{5.37}$$

In the cholesteric phase, as illustrated in Fig. 5.1, the director field varies smoothly along the z-axis, while remaining uniform in the x and y directions: $\mathbf{n} = (n_x(z), n_y(z), 0)$, where n_x and n_y are the components of the tensor order parameter defined in Eq. (2.37). Consequently, gradient terms involving ∂_κ need only account for derivatives with respect to z. Therefore, we may set $\alpha = \kappa = z$. Under this condition, the first term in Eq. (5.37) is given by

$$\begin{aligned}\epsilon_{\alpha\beta\gamma}Q_{\beta\mu}(\mathbf{r})Q_{\gamma\mu}(\mathbf{r}) &= \epsilon_{zxy}\, Q_{x\mu}(\mathbf{r})Q_{y\mu}(\mathbf{r}) + \epsilon_{zyx}\, Q_{y\mu}(\mathbf{r})Q_{x\mu}(\mathbf{r}) \\ &= Q_{x\mu}(\mathbf{r})Q_{y\mu}(\mathbf{r}) - Q_{x\mu}(\mathbf{r})Q_{y\mu}(\mathbf{r}) \\ &= 0, \end{aligned} \tag{5.38}$$

and the third term in Eq. (5.37) is given by

$$\begin{aligned}\epsilon_{\alpha\beta\gamma}Q_{\beta\mu}(\mathbf{r})\partial_\kappa\partial_\kappa Q_{\gamma\mu}(\mathbf{r}) &= \epsilon_{zxy}\, Q_{x\mu}\partial_z^2 Q_{y\mu} + \epsilon_{zyx}\, Q_{y\mu}\partial_z^2 Q_{x\mu} \\ &= (Q_{xx} - Q_{yy})\partial_z^2 Q_{xy} - Q_{xy}(\partial_z^2 Q_{yy} - \partial_z^2 Q_{xx}) \\ &= 0, \end{aligned} \tag{5.39}$$

leaving only the second term in Eq. (5.37) to represent the chiral interaction. Then the chiral interaction term Eq. (5.37) becomes

$$\beta F_1 = \frac{2}{9}\rho^2 C_L \int \epsilon_{z\beta\gamma}Q_{\beta\mu}(\mathbf{r})\partial_z Q_{\gamma\mu}(\mathbf{r})d\mathbf{r}, \tag{5.40}$$

where the constant C_L is defined as

$$C_L = -\int \beta u_1(r_{12})\frac{r_{12,z}^2}{r_{12}} dr_{12}. \tag{5.41}$$

Similarly, substituting Eq. (5.36) into Eq. (5.35), we obtain

$$\begin{aligned}\beta F_2 &= \frac{1}{2}\rho^2 A_L \int \frac{2}{3} Q_{\alpha\beta}(\mathbf{r})Q_{\alpha\beta}(\mathbf{r})d\mathbf{r} \\ &\quad + \frac{1}{2}\rho^2 B_L \int \frac{1}{3} Q_{\alpha\beta}(\mathbf{r})\partial_z\partial_z Q_{\alpha\beta}(\mathbf{r})d\mathbf{r}, \end{aligned} \tag{5.42}$$

where the coefficients A_L and B_L are defined by

$$A_L = -\int \beta\tilde{u}_2(r_{12})dr_{12}, \tag{5.43}$$

$$B_L = -\int \beta\tilde{u}_2(r_{12})r_{12,z}^2 dr_{12}, \tag{5.44}$$

and we have used the relation $\int Q_{\alpha\beta}\partial_z Q_{\alpha\beta}d\mathbf{r} = 0$. Note that the interaction term $\beta\tilde{u}_2(r_{12})$ is given by Eq. (5.21). The intermolecular potentials u_1 and u_2 in Eq. (5.21) are assumed to be short-range interactions with a characteristic length scale on the order of d_0, as illustrated in Fig. 5.4. Therefore, we set $r_{12,z} = d_0$, and introduce two position-independent interaction parameters $\nu_M = \beta u_2 = \beta U_M$ and $c_L = \beta u_1$. The three constants are then given by

$$A_L = -v_{rr}(\frac{5}{4} + \nu_M), \tag{5.45}$$

$$B_L = -v_{rr}(\frac{5}{4} + \nu_M)d_0^2/3, \tag{5.46}$$

$$C_L = -v_{rr}c_L d_0, \tag{5.47}$$

where $\int d r_{12} = (\pi/4)L^2D (= v_{rr})$ is the average excluded volume between rod-like molecules and we have used $r_{12,z}^2 \simeq d_0^2/3$ in Eq. (5.44) and $r_{12,z}^2/r_{12} \simeq d_0$ in Eq. (5.41). The interaction parameter $\nu_M (= U_M/(k_BT) > 0)$ represents the attractive intermolecular interaction for rod-like molecules, as treated in the Maier-Saupe theory for the nematic phase (see Section 4.2). The 5/4 term arises from the excluded volume interactions between rod-like molecules (Eq. (4.29)). The interaction parameter $c_L = u_1/(k_BT)$ is a chiral parameter (pseudoscalar) that characterizes the strength of the twisting interaction between rod-like molecules. As c_L increases, the twisting becomes stronger, resulting in a shorter helical pitch.

Using Eqs. (5.45)–(5.47), we can calculate F_1 and F_2. The free energy Eq. (5.26) can be summarized as follows:

$$F = F_{iso} + F_{nem} + F_{dis}. \tag{5.48}$$

The first term shows the free energy of the isotropic phase:

$$b^3\beta F_{iso}/V = \frac{\phi}{n_r}(\beta\mu_r^\circ - 1 + \ln\phi + n_r\phi), \tag{5.49}$$

and the second term is the free energy of the nematic phase:

$$\begin{aligned} b^3\beta F_{nem}/V &= \frac{\phi}{n_r}\int f_d(\mathbf{n}(\mathbf{r})\cdot\mathbf{\Omega})\ln\left[4\pi f_d(\mathbf{n}(\mathbf{r})\cdot\mathbf{\Omega})\right]d\mathbf{\Omega} \\ &\quad -\frac{1}{2}(\frac{5}{4} + \nu_M)\phi^2\frac{2}{3}Q_{\alpha\beta}Q_{\alpha\beta}, \end{aligned} \tag{5.50}$$

and the third term shows the distortion-free energy for the cholesteric phase:

$$\begin{aligned} b^3\beta F_{dis} &= -\frac{1}{2}(\frac{5}{4} + \nu_M)\phi^2\frac{1}{9}d_0^2\int Q_{\alpha\beta}\partial_z\partial_z Q_{\alpha\beta}d\mathbf{r} \\ &\quad -\frac{2}{9}c_L\phi^2 d_0\int \epsilon_{z\beta\gamma}Q_{\beta\mu}\partial_z Q_{\gamma\mu}d\mathbf{r}, \end{aligned} \tag{5.51}$$

where we define the axial ratio $n_r = L/D$, the molecular volume unit $b^3 = (\pi/4)D^3$, and the volume of a single rod-like molecule as $v_r = (\pi/4)D^2L = b^3n_r$. The volume fraction (or concentration) of rod-like molecules is given by $\phi = v_r\rho$. Equations (5.50) and (5.51) describe the free energy contributions in terms of a general tensor order parameter.

5.1.4 Distortion free energy

We calculate the distortion free energy F_{dis} associated with the twisting of the director in the cholesteric phase. As shown in Eq. (5.1), consider the case where the director in the cholesteric phase twists along the z-axis with a pitch length $p(= 2\pi/|q|)$. The tensor order parameter is given by

$$Q_{\alpha\beta}(\mathbf{r}) = S(\frac{3}{2}n_\alpha(\mathbf{r})n_\beta(\mathbf{r}) - \frac{1}{2}\delta_{\alpha\beta}), \tag{5.52}$$

and the scalar order parameter:

$$S = \int P_2(\mathbf{n}(\mathbf{r}) \cdot \mathbf{\Omega}) f_d(\mathbf{n}(\mathbf{r}) \cdot \mathbf{\Omega}) d\mathbf{\Omega}. \tag{5.53}$$

We consider here that the orientational order parameter S is assumed to be independent of position $\mathbf{r}$ throughout the system. However, we consider the case where the director $\mathbf{n}$ is twisted. (When considering the effects of interfaces, the value of S would typically depend on the position, but here, we focus on the bulk properties, assuming no interface effects.)

Using Eqs. (5.1) and (5.52), we can calculate the second term Eq. (5.51) as follows:

$$\begin{aligned} \epsilon_{z\beta\gamma} Q_{\beta\mu}(\mathbf{r}) \partial_z Q_{\gamma\mu}(\mathbf{r}) &= \epsilon_{zxy} Q_{x\mu}(\mathbf{r}) \partial_z Q_{y\mu}(\mathbf{r}) + \epsilon_{zyx} Q_{y\mu}(\mathbf{r}) \partial_z Q_{x\mu}(\mathbf{r}) \\ &= Q_{xx}\partial_z Q_{yx} + Q_{xy}\partial_z Q_{yy} + Q_{xz}\partial_z Q_{yz} - Q_{yx}\partial_z Q_{xx} \\ &\quad - Q_{yy}\partial_z Q_{xy} - Q_{yz}\partial_z Q_{xz}, \\ &= \frac{9}{4} S^2 q, \end{aligned} \tag{5.54}$$

where we have used $dn_x/dz = -qn_y$ and $dn_y/dz = qn_x$. Similarly, we also have

$$Q_{\alpha\beta}(\mathbf{r}) Q_{\alpha\beta}(\mathbf{r}) = \frac{3}{2} S^2, \tag{5.55}$$

$$Q_{\alpha\beta}(\mathbf{r}) \partial_z Q_{\alpha\beta}(\mathbf{r}) = 0, \tag{5.56}$$

$$Q_{\alpha\beta}(\mathbf{r}) \partial_z \partial_z Q_{\alpha\beta}(\mathbf{r}) = -\frac{9}{2} S^2 q^2, \tag{5.57}$$

where we use

$$n_\alpha \partial_\kappa n_\alpha = \frac{1}{2} \partial_\kappa n_\alpha^2 = \frac{1}{2} \partial_\kappa (1) = 0, \tag{5.58}$$

and

$$\delta_{\alpha\beta} \partial_\kappa n_\alpha n_\beta = \partial_\kappa n_\alpha^2 = 0. \tag{5.59}$$

Using Eqs.(5.54)–(5.57), the free energy for the nematic phase Eq. (5.50) is given by

$$\begin{aligned} b^3 \beta F_{nem}/V &= \frac{\phi}{n_r} \int f_d(\mathbf{n}(\mathbf{r}) \cdot \mathbf{\Omega}) \ln \left[4\pi f_d(\mathbf{n}(\mathbf{r}) \cdot \mathbf{\Omega}) \right] d\mathbf{\Omega} \\ &\quad - \frac{1}{2} (\frac{5}{4} + \nu_M) S^2 \phi^2, \end{aligned} \tag{5.60}$$

and the distortion free energy Eq. (5.51) is given by

$$b^3 \beta F_{dis}/V = -\frac{1}{2} (\frac{5}{4} + \nu_M) S^2 \phi^2 g(Q), \tag{5.61}$$

where the distortion function is given by

$$g(Q) = \frac{c_L}{\frac{5}{4} + \nu_M} Q - \frac{1}{2} Q^2, \tag{5.62}$$

with the dimensionless wavenumber $Q = d_0 q$. The constants c_L and Q are pseudoscalar quantities. When $c_L > 0$ and $Q > 0$, they correspond to a right-handed helix, while when $c_L < 0$ and $Q < 0$, they correspond to a left-handed helix. The product of these two pseudoscalar quantities is a true scalar, which matches the scalar nature of the free energy.

The wavenumber (Q) dependence of the distortion free energy (Eq. (5.61)) exhibits a minimum at the wavenumber Q_0, or q_0, as shown in Fig. 2.6. The equilibrium pitch wavenumber (Q_0) is obtained by differentiating Eq. (5.61) with respect to Q: $(\partial F_{dis}/\partial Q)_T = 0$,

$$Q_0 = d_0 q_0 = \frac{c_L}{\frac{5}{4} + \nu_M}. \tag{5.63}$$

An important conclusion from the mean field theory is that the equilibrium pitch wavenumber is given by the ratio of the chiral interaction (chirality parameter c_L) to the nematic interaction $(5/4 + \nu_M)$. For example, when the chiral interaction c_L increases, the twisting force strengthens, leading to an increase in the wavenumber Q_0. Cholesteric liquid crystals are observed when the liquid crystalline molecules themselves possess chirality. However, it is also possible to induce a cholesteric phase by adding a chiral dopant to an achiral nematic liquid crystal. Even a small amount of chiral dopant can introduce a twist, increasing the chiral interaction c_L and resulting in the appearance of the cholesteric phase. For a typical low-molecular-weight liquid crystal with $d_0 \simeq 3$ nm and a cholesteric pitch length of $p = 300$ nm, the equilibrium wavenumber is given by $Q_0 = 2\pi d_0/p_0 \simeq 0.06$. From Eq. (5.63), the chiral interaction parameter can be estimated as

$$c_L \simeq 0.06(\frac{5}{4} + \nu_M). \tag{5.64}$$

If the intermolecular interaction is comparable to the thermal energy, we can assume $\nu_M \simeq 1$, which yields $c_L \simeq 0.1$. In any case, the theoretical framework is expected to remain valid within the range where $Q_0 < 1$. The interaction parameters c_L and ν_M are inversely proportional to temperature, so the magnitude of $|Q_0|$ in Eq. (5.63) decreases as temperature increases. This means that as the temperature rises, the pitch length increases. If the excluded volume term (the 5/4 term) in Eq. (5.63) is neglected, Q_0 simplifies to a constant: $Q_0 = c_L/\nu_M$.

Equation (5.62) can be rewritten using Q_0 as

$$g(Q) = (Q_0 - \frac{1}{2}Q)Q. \tag{5.65}$$

When $Q = Q_0$, we obtain

$$g(Q_0) = \frac{1}{2}Q_0^2, \tag{5.66}$$

Thus, the distortion free energy in thermal equilibrium is given by

$$b^3 \beta F_{dis}/V = -\frac{1}{4}(\frac{5}{4} + \nu_M)S^2\phi^2 Q_0^2. \tag{5.67}$$

Since the wavenumber Q_0 is nonzero, the distortion energy $F_{\rm dis}$ is always negative. This indicates that the formation of the cholesteric phase reduces the free energy compared to the nematic phase. As a result, there is no phase transition between the nematic and cholesteric phases; instead, the only true phase transition occurs between the isotropic and cholesteric phases. This implies that in any system with even a small degree of molecular twist (i.e., any nonzero value of the chiral interaction parameter c_L), the cholesteric phase will spontaneously emerge. In this sense, molecular chirality is a ubiquitous phenomenon in soft matter systems. For instance, the well-known double-helix structure of DNA reflects the intrinsic chirality of the constituent molecules.

5.1.5 Distribution function of the cholesteric phase

In this section, we derive the orientational distribution function $f_d(\mathbf{n}(\mathbf{r}) \cdot \boldsymbol{\Omega})$. The equilibrium orientational distribution function is determined by minimizing the total free energy

Eq. (5.48). This condition is given by

$$\left(\frac{\delta F}{\delta f_d(\mathbf{n}(\mathbf{r})\cdot\mathbf{\Omega})}\right)_T = 0. \tag{5.68}$$

Differentiating Eqs. (5.60) and (5.61) with respect to the distribution function $f_d(\mathbf{n}(\mathbf{r})\cdot\mathbf{\Omega})$, while noting that the orientational order parameter S in Eq. (5.53) also depends on $f_d(\mathbf{n}(\mathbf{r})\cdot\mathbf{\Omega})$, we obtain

$$\begin{aligned}\left(\frac{\delta F_{nem}}{\delta f_d(\cdots)}\right)_T &= \frac{\phi}{n_r}\int \ln[4\pi f_d(\cdots)+1]d\mathbf{\Omega} \\ &\quad -(\frac{5}{4}+\nu_M)\phi^2 S\int P_2(\cdots)d\mathbf{\Omega},\end{aligned} \tag{5.69}$$

and

$$\left(\frac{\delta F_{dis}}{\delta f_d(\cdots)}\right)_T = -\frac{1}{2}(\frac{5}{4}+\nu_M)\phi^2 g(Q_0)S\int P_2(\cdots)d\mathbf{\Omega}. \tag{5.70}$$

Then, the distribution function is given by

$$f_d(\mathbf{n}(\mathbf{r})\cdot\mathbf{\Omega}) = \frac{1}{Z_0}\exp\left[\Gamma_0\big(1+g(Q_0)\big)SP_2(x)\right], \tag{5.71}$$

where $x = \cos\theta$ and

$$\Gamma_0 = n_r(\frac{5}{4}+\nu_M)\phi. \tag{5.72}$$

The constant Z_0 is determined using the normalization condition for the distribution function (Eq. (5.25)):

$$Z_0 = 4\pi\int_0^1 \exp\left[\Gamma_0\big(1+g(Q_0)\big)SP_2(x)\right]dx, \tag{5.73}$$

Substituting Eq. (5.71) into Eq. (5.53), we obtain

$$S = I_1(\Gamma_0 S)/I_0(\Gamma_0 S), \tag{5.74}$$

where we define

$$I_m(\Gamma_0 S) = \int_0^1 [P_2(x)]^m \exp\left[\Gamma_0\big(1+g(Q_0)\big)SP_2(x)\right]dx, \tag{5.75}$$

with $m = 0, 1, 2$, and $Z_0 = 4\pi I_0(\Gamma_0 S)$. Thus, the orientational order parameter S is determined as the solution to the self-consistent Eq. (5.74). In the isotropic phase where $S = 0$, we have $I_1(0) = 0$ and $I_0(0) = 1$.

We define the free energy of the cholesteric phase, which includes the contribution from the nematic phase, as

$$F_{ch} = F_{nem} + F_{dis}, \tag{5.76}$$

so that the total free energy ($F = F_{iso} + F_{ch}$) is given by

$$v_r\beta F/V = f_{iso} + f_{ch}, \tag{5.77}$$

where $v_r = b^3 n_r$. The first term is the isotropic free energy (Eq. (5.49)):

$$f_{iso} = v_r\beta F_{iso}/V = \phi(\beta\mu_r^\circ - 1 + \ln\phi + n_r\phi), \tag{5.78}$$

and the second term shows the free energy of the cholesteric phase (Eqs. (5.60) and (5.61)):

$$f_{ch} = v_r\beta F_{ch}/V = \phi(\frac{1}{2}\Gamma_0 S^2\big(1+g(Q_0)\big) - \ln I_0(\Gamma_0 S)\big). \tag{5.79}$$

When $g(Q_0) = 0$, the free energy reduces to that of the nematic phase Eq. (4.38).

5.2 Phase transitions of the cholesteric phase

The cholesteric phase is defined by the conditions $S \neq 0$ and $Q_0 \neq 0$ (Eq. (5.63)). By numerically solving Eq. (5.74) for each temperature and concentration, one can evaluate the behavior of the orientational order parameter S.

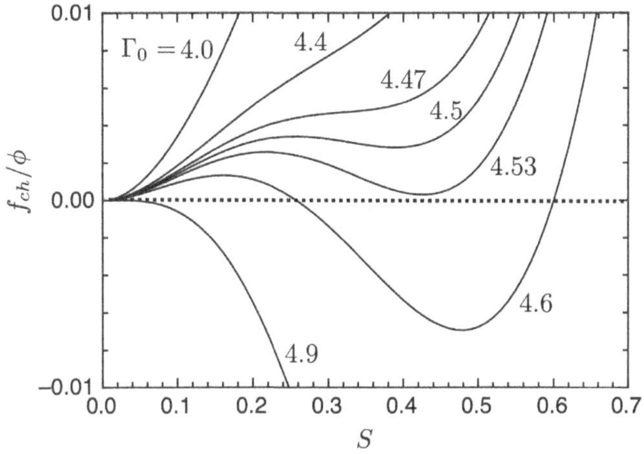

FIGURE 5.5
Free energy of the cholesteruc phase (Eq. (5.79)) plotted against the orientational order parameter S for various values of Γ_0 with $Q_0 = 0.06$.

Figure 5.5 shows the free energy of the cholesteric phase (Eq. (5.79)) plotted as a function of the orientational order parameter S for various values of Γ_0, with $Q_0 = 0.06$. For $\Gamma_0 < \Gamma_0^c (= 4.47)$, the free energy exhibits a single minimum at $S = 0$, corresponding to the isotropic phase. When $\Gamma_0^c < \Gamma_0 < \Gamma_0^{CI} (= 4.532)$, a second local minimum appears at $S > 0$ in addition to the one at $S = 0$, indicating the emergence of a metastable cholesteric phase. At $\Gamma_0 = \Gamma_0^{CI}$, the free energies at $S = 0$ and $S \approx 0.43$ become equal (both zero), leading to a cholesteric-isotropic phase transition (CIT). At this point, the orientational order parameter S exhibits a discontinuous jump, indicating that the transition is first-order. For $\Gamma_0 > \Gamma_0^{CI}$, the local minimum at $S > 0$ becomes lower than that at $S = 0$, and the cholesteric phase becomes thermodynamically more stable than the isotropic phase.

Figure 5.6 shows the orientational order parameter S plotted as a function of Γ_0 for $Q_0 = 0, 0.1$, and 0.5, obtained by numerically solving Eq. (5.74). Since Γ_0 is proportional to the concentration ϕ of rod-like molecules, the horizontal axis can be interpreted as a measure of concentration. As Γ_0 increases, S exhibits a discontinuous jump from zero to $S \approx 0.43$ at a CIT. The value of Γ_0^{CI} decreases with increasing Q_0; that is, the CIT occurs at lower concentrations when the pitch length is shorter (i.e., Q_0 is larger). Beyond the transition point, S continues to increase with Γ_0, reflecting stronger orientational ordering at higher concentrations. For reference, in the absence of twist ($Q_0 = 0$), the system undergoes an isotropic-nematic phase transition at $\Gamma_0^{NI} \approx 4.55$ (see Fig. 4.4). When Q_0 is small, the effect of twist on the transition is of the order of Q_0^2, implying a weak influence on the transition point.

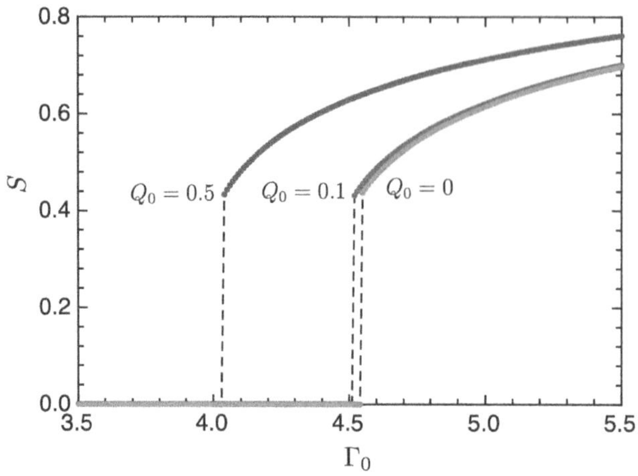

FIGURE 5.6
Orientational order parameter S plotted aginst $\Gamma_0(= n_r(5/4 + \nu_M)\phi)$ for various values of the wavenumber Q_0.

5.2.1 Phase separation between cholesteric and isotropic phases

Figure 5.7 shows the free energy (Eq. (5.77)) as a function of the volume fraction ϕ of rod-like molecules for the case of $n_r = 100$, $\nu_M = 0$, and $Q_0 = 0.06$. The free energy of the nematic phase is shown in Fig. 4.8, where only excluded volume effects were considered (i.e., $\nu_M = 0$). The two curves represent the isotropic free energy f_{iso} with $S = 0$ and the total free energy $f = f_{iso} + f_{ch}$ of the cholesteric phase ($S \neq 0$). For concentrations $\phi < \phi_{CI}$, the isotropic phase has a lower free energy and is thus thermodynamically stable. For $\phi > \phi_{CI}$, the cholesteric phase becomes more stable, having the lower free energy. At the CIT concentration $\phi = \phi_{CI}$, a discontinuous change in the slope of the free energy curve occurs, indicating a first-order phase transition. As shown by the dashed line in the inset, a common tangent can be drawn between the two free energy curves, signaling the coexistence of isotropic and cholesteric phases: I+Ch phase separation. The points of tangency correspond to the concentrations ϕ_I and ϕ_{ch} of the rod-like molecules in the isotropic and cholesteric phases, respectively, under phase separation.

The concentrations of the coexisting phases can be determined more analytically. As discussed in Fig. 4.8, applying the common tangent construction is equivalent to imposing the conditions for phase equilibrium: the equality of chemical potential and osmotic pressure between the coexisting phases. The osmotic pressure is defined as $\Pi = -\big(\partial F/\partial V\big)_N$, and using Eq. (5.77) ,we have the dimensionless osmotic pressure:

$$\begin{aligned}\tilde{\Pi}(\phi, S) &= v_r\beta\Pi = -v_r\beta\big(\partial F/\partial V\big)_N \\ &= \phi + n_r\phi^2 \\ &\quad -\frac{1}{2}\Big(\frac{5}{4} + \nu_M\Big)n_r\phi^2 S^2\big(1 + g(Q_0)\big).\end{aligned} \tag{5.80}$$

The chemical potential of the rod-like molecules is given by

$$\begin{aligned}\beta\mu &= \beta\big(\partial F/\partial N\big)_V, \\ &= \beta\mu_r^\circ + \ln\phi + 2n_r\phi - \ln I_0(\Gamma_0 S).\end{aligned} \tag{5.81}$$

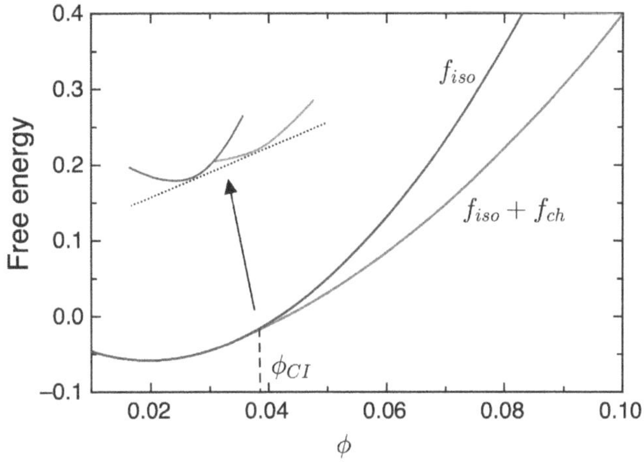

FIGURE 5.7
Free energy as a function of volume fraction ϕ (concentration) for the case of $n_r = 100$, $\nu_M = 0$, and $Q_0 = 0.06$.

To determine the concentrations of the coexisting isotropic and cholesteric phases (ϕ_I, ϕ_{Ch}), the chemical potential and osmotic pressure of the molecules must be equal in both phases. This condition is expressed as

$$\Pi(\phi_I, 0) = \Pi(\phi_{Ch}, S), \tag{5.82}$$

$$\mu(\phi_I, 0) = \mu(\phi_{Ch}, S), \tag{5.83}$$

From Eq. (5.82), the concentration ϕ_I can be expressed as a function of ϕ_{Ch}:

$$\begin{aligned} x_I &= \frac{1}{2}\Bigg[-1 + \Bigg(1 + 4\big[x_{Ch} + x_{Ch}^2 \\ &\quad - \frac{1}{2}\Big(\frac{5}{4} + \nu_M\Big)x_{Ch}^2 S^2 \big(1 + g(Q_0)\big)\big]\Bigg)^{1/2}\Bigg], \end{aligned} \tag{5.84}$$

where $x_I = n_r\phi_I$, $x_{Ch} = n_r\phi_{Ch}$. Equation (5.83) can be rewritten as

$$\ln\frac{x_{Ch}}{x_I} + 2(x_{Ch} - x_I) - \ln I_0(\Gamma_0 S) = 0. \tag{5.85}$$

Thus, using Eq. (5.84), we can solve numerically for x_{Ch} that satisfies Eq. (5.85). For the case $Q_0 = 0.06$ and $\nu_M = 0$, the solution is

$$x_I = 3.506,\ x_{Ch} = 3.880,\ S = 0.577, \tag{5.86}$$

and the volume fraction gives

$$\phi_I = 3.506/n_r,\ \phi_{Ch} = 3.88/n_r. \tag{5.87}$$

Figure 5.8 shows the phase diagram plotted on the concentration (ϕ) versus axial ratio (L/D) plane, calculated from Eq. (5.87). The two curves represent the coexistence boundary between the cholesteric and isotropic phases. The region enclosed by these two curves indicates the I+Ch phase separation. As the axial ratio L/D increases, i.e., as the rod-like

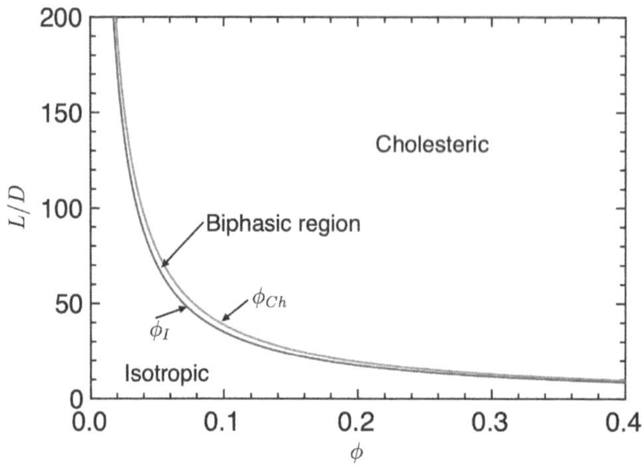

FIGURE 5.8
Phase diagram for $Q_0 = 0.06$ and $\nu_M = 0$.

molecules become longer relative to their diameter, the concentration required for phase separation decreases, shifting the coexistence region toward lower concentrations.

When $Q_0 = 0.5$, the coexistence values are obtained as

$$x_I = 3.453, \ x_{Ch} = 3.830, \ S = 0.579. \tag{5.88}$$

Compared with the results in Eq. (5.86), these values are slightly smaller in the second decimal place. This indicates that, as the pitch wavenumber Q_0 increases, the coexistence concentrations shift toward lower values. In other words, the presence of a molecular twist slightly reduces the concentrations at which the cholesteric and isotropic phases can coexist.

5.2.2 Temperature-dependent phase separation

In the previous section, the results were presented under the assumption that the intermolecular attractive (Maier-Saupe) interaction is absent ($\nu_M = 0$). In this section, we extend the analysis by incorporating the effects of attractive interactions. When the ratio of the chiral interaction strength c_L to the nematic interaction strength ν_M is defined as $\gamma_L = c_L/\nu_M$, the pitch wavenumber in Eq. (5.63) can be rewritten as

$$Q_0 = \frac{\gamma_L}{1 + 1.25\tau}, \tag{5.89}$$

where we introduce the dimensionless temperature parameter $\tau = 1/\nu_M = k_BT/U_M$. As the temperature τ increases, the pitch wave number Q_0 in the thermodynamic equilibrium state decreases. This expression explicitly shows the influence of temperature on the cholesteric phase: higher temperatures lead to a larger effective thermal motion (smaller ν_M), thereby reducing the twist deformation and resulting in a longer pitch (smaller Q_0).

The CIT occurs when the condition $\Gamma_0(1 + g(Q_0)) = 4.55$ is satisfied in Eq. (5.71). Notably, this critical value 4.55 is identical to that used in the nematic phase transition discussed in Section 4.1.4. Substituting Eq. (5.66) into this condition, the CIT point is given by

$$n_r(\frac{5}{4} + \nu_M)\phi(1 + \frac{1}{2}Q_0^2) = 4.55. \tag{5.90}$$

This equation describes the relationship between the concentration ϕ, the pitch wave number Q_0, and the temperature-dependent parameters that determine the phase transition. Considering $\nu_M = U_M/(k_B T)$, the CIT temperature T_{CI} as a function of concentration ϕ is given by

$$\tau_{CI} = \frac{k_B T_{CI}}{U_M} = \frac{n_r(1 + \frac{1}{2}Q_0^2)\phi}{4.55 - \frac{5}{4}n_r(1 + \frac{1}{2}Q_0^2)\phi}. \tag{5.91}$$

This expresses the CIT temperature T_{CI} in terms of the concentration ϕ and the pitch wave number Q_0, illustrating how the transition temperature varies with these parameters. Notably, the CIT temperature diverges when the denominator of Eq. (5.91) approaches zero. This divergence corresponds to a critical concentration ϕ^*, at which the CIT temperature becomes infinite. The critical value ϕ^* is given by

$$\phi^* = \frac{4.55}{\frac{5}{4}n_r(1 + \frac{1}{2}Q_0^2)}, \tag{5.92}$$

representing the upper bound of the concentration range where the CIT can occur. Beyond this concentration, the system remains in the cholesteric phase for all temperatures. The CIT temperature T_{CI} increases with the concentration ϕ and diverges at the critical concentration $\phi = \phi^*$, as given in Eq. (5.92). The critical value ϕ^* decreases with increasing pitch wavenumber Q_0. In other words, as the helical pitch shortens, corresponding to a stronger twisting force, the divergence point ϕ^* shifts toward lower concentrations, and the CIT temperature increases accordingly.

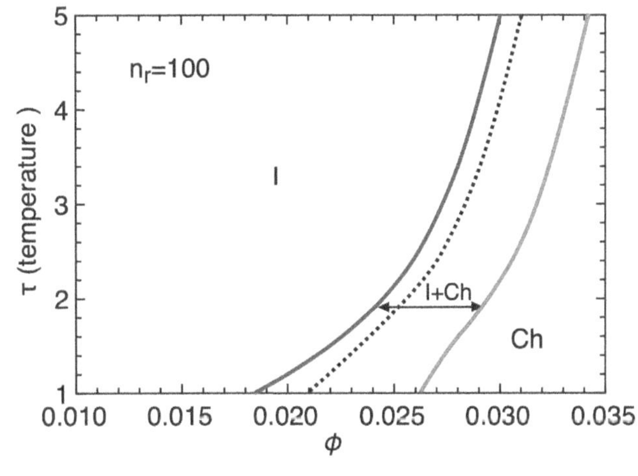

FIGURE 5.9
Phase diagram on the temperature-concentration (T-ϕ) plane for $n_r = 100$, showing the coexistence curves (I+Ch). The coexistence region lies between the two solid curves, and the dotted line shows the first-order CIT. As the temperature increases, the CIT temperature shifts to higher concentrations.

Figure 5.9 shows the phase diagram in the (ϕ, T) plane for $\gamma_L = 1$ and $n_r = 100$, calculated from Eqs. (5.82) and (5.83). The left curve represents the coexistence curve of the isotropic (I) phase (ϕ_I), while the right curve represents the coexistence curve of the cholesteric (Ch) phase (ϕ_{Ch}). The dashed line indicates the CIT point (ϕ, T_{CI}) given by Eq. (5.91). The region enclosed by the coexistence curves corresponds to the I+Ch phase separation region. As temperature decreases, the attractive interactions between rod-like

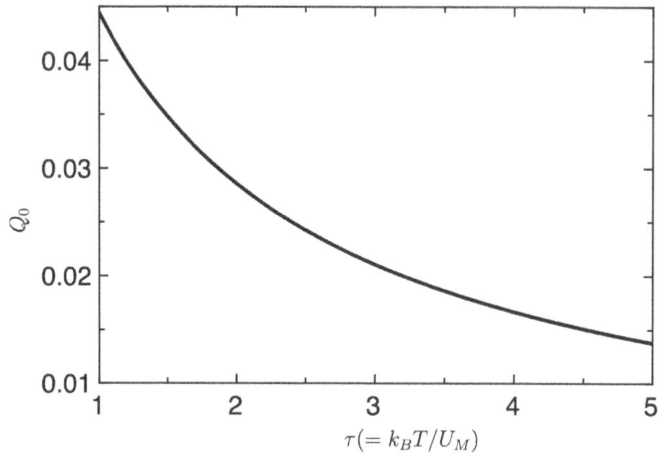

FIGURE 5.10
Temperature dependence of the pitch wave number Q_0 along the the cholesteric concentration of the coexistence curve ϕ_{Ch} in Fig. 5.9.

molecules become more pronounced, shifting the coexistence region toward lower concentrations. This behavior qualitatively agrees with experimental observations in solutions of rod-like molecules.

Figure 5.10 shows the temperature dependence of the pitch wave number $Q_0(\propto 1/p)$ along the cholesteric branch ϕ_{Ch} of the coexistence curve in Fig. 5.9. As the temperature decreases, the pitch wave number increases, indicating a shorter helical pitch.

The theory introduced here describes that the pitch wave number Q_0, as given by Eq. (5.89), depends solely on temperature. However, experimental observations have revealed more diverse behaviors: in some systems, the pitch length increases or decreases monotonically with temperature, while in others it exhibits a maximum or minimum. Moreover, Eq. (5.89) does not incorporate any dependence on concentration. These limitations can be addressed by considering chiral coupling between rod-like molecules and solvent molecules [69]. In addition, helical inversion phenomena—where the handedness of the cholesteric twist reverses with temperature or concentration—can also be captured by an extended theoretical framework [70].

In earlier theoretical studies [65,71], the orientational distribution function was expanded to include terms up to $P_4(\cos\theta)$, allowing for a more detailed description of the temperature dependence of pitch length. These higher-order terms may enable the modeling of more complex thermal behaviors. Spontaneous mirror symmetry breaking accompanied by chirality is a key research topic in the fields of liquid crystals and soft matter, including bio-soft matter [72].

5.3 Twist elastic constant

Finally, we will examine how the results of mean field theory relate to the continuum theory of liquid crystals, specifically the Frank elastic theory. The elastic free energy density (per unit volume) of Frank elasticity (see Section 3.1.1), which describes the spatial distortion of the director $\mathbf{n}$, is given by

$$f_F = F_{dis}/V = \frac{1}{2}K_1(\nabla\cdot\mathbf{n})^2 + \frac{1}{2}K_2(\mathbf{n}\cdot(\nabla\times\mathbf{n}) + q_0)^2 + \frac{1}{2}K_3(\mathbf{n}\times(\nabla\times\mathbf{n}))^2, \tag{5.93}$$

where K_1, K_2, and K_3 are the elastic constants corresponding to splay, twist, and bend deformations, respectively. Using the director for the cholesteric phase (Eq. (5.1)), we have

$$\nabla\cdot\mathbf{n} = 0, \tag{5.94}$$

$$\nabla\times\mathbf{n} = -(d\omega/dz)\mathbf{n}. \tag{5.95}$$

Thus, the second term (twist) in Eq. (5.93) becomes

$$\mathbf{n}\cdot(\nabla\times\mathbf{n}) = -(d\omega/dz) = -q, \tag{5.96}$$

and the third term (bend) is

$$\mathbf{n}\times(\nabla\times\mathbf{n}) = 0. \tag{5.97}$$

Frank elastic energy receives contribution only from the twist term K_2.

Rewriting the distortion function $g(Q)$ (Eq. (5.62)) in terms of the q^2 term, we obtain

$$g(Q) = -\frac{1}{2}d_0^2(\mathbf{n}\cdot(\nabla\times\mathbf{n}) + q_0)^2, \tag{5.98}$$

and the twist elastic energy (Eq. (5.61)) can be written as

$$b^3\beta F_{dis}/V = \frac{1}{2}(\frac{5}{4} + \nu_M)S^2\phi^2\frac{d_0^2}{2}(\mathbf{n}\cdot(\nabla\times\mathbf{n}) + q_0)^2. \tag{5.99}$$

If we assume the length scale $b \simeq d_0 \simeq D$, Eq. (5.99) can be rewritten in the form of Eq. (5.93) and the twist elastic constant K_2 is given by

$$K_2 = \frac{1}{2}k_BT(\frac{5}{4} + \nu_M)S^2\phi^2/D. \tag{5.100}$$

For a typical liquid crystal molecule, 5CB, K_2 takes a value of approximately 0.23×10^{-6} dyn. Using Eq. (5.100) with $S = 1$, $\phi = 1$, $\nu_M = 1$, $D = 5$ nm, and the thermal energy at room temperature $k_BT \simeq 4 \times 10^{-21}$ J, we obtain $K_2 \simeq 10^{-7}$ dyn, which is a reasonable value. In the continuum theory of typical liquid crystals, the elastic constant is treated as a constant. However, from the conclusion of our mean field theory, we can see that the elastic constant K_2 is proportional to the intermolecular interactions $5/4 + \nu_M$ and the orientational order parameter S^2.

6

Liquid Crystals Under an External Field

Liquid crystalline molecules possess anisotropies in both magnetic susceptibility and dielectric constant. When external fields such as an electric field **E** or magnetic field **H** are applied to these liquid crystalline molecules, the interaction between the external field and the anisotropy of the dielectric constant or magnetic susceptibility causes the orientation of the liquid crystalline molecules to change, leading to various phase transitions. The effects of external fields are one of the important subjects in liquid crystal systems. In this chapter, we introduce mean field theories describing nematic and cholesteric phases under external fields.

6.1 Dielectric anisotropy

The dielectric constant ϵ indicates the polarization of charge within a molecule. As shown in Fig. 6.1, when a liquid crystal molecule is divided into its long and short axes, the dielectric constants in these directions, $\epsilon_{\parallel}$ and $\epsilon_{\perp}$, can be defined. The dielectric anisotropy is defined as $\Delta\epsilon = \epsilon_{\parallel} - \epsilon_{\perp}$. When the dielectric constants in the long axis and short axis directions are the same, $\Delta\epsilon = 0$, the material is dielectric isotropic. On the other hand, when the dielectric constants in these directions differ, the material exhibits dielectric anisotropy. Suppose the dielectric constant along the long axis is larger, $\Delta\epsilon > 0$. In that case, the material has positive dielectric anisotropy, and if the dielectric constant along the short axis is larger, $\Delta\epsilon < 0$, the material has negative dielectric anisotropy. The charge polarization is a result of the electron distribution within the molecular structure, and the difference in the sign of dielectric anisotropy depends on the structure of the liquid crystalline molecule. When an electric field is applied to such a liquid crystalline molecule, polarization occurs in the direction of the electric field due to Coulomb forces, and the molecule tends to align with the direction of the electric field. For liquid crystalline molecules with $\Delta\epsilon > 0$, the long axis of the molecule tends to align with the direction of the external field, whereas for liquid

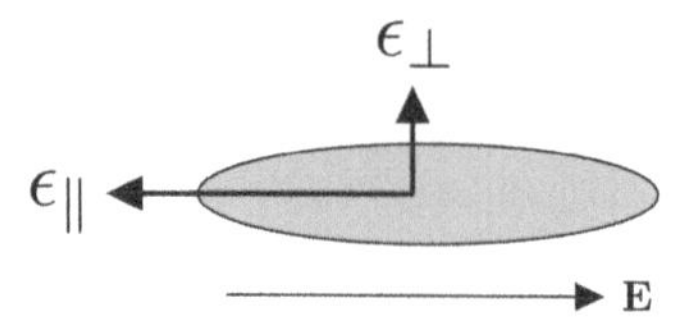

FIGURE 6.1
Dielectric anisotropy is defined as $\Delta\epsilon = \epsilon_{\parallel} - \epsilon_{\perp}$.

DOI: 10.1201/9781003737711-6

crystalline molecules with $\Delta\epsilon < 0$, the short axis tends to align with the direction of the external field.

Many liquid crystalline molecules that form the nematic phase possess a permanent dipole. However, in the absence of an external field, the dipoles appear with equal probability in the upward and downward directions. When an electric field $\mathbf{E}$ is applied to this nematic phase, it induces a polarization $\mathbf{p}$. The magnitude of the polarization is proportional to the electric field and is given by

$$\mathbf{p} = \epsilon_\perp \mathbf{E} + \Delta\epsilon(\mathbf{E}\cdot\mathbf{n})\mathbf{n}, \tag{6.1}$$

where $\mathbf{n}$ denotes the nematic director. If $\mathbf{E}$ is parallel to $\mathbf{n}$, we have $\mathbf{p} = \epsilon_\parallel \mathbf{E}$, whereas if $\mathbf{E}$ is perpendicular to $\mathbf{n}$, then $\mathbf{p} = \epsilon_\perp \mathbf{E}$. The free energy per unit volume under an applied electric field is given by

$$\begin{aligned} f_{elec} &= -\int_0^E \mathbf{p}\cdot d\mathbf{E} \\ &= -\frac{1}{2}\epsilon_\perp E^2 - \frac{1}{2}\Delta\epsilon(\mathbf{E}\cdot\mathbf{n})^2, \end{aligned} \tag{6.2}$$

where the first term is a constant independent of molecular orientation, while the second term depends on the molecular alignment.

A similar discussion applies to magnetic susceptibility anisotropy. Let $\chi_\parallel$ and $\chi_\perp$ denote the magnetic susceptibility along the long and short axes of the molecule, respectively, and define the susceptibility anisotropy as $\Delta\chi = \chi_\parallel - \chi_\perp$. When a magnetic field is applied, it interacts with this susceptibility anisotropy, causing the molecules to reorient. When a magnetic field $\mathbf{H}$ is applied at an arbitrary angle to the director $\mathbf{n}$, the magnetization of the material is given

$$\mathbf{M} = \mu_0\chi_\perp \mathbf{H} + \mu_0\Delta\chi(\mathbf{H}\cdot\mathbf{n})\mathbf{n}, \tag{6.3}$$

where μ_0 is the permeability of free space. The free energy per unit volume associated with the interaction between the liquid crystalline molecules and the magnetic field is given by

$$\begin{aligned} f_{mag} &= -\int_0^H \mathbf{M}\cdot d\mathbf{H} \\ &= -\frac{1}{2}\mu_0\chi_\perp H^2 - \frac{1}{2}\mu_0\Delta\chi(\mathbf{H}\cdot\mathbf{n})^2, \end{aligned} \tag{6.4}$$

where the first term is a constant independent of molecular orientation, while the second term represents the interaction between the magnetic field and the director. Clearly, when $\Delta\chi > 0$, the director aligns parallel to the magnetic field to lower the free energy. Using the concepts of dielectric and susceptibility anisotropy introduced above, the following sections will examine the effects of applying external fields to the nematic and cholesteric phases.

6.2 Nematic phase under an external field

The change in free energy due to the external field $\mathbf{H}$ can be written using the second term in Eq. (6.4), expressed as the square of the inner product between the external field $\mathbf{H}$ and the director $\mathbf{n}$. Using the tensor order parameter $Q_{\alpha\beta}$ (Eq. (2.37)), the free energy of the external field, representing the coupling between the external field and liquid crystalline molecules, can be given by

$$F_{ext} = -\phi\mu_0\Delta\chi \int H_\alpha Q_{\alpha\beta}(\mathbf{r})H_\beta d\mathbf{r}, \tag{6.5}$$

where H_α shows the $\alpha(= x, y, z)$ components of the external field vector (magnetic field $\mathbf{H}$ or electric field $\mathbf{E}$), $\phi = b^3 n_r \rho$ is the volume fraction of rods with $n_r = L/D$, and $\rho = N/V$ is the number density of rods. We consider the magnetic field $\mathbf{H}$ as the external field, but for the electric field $\mathbf{E}$, $\mu_0 \Delta\chi$ changes to $\Delta\epsilon$.

The total free energy of the nematic phase under an external field is given by

$$F = F_{nem} + F_{ext}, \tag{6.6}$$

where the first term is the free energy (Eq. (4.64)) of the nematic phase, and the second term represents the contribution from the external field (Eq. (6.5)).

Let the external magnetic field be applied parallel to the z-direction:

$$\mathbf{H} = (0, 0, H). \tag{6.7}$$

Since the director $\mathbf{n}$ aligns parallel to the external field $\mathbf{H}$, we can set

$$\mathbf{n} = (0, 0, 1). \tag{6.8}$$

Thus, using

$$\begin{aligned} H_\alpha Q_{\alpha\beta}(\mathbf{r}) H_\beta &= H_z Q_{zz} H_z \\ &= H^2 S\big(\frac{3}{2} n_z^2 - \frac{1}{2}\big) = H^2 S, \end{aligned} \tag{6.9}$$

the external field contribution to the free energy (Eq. (6.5)) is given by

$$b^3 \beta F_{ext}/V = -\phi S \tilde{H}, \tag{6.10}$$

where the dimensionless parameter

$$\tilde{H} = b^3 \beta \mu_0 \Delta\chi H^2, \tag{6.11}$$

represents the dimensionless strength of the external field.

The total free energy (Eq. (6.6)) is given by

$$b^3 \beta F/V = \frac{\phi}{n_r} \int f_d(\theta) \ln\left[4\pi f_d(\theta)\right] d\mathbf{\Omega} - \frac{1}{2}\nu_M \phi^2 S^2 - \phi S \tilde{H}, \tag{6.12}$$

where we have used Maier-Saupe model (Eq. (4.64)) for the nematic free energy F_{nem}. The orientational distribution function is given by $(\delta F/\delta f_d(\theta))_T = 0$, under the normalization condition:

$$\int f_d(\theta) d\mathbf{\Omega} = 1. \tag{6.13}$$

We then obtain

$$f_d(\theta) = \frac{1}{Z_0} \exp\left(n_r\big(\nu_M S\phi + \tilde{H}\big) P_2(\cos\theta)\right). \tag{6.14}$$

The constant Z_0 is determined by the normalization condition as $Z_0 = 4\pi I_0(S)$ and the function I_0 is defined as

$$I_m(S) = \int_0^1 \left[P_2(x)\right]^m \exp\left[n_r(\nu_M S\phi + \tilde{H}) P_2(x)\right] dx, \tag{6.15}$$

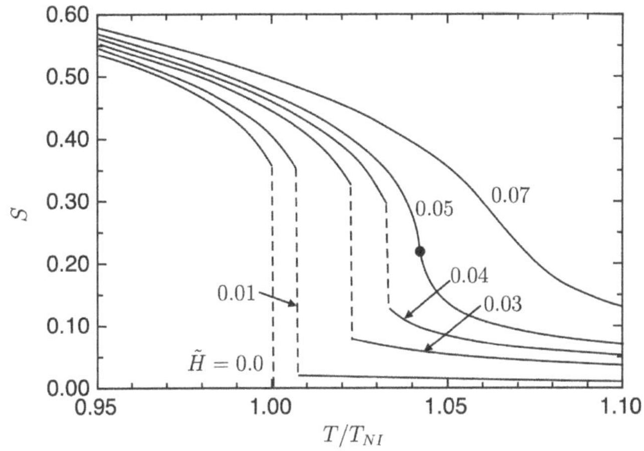

FIGURE 6.2
Orientational order parameter S plotted against the temperature T/T_{NI} for various values of the external field $\tilde{H}$.

for $m = 0, 1, 2$. The orientational order parameter (Eq. (4.21)) is determined by

$$S = I_1(S)/I_0(S). \tag{6.16}$$

Substituting Eq. (6.14) into Eq. (6.12), we obtain the nematic free energy under the external field:

$$b^3\beta F/V \quad = \quad \frac{1}{2}\nu_M \phi^2 S^2 - \frac{\phi}{n_r} \ln I_0(S) \tag{6.17}$$

Although the explicit linear field term does not appear in Eq. (6.17), the effect of the external field is implicitly included in the function $I_0(S)$ through Eq. (6.15), since the argument of the exponential contains $\tilde{H}$.

Figure 6.2 shows the orientational order parameter S plotted as a function of the reduced temperature T/T_{NI} for various values of the external field $\tilde{H}$, where T_{NI} denotes the NIT temperature in the absence of an external field ($\tilde{H} = 0$). As the external field increases, the transition temperature increases. For instance, at $\tilde{H} = 0.01$, the system remains in a paranematic phase, exhibiting a small but finite orientational order even at temperatures above T_{NI}. With a further increase in the external field, the order parameter varies continuously with temperature, and at a critical field $\tilde{H}_c$, the transition becomes second-order. The black circle indicates this critical point, characterized by $dS/dT = \infty$, where the first-order phase transition changes into a continuous second-order transition. Such field-induced nematic transitions have been widely observed in experiments [73].

6.3 Twist-untwist transition in the cholesteric phase

Let us consider the case where an external field is applied perpendicular to the pitch axis of a cholesteric (Ch) phase. As illustrated in Fig. 6.3, a magnetic field **H** (or electric field **E**)

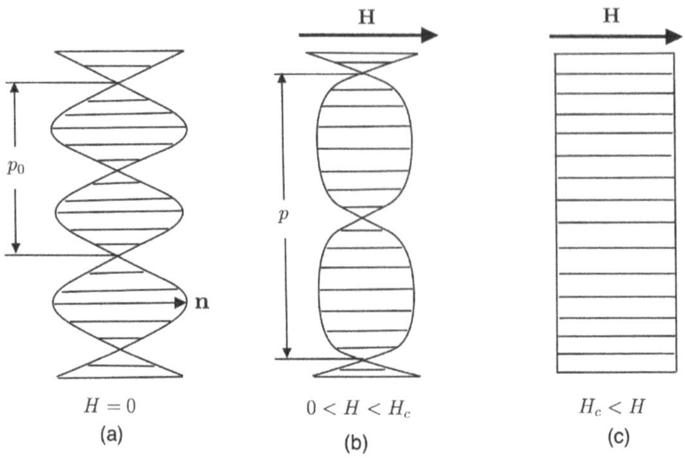

FIGURE 6.3
When a magnetic field **H** (or an electric field **E**) is applied perpendicular to the pitch axis of the cholesteric phase, the cholesteric phase undergoes a transition to the nematic phase, known as the twist-untwist phase transition.

is applied in the direction perpendicular to the cholesteric pitch axis. When the magnetic susceptibility anisotropy $\Delta\chi$ is positive ($\Delta\chi > 0$), the director $\mathbf{n}$ of the rod-like molecules tends to align with the direction of the external field. The pitch length p of the cholesteric helix increases with increasing field strength H, as shown in Fig. 6.3(a) and (b). The pitch length in the absence of the field ($H = 0$) is denoted by p_0. At a certain critical field H_c, the pitch diverges to infinity (Fig. 6.3(c)). For $H > H_c$, the entire system aligns uniformly along the external field direction, resulting in a transition to the nematic phase. This transition from the cholesteric to the nematic phase induced by an external field is known as the twist-untwist transition. It was theoretically predicted by de Gennes's continuum theory [74] and later confirmed experimentally by Meyer [75]. When the external field is removed, the system reverts to the original cholesteric phase. The ability to control molecular alignment in liquid crystals with external fields underlies liquid crystal display technology and its various applications. In this Section, we will develop the theory based on the free energy of the cholesteric phase, as discussed in Chapter 5.

6.3.1 Free energy under an external field

The total free energy of the cholesteric phase under the external field system is given by

$$F = F_{iso} + F_{nem} + F_{dis} + F_{ext}. \tag{6.18}$$

The first term shows the isotropic free energy (Eq. (5.49)):

$$b^3\beta F_{iso}/V = \frac{\phi}{n_r}(\beta\mu_r^\circ - 1 + \ln\phi + n_r\phi), \tag{6.19}$$

and the second term shows the nematic free energy (Eq. (5.50)):

$$\begin{aligned} b^3\beta F_{nem}/V &= \frac{\phi}{n_r}\int f_d(\mathbf{n}(\mathbf{r})\cdot\mathbf{\Omega})\ln\left[4\pi f_d(\mathbf{n}(\mathbf{r})\cdot\mathbf{\Omega})\right]d\mathbf{\Omega} \\ &\quad -\frac{1}{2}\left(\frac{5}{4}+\nu_M\right)\phi^2\frac{2}{3}Q_{\alpha\beta}Q_{\alpha\beta}, \end{aligned} \tag{6.20}$$

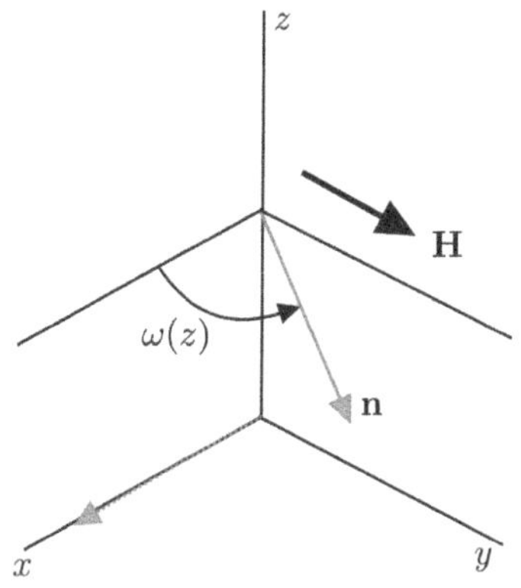

FIGURE 6.4
Schematic illustration of the rotation of the director $\mathbf{n}(z)$ under the external field $\mathbf{H}$ applied perpendicular to the pitch axis (z-axis), along the y-axis.

and the third term is the distortion free energy (Eq. (5.51)):

$$\begin{aligned} b^3\beta F_{dis} &= -\frac{1}{2}(\frac{5}{4}+\nu_M)\phi^2\frac{1}{9}d_0^2\int Q_{\alpha\beta}\partial_z^2 Q_{\alpha\beta}d\mathbf{r}, \\ &\quad -\frac{2}{9}c_L\phi^2 d_0\int \epsilon_{z\beta\gamma}Q_{\beta\mu}\partial_z Q_{\gamma\mu}d\mathbf{r}. \end{aligned} \tag{6.21}$$

As shown in Fig. 6.4, consider a cholesteric phase with a pitch length p and a pitch axis aligned along the z-axis. When an external field $\mathbf{H}$ is applied along the y-axis, the external field is expressed as

$$\mathbf{H} = (0, H, 0), \tag{6.22}$$

where H represents the magnitude of the external field. The director $\mathbf{n}$ rotates in a helical manner along the z-axis. Let the angle between the director and the x-axis at position z be denoted by $\omega(z)$. At any given position z, the director is given by

$$\mathbf{n} = (\cos\omega(z), \sin\omega(z), 0). \tag{6.23}$$

In equilibrium, the function $\omega(z)$ is determined to minimize the free energy of the system. Using Eq. (6.23), we have

$$\epsilon_{z\beta\gamma}Q_{\beta\mu}(z)\partial_z Q_{\gamma\mu}(z) = \frac{9}{4}S^2\left(\frac{\partial\omega}{\partial z}\right), \tag{6.24}$$

$$\epsilon_{z\beta\gamma}Q_{\beta\mu}(z)\partial_z^2 Q_{\gamma\mu}(z) = 0, \tag{6.25}$$

$$Q_{\alpha\beta}(z)Q_{\alpha\beta}(z) = \frac{3}{2}S^2, \tag{6.26}$$

$$Q_{\alpha\beta}(z)\ \partial_z^2 Q_{\alpha\beta}(z) = -\frac{9}{2}S^2\left(\frac{\partial\omega}{\partial z}\right)^2, \tag{6.27}$$

and then Eq. (6.21) is given by

$$\begin{aligned} b^3\beta F_{dis} &= \frac{1}{2}(\frac{5}{4}+\nu_M)\phi^2 S^2 \int_0^p \frac{1}{2}\left(\frac{\partial \omega}{\partial z}\right)^2 d_0^2 dz V/p \\ &\quad -\frac{1}{2}c_L\phi^2 S^2 \int_0^p \left(\frac{\partial \omega}{\partial z}\right) d_0 dz V/p, \\ &= \frac{1}{2}(\frac{5}{4}+\nu_M)\phi^2 S^2 QV \int_0^{1/Q} \frac{1}{8\pi^2}\left(\frac{\partial \omega}{\partial \tilde{z}}\right)^2 d\tilde{z} \\ &\quad -\frac{1}{2}c_L\phi^2 S^2 QV \int_0^{1/Q} \frac{1}{2\pi}\left(\frac{\partial \omega}{\partial \tilde{z}}\right) d\tilde{z}, \end{aligned} \tag{6.28}$$

where we define $\tilde{z} = z/(2\pi d_0)$, the dimensionless pitch wave number $Q = qd_0 = 2\pi d_0/p$, and the integration variable is changed from z to $\tilde{z}$.[1] We also have

$$\begin{aligned} H_\alpha Q_{\alpha\beta} H_\beta &= H_y Q_{yy} H_y \\ &= H^2 S\frac{3}{2}\big(\sin^2\omega(z) - \frac{1}{3}\big), \end{aligned} \tag{6.29}$$

and the external free energy (Eq. (6.5)) is given by

$$\begin{aligned} b^3\beta F_{ext} &= -b^3\beta\phi\mu_0\Delta\chi \int_0^p H_y Q_{yy} H_y dz V/p \\ &= -\frac{1}{2}\phi S(3b^3\mu_0\beta\Delta\chi H^2)QV \int_0^{1/Q} \big(\sin^2\omega(\tilde{z}) - \frac{1}{3}\big) d\tilde{z}. \end{aligned} \tag{6.30}$$

Using Eqs. (6.28) and (6.30), we obtain

$$b^3\beta(F_{dis}+F_{ext})/V = -\frac{1}{2}(\frac{5}{4}+\nu_M)\phi^2 S^2 g_1(Q), \tag{6.31}$$

where the function $g_1(Q)$ is given by

$$\begin{aligned} g_1(Q) &= -\frac{Q}{8\pi^2}\int_0^{1/Q}\left[\left(\frac{\partial \omega}{\partial \tilde{z}}\right)^2 - 4\pi Q_0\left(\frac{\partial \omega}{\partial \tilde{z}}\right)\right. \\ &\quad \left. -\big(\frac{2\pi}{\xi}\big)^2(\sin^2\omega - \frac{1}{3})\right] d\tilde{z}, \end{aligned} \tag{6.32}$$

the coherence length is defined as

$$\xi = \sqrt{\frac{\phi S}{6\tilde{H}}(\frac{5}{4}+\nu_M)}, \tag{6.33}$$

and

$$Q_0 (= 2\pi d_0/p_0) = \frac{c_L}{\frac{5}{4}+\nu_M}, \tag{6.34}$$

shows the pitch wave number in the absence of the external field. As a dimensionless parameter characterizing the strength of the external field, we define

$$\tilde{H} = b^3\beta\mu_0\Delta\chi H^2. \tag{6.35}$$

The value of ξ decreases with increasing the external field.

[1] Using $\int d\mathbf{r} = \int dxdydz = (V/p)\int_0^p dz$.

6.3.2 Pitch length under an external field

The pitch length in the equilibrium state under the external field is determined by minimizing the free energy (Eq. (6.31)). The Euler-Lagrange equation for the angle $\omega(z)$, which minimizes Eq. (6.31), is given by

$$\Big(\frac{\xi}{2\pi}\Big)^2\Big(\frac{\partial\omega}{\partial\tilde{z}}\Big)^2 + \sin^2\omega = c, \tag{6.36}$$

where c is a constant determined by the condition for minimizing the free energy. The cholesteric pitch length p is given by

$$p = \int_0^p dz = \int_0^{1/Q} 2\pi d_0\Big(\frac{\partial\tilde{z}}{\partial\omega}\Big)d\omega. \tag{6.37}$$

Using Eq. (6.36), the pitch wavenumber number Q is given by

$$\frac{1}{Q} = \Big(\frac{2}{\pi}\Big)\xi K(c), \tag{6.38}$$

where $K(c)$ is the complete elliptic integral of the first kind:

$$K(c) = \int_0^{\pi/2}\frac{d\omega}{\sqrt{c-\sin^2\omega}}. \tag{6.39}$$

Substituting Eqs. (6.36) and (6.39) into Eq. (6.32), the function $g_1(Q)$ can be rewritten as

$$g_1(Q) = \frac{1}{2\xi^2}\Big(c-\frac{1}{3}\Big) - Q\Big(-Q_0 + \frac{2}{\pi\xi}E(c)\Big), \tag{6.40}$$

where $E(c)$ is the elliptic integral of the second kind, defined as

$$E(c) = \int_0^{\pi/2}\sqrt{c-\sin^2\omega}\; d\omega. \tag{6.41}$$

The constant c is determined from Eq. (6.40) by solving the condition $dg_1(Q)/dc = 0$, which yields

$$E(c) = \frac{\pi}{2}\xi Q_0, \tag{6.42}$$

where note that Q is a function of c and $dE(c)/dc = \pi/(4Q\xi)$ with Eq. (6.38). The ratio of the pitch length p_0 in the absence of an external field to the pitch length p in the presence of an external field can be obtained by eliminating ξ using Eqs. (6.38) and (6.42):

$$\frac{p}{p_0} = \frac{Q_0}{Q} = \frac{4}{\pi^2}K(c)E(c). \tag{6.43}$$

Substituting Eq. (6.42) into Eq. (6.40), the function $g_1(Q)$ that minimizes the distortion free energy is given by

$$g_1(Q) = \frac{1}{2\xi^2}\Big(c-\frac{1}{3}\Big). \tag{6.44}$$

The dependence of the first-kind elliptic integral $K(c)$, the second-kind elliptic integral $E(c)$, and the normalized pitch length p/p_0 on the constant c is shown in Fig. 6.5. As the

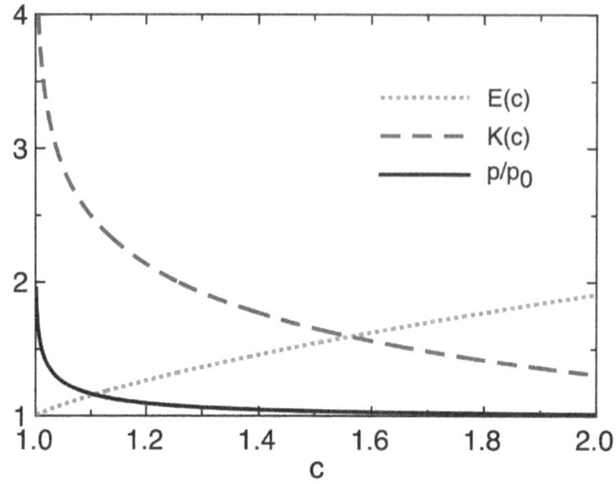

FIGURE 6.5
The dependence of the first-kind elliptic integral $K(c)$, the second-kind elliptic integral $E(c)$, and the normalized pitch length p/p_e on the constant c.

constant c approaches 1 ($c \to 1$), the function $K(c)$ diverges to infinity (i.e., the pitch length diverges), and the pitch wavenumber Q continuously approaches zero. The second-kind elliptic integral $E(c)$ is greater than 1 and decreases as c decreases toward 1. For large c, it increases proportionally to $\sqrt{c}$: $E(c) \propto \sqrt{c}$.

The determination of the constant c follows two possible cases based on Eq. (6.42):

1. When $(\pi/2)\xi Q_0 > 1$, the value of c is determined to satisfy Eq. (6.42), and the pitch wavenumber of the cholesteric phase is given by Eq. (6.43) (see Fig.6.3(b)). The cholesteric phase is characterized by $c > 1$ or $Q > 0$, $S \neq 0$.
2. When $(\pi/2)\xi Q_0 < 1$, no value of c satisfies Eq. (6.42). In this case, from Eq. (6.36), we obtain $\omega = \pi/2$ and $c = 1$. This means that the director aligns parallel to the external field, corresponding to the nematic phase (see Fig. 6.3(c)). As shown in Fig. 6.5, when $c = 1$, the pitch length becomes infinite ($Q = 0$), which corresponds to the nematic phase ($Q = 0$, $S \neq 0$). The isotropic phase is defined by $Q = S = 0$.

As the external field H increases, a phase transition from the cholesteric phase to the nematic phase occurs at a critical field. This critical field (H_c, also referred to as the critical electric field) is determined by the condition $E(c = 1) = 1$. From Eq. (6.42), the critical field is expressed as

$$H_c = \frac{\pi}{2} Q_0 \sqrt{\frac{\left(\frac{5}{4} + \nu_L\right) \phi S}{6\beta \Delta\chi}} = \frac{\pi^2}{p_0} \sqrt{\frac{K_2}{3\mu_0 \Delta\chi \phi S}}. \tag{6.45}$$

The latter expression is derived using the twist elastic constant K_2 (Eq. (5.100)) and Eq. (6.34). For $H < H_c$, the pitch length of the cholesteric phase increases with increasing field strength. For $H > H_c$, the pitch length diverges, resulting in a transition to the nematic phase (see Fig.6.3(c)) [1]. As shown in Fig.6.5, the pitch length ratio p/p_0 (Eq. (6.43)) increases as the constant c decreases, and diverges as c approaches 1.

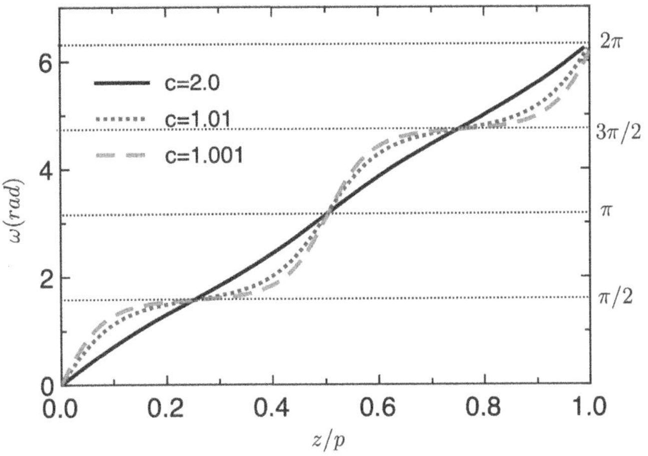

FIGURE 6.6
Distribution of the twist angle ω along the pitch direction z/p for various values of the constant c.

Using Eq. (6.36), the distribution of the twist angle ω along the pitch direction z is given by

$$\frac{z}{p} = \frac{1}{4K(c)} \int_0^{\omega} \frac{d\omega}{\sqrt{c - \sin^2 \omega}}. \tag{6.46}$$

Figure 6.6 shows the distribution of the twist angle $\omega(z)$ for various values of the constant c. When $c = 2$, corresponding to a relatively weak external field, the twist angle changes linearly along the pitch axis (z-axis). However, near $c \simeq 1$, where the external field is strong, there are regions around $\theta = \pi/2, 3\pi/2$, where the twist angle remains nearly constant. This situation resembles that shown in Fig. 6.3(b). Near $\theta = 0, \pi, 2\pi$, the twist angle changes sharply. At $c = 1$, the entire system undergoes a phase transition to the nematic phase.

6.3.3 Cholesteric-nematic phase transitions induced by an external field

The orientational distribution function in thermal equilibrium is determined by minimizing the free energy (Eq. (6.18)):

$$\left(\frac{\delta F}{\delta f_d(\mathbf{n}(\mathbf{r}) \cdot \mathbf{\Omega})} \right)_T = 0. \tag{6.47}$$

This yields

$$f_d(x) = \frac{1}{Z_0} \exp\left[\Gamma_0 S P_2(x)\right], \tag{6.48}$$

and

$$\Gamma_0 = n_r\left(\frac{5}{4} + \nu_M\right)\phi\left(1 + g_1(Q)\right), \tag{6.49}$$

with the normalization condition $\int f_d(\mathbf{\Omega}) d\mathbf{\Omega} = 1$ (Eq. (4.11)): $Z_0 = 4\pi I_0(\Gamma_0 S)$, where we define the function

$$I_m(\Gamma_0 S) \quad = \quad \int_0^1 \left[P_2(x)\right]^m \exp\left[\Gamma_0 S P_2(x)\right] dx, \tag{6.50}$$

for $m = 0, 1, 2$. The orientational order parameter S (Eq. (4.21)) is determined by the self-consistent equation:

$$S = \frac{I_1(\Gamma_0 S)}{I_0(\Gamma_0 S)}. \tag{6.51}$$

The total free energy is given by the sum of Eqs. (6.19), (6.20), and (6.31). Applying the distribution function (Eq. (6.48)) to the nematic free energy (Eq. (6.20)), the free energy is expressed as

$$\begin{aligned} b^3\beta F/V &= \frac{\phi}{n_r}(\beta\mu_r^\circ - 1 + \ln\phi + n_r\phi) \\ &+\frac{1}{2}(\frac{5}{4} + \nu_M)\phi^2 S^2(1+g_1(Q)) - \frac{\phi}{n_r}\ln I_0(\Gamma_0 S). \end{aligned} \tag{6.52}$$

The only difference from Eq. (5.79) in Chapter 5 is that the function $g_1(Q)$ is now given by Eq. (6.44). (Note that $v_r = b^3 n_r$.) The expressions for the osmotic pressure Π and the chemical potential μ under an external field remain the same as those derived in Chapter 5. When the volume fraction is $\phi = 1$, the free energy reduces to that of a pure thermotropic liquid crystal.

We summarize below the procedure for calculating the order parameters S and Q:

1. Specify the concentration ϕ, temperature τ (where $\tau = 1/\nu_M$), external field strength $\tilde{H}$, and the parameter Q_0.
2. Set an initial value for S and calculate the coherence length ξ using Eq. (6.33).
3. If $(\pi/2)\xi Q_0 > 1$, determine the constant c such that Eq. (6.42) is satisfied. Otherwise, set $c = 1$ and $Q = 0$.
4. Substitute the value of c into the function $g_1(Q)$ using Eq. (6.44).
5. Since ξ depends on S via Eq. (6.33), check whether the current value of S satisfies Eq. (6.51).
6. Iterate steps 2 through 5 until a self-consistent value of S that satisfies Eq. (6.51) is obtained.

For example, we consider the case where the external field $\tilde{H} = 2.5 \times 10^{-5}$, the temperature $\tau(= 1/\nu_M) = 2$, $n_r = 100$, and $Q_0 = 0.06$. Figure 6.7(a) shows the constant c and the orientational order parameter S, and (b) shows the pitch wave number Q, as functions of the concentration ϕ of rod-like molecules. At low concentrations, due to the influence of the external field, S takes a small but nonzero value, indicating a nearly isotropic (I) phase. This phase is referred to as the paranematic (pN) phase. The prefix para- here implies "pseudo" (Fig. 6.2), indicating that the order parameter S remains small. As the concentration increases, the value of S exhibits a discontinuous jump at the pN-N phase transition concentration ϕ_{pN}. With further increase in concentration, a continuous transition from the N to the Ch phase occurs at ϕ_{NC}, where the value of c begins to increase continuously from 1. As shown in Fig. 6.7(b), the wavenumber Q also starts to increase continuously at ϕ_{NC}. For $\phi > \phi_{NC}$, the cholesteric phase becomes stable. In the intermediate region $\phi_{pN} < \phi < \phi_{NC}$, we have $S > 0$ and $c = 1$, indicating the nematic phase. In the absence of an external field ($\tilde{H} = 0$), the I-Ch phase transition occurs at approximately $\phi \simeq 0.025$ (see Fig.5.9). As the external field strength increases, this transition shifts to lower concentrations.

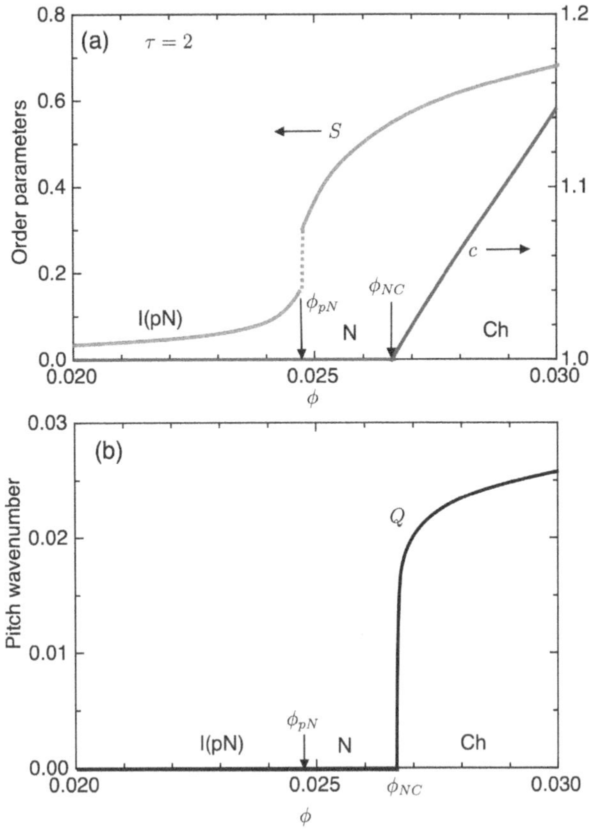

FIGURE 6.7
(a) Orientational order parameter S, the constant c, and (b) the pitch wavenumber Q (b), plotted against ϕ for $\tilde{H} = 2.5 \times 10^{-5}$ and $\tau = 2$.

Thus, in the presence of an external field, the Ch-N phase transition occurs as a function of the concentration of rod-like molecules. As discussed in Chapter 5, in the absence of an external field, the cholesteric phase is always more stable than the nematic phase, and therefore the Ch-N transition does not occur. However, applying an external field induces a Ch-N phase transition.

Using Eq. (6.52), we analyze the phase separation behavior. Figure 6.8 shows the phase diagram in the temperature-concentration (τ-ϕ) plane for $n_r = 100$ under the external field of $\tilde{H} = 1 \times 10^{-6}$. The solid line represents the coexistence curve corresponding to the phase separation between the para-nematic (pN) and cholesteric (Ch) phases (pN+Ch). The dashed line indicates the pN-N phase transition concentration ϕ_{pN}, while the dotted line represents the N-Ch transition concentration ϕ_{NC}. Compared to the phase diagram without an external field (Fig. 5.9), the coexistence curve (solid line) shifts slightly toward lower concentrations. In solutions of rod-like molecules, the pN-N and Ch-N phase transitions take place within the binodal lines, indicating that the nematic phase is metastable throughout this region of the phase diagram.

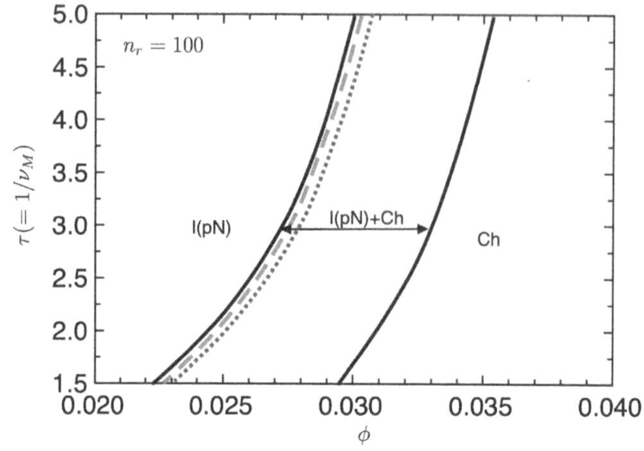

FIGURE 6.8
Phase diagram under the external field $\tilde{H} = 1 \times 10^{-6}$ for $n_r = 100$. The solid line shows the coexistence curve.

6.4 Heliconical cholesteric phase under an external field

6.4.1 Heliconical cholesteric phase

In this section, we consider the case where an external field $\mathbf{E}$ is applied parallel to the pitch axis (i.e., the z-axis) of the cholesteric phase:

$$\mathbf{E} = (0, 0, E). \tag{6.53}$$

Under this condition, the director $\mathbf{n}$ of the cholesteric phase tilts by an angle ϵ, referred to as the conical angle, from the pitch axis, resulting in a heliconical cholesteric phase (Ch_h) as shown in Fig. 6.9 [75].

Figure 6.9 shows the behavior of the director $\mathbf{n}$ in the Ch_h phase with a pitch length $p(= 2\pi/|q|)$. As it progresses in the positive z-axis direction, the director maintains a tilt angle ϵ and exhibits a right-handed heliconical structure (a left-handed helix is also possible). Since the director $\mathbf{n}$ undergoes helical rotation along the z-axis, its azimuthal angle can be expressed as $\omega(z)$. At a given position z, the director is given by

$$\mathbf{n} = (\sin\epsilon \cos\omega(z), \sin\epsilon \sin\omega(z), \cos\epsilon). \tag{6.54}$$

The azimuthal angle ω is a function of the coordinate z and is given by $\omega(z) = qz$. When $\epsilon = 0$, the phase corresponds to the nematic phase, while for $\epsilon = \pi/2$, it shows the conventional cholesteric phase. As the external field strength E increases, the phase transition from the Ch to the Ch_h phase, and eventually to the N phase, is expected.

6.4.2 Free energy of the heliconical cholesteric phase

The free energy of liquid crystal phases, including N, Ch, and Ch_h phases, under the external field is given by [76]

$$F_{LC} = F_{nem} + F_{dis} + F_{ext}, \tag{6.55}$$

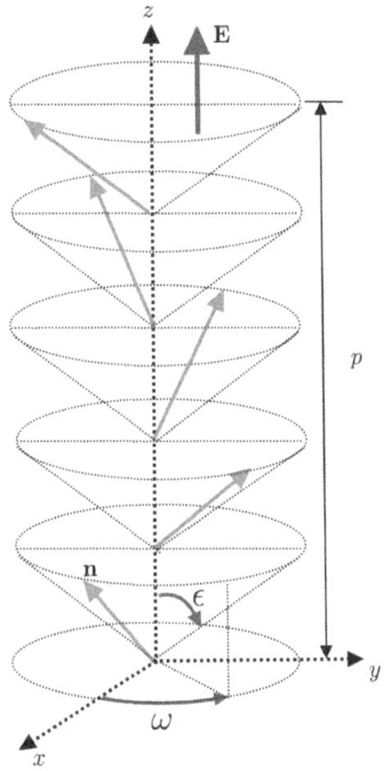

FIGURE 6.9
When an external field **E** is applied parallel to the pitch axis (z-axis) of the cholesteric phase, the director **n** exhibits a heliconical cholesteric (Ch_h) structure.

where the first term is the free energy (Maier-Saupe model) of the nematic phase in the absence of the external field (Eq. (5.50)):

$$\begin{aligned} b^3\beta F_{nem}/V &= \frac{\phi}{n_r}\int f_d(\mathbf{n}(\mathbf{r})\cdot\mathbf{\Omega})\ln[4\pi f_d(\mathbf{n}(\mathbf{r})\cdot\mathbf{\Omega})]d\mathbf{\Omega} \\ &\quad -\frac{1}{2}\nu_M\phi^2S^2, \end{aligned} \tag{6.56}$$

and the second term shows the distortion free energy for the cholesteric phase (Eq. (5.51)):[2]

$$\begin{aligned} b^3\beta F_{dis} &= \frac{1}{2}\nu_M\phi^2\frac{1}{9}d_0^2\int \partial_\gamma Q_{\alpha\beta}(\mathbf{r})\partial_\gamma Q_{\alpha\beta}(\mathbf{r})d\mathbf{r}, \\ &\quad -\frac{2}{9}c_L\phi^2 d_0\int \epsilon_{\alpha\beta\gamma}Q_{\mu\beta}(\mathbf{r})\partial_\alpha Q_{\mu\gamma}(\mathbf{r})d\mathbf{r}, \end{aligned} \tag{6.57}$$

where $\phi(= b^3n_r\rho)$ is the volume fraction of rods. In this section, we focus on short rod-like molecules and neglect the excluded volume term $5/4$ in Eqs. (5.50) and (5.51). The last term F_{ext} in Eq. (6.55) shows the free energy of the external field. When the external electric field **E** is applied to the liquid crystalline molecules, having a dielectric anisotropy $\Delta\epsilon$, the external free energy, given by Eq. (6.5) and using Eqs. (6.53) and (6.54), is expressed as

$$\begin{aligned} F_{ext} &= -\phi\Delta\epsilon\int E_\alpha Q_{\alpha\beta}(\mathbf{r})E_\beta d\mathbf{r} \\ &= -\phi\Delta\epsilon E_z^2 S\left(\frac{3}{2}n_z^2-\frac{1}{2}\right)V \\ &= -\phi\Delta\epsilon E^2 S\left(1-\frac{3}{2}\sin^2\epsilon\right)V. \end{aligned} \tag{6.58}$$

[2] It is useful to note that $\int Q_{\alpha\beta}\partial_\gamma\partial_\gamma Q_{\alpha\beta}d\mathbf{r} = -\int(\partial_\gamma Q_{\alpha\beta})^2d\mathbf{r}$, which follows from integration by parts with the assumption that the surface term vanishes.

We now derive the distortion free energy. Using the tensor order parameter given in Eq. (5.52) together with Eqs. (3.2) and (3.7), we obtain

$$(\partial_\gamma Q_{\alpha\beta})^2 = \frac{9}{2} S^2 \partial_\alpha n_\beta \partial_\alpha n_\beta, \tag{6.59}$$

where we assume that the orientational order parameter (S) is spatially uniform. From Eq. (3.10), we also have

$$\partial_\alpha n_\beta(\mathbf{r}) \partial_\alpha n_\beta(\mathbf{r}) = \big(\nabla \cdot \mathbf{n}(\mathbf{r})\big)^2 + \big(\mathbf{n}(\mathbf{r}) \cdot \nabla \times \mathbf{n}(\mathbf{r})\big)^2 + \big(\mathbf{n}(\mathbf{r}) \times \nabla \times \mathbf{n}(\mathbf{r})\big)^2, \tag{6.60}$$

where we neglect the surface contribution to the free energy. Substituting Eq. (6.60) into Eq. (6.59), we obtain

$$\begin{aligned}(\partial_\gamma Q_{\alpha\beta})^2 &= \frac{9}{2} S^2 \Big[\tilde{K}_1 \big(\nabla \cdot \mathbf{n}(\mathbf{r})\big)^2 + \tilde{K}_2 \big(\mathbf{n}(\mathbf{r}) \cdot \nabla \times \mathbf{n}(\mathbf{r})\big)^2 \\ &\quad + \tilde{K}_3 \big(\mathbf{n}(\mathbf{r}) \times \nabla \times \mathbf{n}(\mathbf{r})\big)^2\Big],\end{aligned} \tag{6.61}$$

where we have added the coefficient $\tilde{K}_i$ ($i = 1, 2, 3$), corresponding to the dimensionless elastic constant of a pure liquid crystal ($\phi = 1$). Using the director Eq. (6.54), we have

$$\big(\nabla \cdot \mathbf{n}(\mathbf{r})\big)^2 = 0, \tag{6.62}$$

$$\big(\mathbf{n}(\mathbf{r}) \cdot \nabla \times \mathbf{n}(\mathbf{r})\big)^2 = (\partial_z \omega)^2 \sin^4 \epsilon = q^2 \sin^4 \epsilon, \tag{6.63}$$

$$\begin{aligned}\big(\mathbf{n}(\mathbf{r}) \times \nabla \times \mathbf{n}(\mathbf{r})\big)^2 &= (\partial_z \omega)^2 \sin^2 \epsilon (1 - \sin^2 \epsilon) \\ &= q^2 \sin^2 \epsilon (1 - \sin^2 \epsilon).\end{aligned} \tag{6.64}$$

Substituting Eqs. (6.62), (6.63), and (6.64) into (6.61), we obtain the first term in Eq. (6.57):

$$\int \nu_M (\partial_\gamma Q_{\alpha\beta})^2 d\mathbf{r} = \nu_M \frac{9}{2} S^2 q^2 \sin^2 \epsilon \big(\tilde{K}_2 \sin^2 \epsilon + \tilde{K}_3 (1 - \sin^2 \epsilon)\big) V, \tag{6.65}$$

where the first term corresponds to the twist distortion and the second is the bend distortion. Similarly, using the director Eq. (6.54), the second term in Eq. (6.57), including the first spatial derivative of the tensor order parameter, is given by

$$\begin{aligned}\int c_L \epsilon_{\alpha\beta\gamma} Q_{\mu\beta} \partial_\alpha Q_{\mu\gamma} d\mathbf{r} &= c_L \epsilon_{z\beta\gamma} Q_{\mu\beta} \partial_z Q_{\mu\gamma} V \\ &= c_L (\epsilon_{zxy} Q_{\mu x} \partial_z Q_{\mu y} + \epsilon_{zyx} Q_{\mu y} \partial_z Q_{\mu x}) V \\ &= V c_L \frac{9}{4} \tilde{K}_2 S^2 q \sin^2 \epsilon,\end{aligned} \tag{6.66}$$

which is proportional to the twist distortion $(\mathbf{n}(\mathbf{r}) \cdot \nabla \times \mathbf{n}(\mathbf{r}))$, where we have added the dimensionless elastic constant $\tilde{K}_2$ (see Chapter 3). Both c_L and q are pseudoscalar quantities that change sign under spatial inversion, whereas their product $c_L q$ remains invariant and thus represents a true scalar. Substituting Eqs. (6.65) and (6.66) into Eq. (6.57), the free energy F_{dis} is then given by

$$b^3 \beta F_{dis}/V = \frac{1}{2} \nu_M \phi^2 S^2 \Big[\frac{1}{2} (\tilde{K}_2 - \tilde{K}_3) Q^2 y^2 + \big(\frac{1}{2} \tilde{K}_3 Q - \tilde{K}_2 Q_0\big) Q y\Big], \tag{6.67}$$

where the heliconical order parameter is defined by

$$y = \sin^2 \epsilon, \tag{6.68}$$

for the Ch_h phase, and the pitch wavenumber $Q = qd_0$. The value $Q_0(= c_L/\nu_M)$ represents the pitch wavenumber of the pure Ch phase in the absence of an external field (see Eq.(5.63)).

Then, the free energy in the presence of the external field is given by

$$F_{LC} = F_{nem} + F_{dis} + F_{ext} = F_N + F_d, \tag{6.69}$$

where we have separated the free energy into two terms for convenience. The nematic free energy under the external field is defined by incorporating the first term of Eq. (6.58) into Eq. (6.56):

$$\begin{aligned} b^3\beta F_N/V &= b^3\beta\big(F_{nem}/V - \Delta\epsilon E^2\phi S\big) \\ &= \frac{\phi}{n_r}\int f_d(\mathbf{n}(\mathbf{r})\cdot\mathbf{\Omega})\ln[4\pi f_d(\mathbf{n}(\mathbf{r})\cdot\mathbf{\Omega})]d\mathbf{\Omega} \\ &\quad -\frac{1}{2}\nu_M\phi^2S^2 - \phi S\tilde{H}_{el}, \end{aligned} \tag{6.70}$$

where we define the dimensionless parameter of the external field

$$\tilde{H}_{el} = b^3\beta\Delta\epsilon E^2. \tag{6.71}$$

The distortion free energy under the external field is defined by incorporating the second term of Eq. (6.58) into Eq. (6.67):

$$\begin{aligned} b^3\beta F_d/V &= b^3\beta\big(F_{dis}/V + \frac{3}{2}\phi\Delta\epsilon E^2 S\sin^2\epsilon\big) \\ &= \frac{1}{2}\nu_M\phi^2S^2 g_d(y,Q), \end{aligned} \tag{6.72}$$

where we define the distortion function

$$g_d(y,Q) = \frac{1}{2}(\tilde{K}_2 - \tilde{K}_3)Q^2y^2 + \left[(\frac{1}{2}\tilde{K}_3Q - \tilde{K}_2Q_0)Q + \eta\right]y, \tag{6.73}$$

and the dimensionless external field

$$\eta = \frac{3\tilde{H}_{el}}{\nu_M\phi S} \propto E^2. \tag{6.74}$$

When $y = 1$, or $\epsilon = \pi/2$, the bend term $\tilde{K}_3$ disappears and the Ch phase appears. When $y = 0$, or $\epsilon = 0$, we have the N phase because of $g_d(y) = 0$. Depending on the strength $\tilde{H}_{el}$ of the external field, we have N, Ch, and Ch_h phases.

6.4.3 Determination of the conical angle

We here derive the equilibrium value y^* and pitch wavenumber Q^*. In a thermal equilibrium state, the conical angle ϵ, or the order parameter y of the Ch_h phase, is determined by $(\partial g_d(y,Q)/\partial y)_{S,Q} = 0$. We then obtain

$$y = \frac{1}{(\tilde{K}_2 - \tilde{K}_3)Q^2}\left[(\tilde{K}_2Q_0 - \frac{1}{2}\tilde{K}_3Q)Q - \eta\right], \tag{6.75}$$

and $(\partial g_d(y,Q)/\partial Q)_{S,y} = 0$ yields

$$Q = \frac{\tilde{K}_2 Q_0}{(\tilde{K}_2 - \tilde{K}_3)y + \tilde{K}_3}. \tag{6.76}$$

Using Eqs. (6.75) and (6.76), we obtain the equilibrium value

$$y^* = \frac{\tilde{K}_2 Q_0/Q^* - \tilde{K}_3}{(\tilde{K}_2 - \tilde{K}_3)}, \tag{6.77}$$

for the conical angle and

$$Q^* = \pm\sqrt{\frac{2\eta}{\tilde{K}_3}}, \tag{6.78}$$

for the pitch wavenumber, where the sign (+) shows the left-handed helix with $Q_0 > 0$ and the sign (−) is the right-handed helix with $Q_0 < 0$. In the Ch phase of chiral systems, either (+) or (−) twist appears in a sample. The heliconical cholesteric state (Ch_h) induced by the external field in the Ch (N^*) phase has the same chirality everywhere in the sample. The pitch length $p(\propto 1/Q)$ is inversely proportional to the external field E.

The values of y^* and Q^* must be a minimum of the free energy (Eq. (6.72)). The minimum of the free energy $\hat{f}_d(= b^3\beta F_d/V)$ occurs when the Hessian matrix is a positive definite matrix:

$$\begin{aligned} |H(y^*,Q^*)| &= \begin{vmatrix} \frac{\partial^2 \hat{f}_d}{\partial y^2} & \frac{\partial^2 \hat{f}_d}{\partial y \partial Q} \\ \frac{\partial^2 \hat{f}_d}{\partial Q \partial y} & \frac{\partial^2 \hat{f}_d}{\partial Q^2} \end{vmatrix} \\ &= \frac{1}{2}\nu_M \phi^2 S^2 \tilde{K}_3(\tilde{K}_2 - \tilde{K}_3) Q^{*2} y^* > 0. \end{aligned} \tag{6.79}$$

The Ch_h phase is stabilized when $\tilde{K}_2$(twist) $> \tilde{K}_3$(bend) > 0, as this condition ensures $y^* > 0$.

6.4.4 Orientational distribution function

The orientational distribution function $f_d(\mathbf{n}(\mathbf{r}) \cdot \mathbf{\Omega})$ of the liquid crystal molecule is determined by the free energy F_{LC} (Eq. (6.69)) with respect to the distribution function: $(\delta F_{LC}/\delta f_d(\Omega))_T = 0$, under the normalization condition:

$$\int f_d(\mathbf{n}(\mathbf{r}) \cdot \mathbf{\Omega}) d\mathbf{\Omega} = 1. \tag{6.80}$$

We then obtain the distribution function of the liquid crystal molecule:

$$f_d(x) = \frac{1}{Z_0} \exp\left[\Gamma_0 P_2(x)\right], \tag{6.81}$$

where we define $x = \cos\theta$ and

$$\Gamma_0 = n_r \left[\nu_M \phi S(1 - g_d(y,Q)) + (1 + \frac{3}{2}y)\tilde{H}_{el}\right]. \tag{6.82}$$

The constant Z_0 is determined by the normalization condition (6.80) as $Z_0 = 4\pi I_0(S)$ and the functions $I_m(S)$ is defined as:

$$I_m(S) = \int_0^1 [P_2(x)]^m \exp\left[\Gamma_0 P_2(x)\right] dx, \tag{6.83}$$

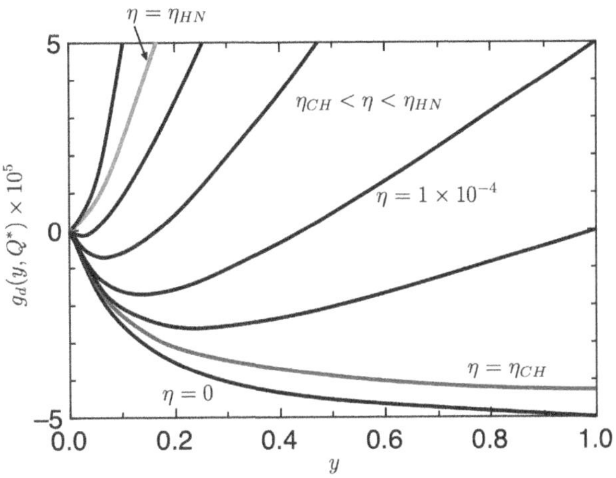

FIGURE 6.10
Distortion function $g_d(y, Q^*)$, Eq. (6.73), plotted against the order parameter y of the Ch_h phase for various values of the external field η. The distortion function g_d is a quadratic $g_d(y) = Ay^2 + By$ and can be easily understood in the framework of a Landau expansion in a series of y of the free energy F_d.

for $m = 0, 1, 2 \cdots$. Using Eq. (6.81), the scalar orientational order parameter $S = \int P_2(x) f_d(\Omega) d\mathbf{\Omega}$ can be determined by the self-consistent equation:

$$S = I_1(S)/I_0(S). \tag{6.84}$$

Using the distribution function, Eq. (6.81), the free energy (Eq. (6.69)) of the liquid crystal phases is given by

$$\begin{aligned} f_{LC} &= b^3 \beta F_{LC}/V \\ &= \frac{1}{2}\nu_M \phi^2 S^2 \left(1 - g_d(y, Q) + \eta y\right) - \frac{\phi}{n_r} \ln I_0(S). \end{aligned} \tag{6.85}$$

6.4.5 Phase transitions between N, Ch, and Ch_h phases

We examine how Eq. (6.72) describes phase transitions between Ch_h, Ch, and N phases. Equation (6.72) is the quadratic function of y: $g_d(y) = Ay^2 + By$, and can be easily understood in the framework of a Landau expansion in a series of y of the free energy F_d.

Figure 6.10 shows the distortion function $g_d(y, Q^*)$ plotted against y for various values of the external field η for $\tilde{K}_2/\tilde{K}_3 = 10$. The value of Q^* is given by Eq. (6.78). We estimate the pitch wavenumber $Q_0 (= 2\pi d_0/p_0) \simeq 0.01$ of the Ch phase in pure liquid crystalline molecules with the typical pitch length $p_0 = 300$ nm and the length $d_0 = 0.5$ nm.

(a) When $0 < \eta < \eta_{CH}$, the function $g_d(y)$ decreases with increasing y and has a minimum at a value of $y = 1$. In this case, the stable state is given by $y > 1$ and we have $Q = Q_0$ from Eq. (6.76). This shows the Ch phase with $Q = Q_0$, $y = 1$ (cone angle: $\epsilon = \pi/2$), and $S > 0$.

(b) When $\eta = \eta_{CH}$, the distortion free energy F_d has a minimum at $y = 1$ and the second-order Ch-Ch_h phase transition appears. Using Eqs. (6.77) and (6.78), the critical value of the Ch-Ch$_h$ transition is given by

$$\eta_{CH} = \frac{Q_0^2}{2}\tilde{K}_3, \tag{6.86}$$

at which we have $Q^*_{CH} = Q_0$.

(c) As increasing the external field $\eta_{CH} < \eta < \eta_{HN}$, the distortion function $g_d(y)$ has a minimum at y^* (Eq. (6.77)). The value of Q^* (Eq. (6.78)) increases with η. In this region, the stable Ch$_h$ phase appears ($0 < y^* < 1$).

(d) When $\eta = \eta_{HN}$, the function $g_d(y)$ has a minimum at $y^* = 0$ ($\epsilon = 0$), where the second-order Ch$_h$-N phase transition takes place. From Eqs. (6.77) and (6.78), the critical value of the Ch$_h$-N phase transition is given by

$$\eta_{HN} = \frac{Q_0^2}{2}\frac{\tilde{K}_2^2}{\tilde{K}_3} = \Big(\frac{\tilde{K}_2}{\tilde{K}_3}\Big)^2 \eta_{CH}, \tag{6.87}$$

where we have $Q^*_{HN} = (\tilde{K}_2/\tilde{K}_3)Q_0$.

(e) When $\eta > \eta_{HN}$, we have the N phase with $y^* = 0$ and $F_d = 0$.

The orientational order parameter S can be numerically solved from Eq. (6.84). The value of η, Eq. (6.74), depends on the concentration ϕ, temperature through $\nu_M(\propto 1/T)$, S, and the external field $\tilde{H}_{el}$.

Figure 6.11 shows the helicoidal order parameter y and the orientational order parameter S plotted against temperature T/T_{CI} for $\tilde{H}_{el} = 4 \times 10^{-6}$ (a), 4×10^{-4} (b), and 5×10^{-4} (c), where T_{CI} shows the CIT temperature of a pure liquid crystal in the absence of the external field. Figure 6.11(a), for a weak external field, shows the first-order phase transition between the Ch phase with $y = 1$, $S > 0$, and the I (para-nematic) phase with $y = 0$, $S \simeq 0$. In Fig. 6.11(b), we find the first-order Ch$_h$-I phase transition, where the value of y discontinuously changes from $y = 0$ to a finite value at $T/T_{CI} \simeq 1$. As increasing the external field, Fig 6.11(c), the second-order Ch$_h$-N transition appears, where the value of y continuously changes at $T/T_{CI} \simeq 0.905$. At $T/T_{CI} \simeq 1$, we have the first-order NIT, where the value of S jumps. As the temperature decreases, we find that the I phase changes to the Ch$_h$ through the N phase. The Ch$_h$ phase appears at the lower temperature side of the N phase. The Ch$_h$-N transition temperature decreases with increasing the external field. There is competition between chirality and uniaxial nematic ordering. The former is strongly affected by the ratio K_2/K_3 of the elastic constants, and the latter needs the external field. For larger values of K_2/K_3, the chirality dominates and then the Ch$_h$ or Ch phase appears below $T/T_{CI} < 1$. On increasing the strength of the external field, the uniaxial nematic phase induced by the external field dominates, and the N phase appears at higher temperatures of the Ch$_h$ phase. In a strong external field, the Ch$_h$ phase disappears, leaving only the NIT.

The external field parameter η (Eq. (6.74)) is proportional to the square of the electric field, E^2. From Eq. (6.87), the ratio of the twist and bend elastic constants is given by

$$\frac{K_2}{K_3} = \sqrt{\frac{\eta_{HN}}{\eta_{CH}}} \simeq \frac{E_{HN}}{E_{CH}} > 1, \tag{6.88}$$

where E_{HN} and E_{CH} denote the critical electric fields at the Ch$_h$-N and Ch-Ch$_h$ phase transitions, respectively. The appearance of the Ch$_h$ phase has been experimentally observed between the N and Ch phases, showing that the Ch$_h$ phase emerges within the range $E_{CH} < E < E_{HN}$ [76, 77]. Equation (6.88) further implies that the stability region of the Ch$_h$ phase expands with increasing K_2/K_3.

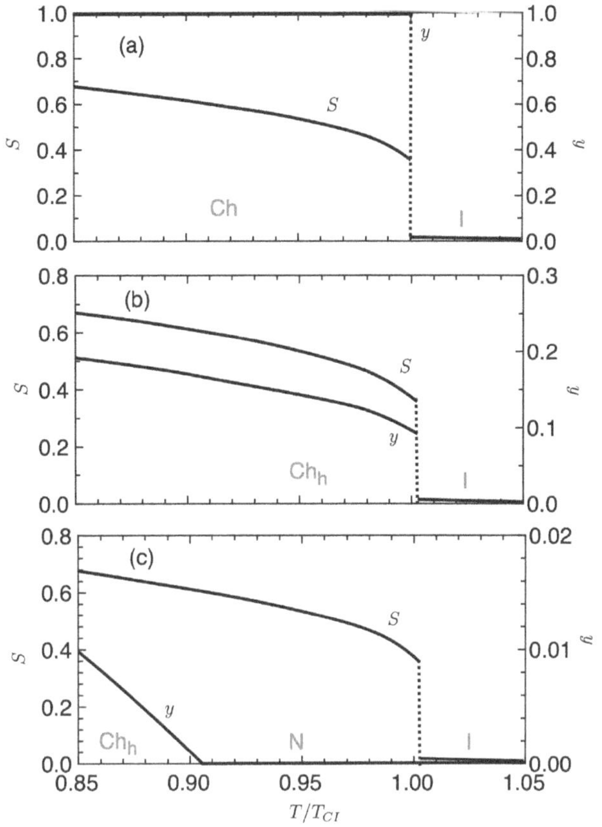

FIGURE 6.11
Heliconical order parameter y and orientational order parameter S plotted against temperature for $\tilde{H}_{el} = 4 \times 10^{-6}$ (a), 4×10^{-4} (b), and 5×10^{-4} (c). The order parameter y of the Ch_h phase discontinuously changes in (a) and (b), but in (c) the value of y continuously changes.

The continuum theory of the Ch_h phase was discussed in de Gennes's textbook (p. 292) [1], where it was noted that this conical phase had not been observed. More recently, however, experiments conducted under applied electric fields have successfully demonstrated a phase transition to the Ch_h phase [77]. Around the same time, studies on banana-shaped liquid crystalline molecules revealed the existence of the twist-bend nematic (N_{TB}) phase [78–80]. The director configuration of the N_{TB} phase is identical to that of the Ch_h phase (Eq. (6.54)), with a pitch length on the order of 10 nm, approximately two orders of magnitude shorter than that of a conventional cholesteric phase. Moreover, while conventional Ch and Ch_h phases exhibit either right- or left-handed helices (and are therefore classified as chiral nematic phases), the N_{TB} phase is unique in that right- and left-handed helices coexist with equal probability within the same system. For readers interested in a more detailed theoretical treatment of the N_{TB} phase, see Refs. [78, 81, 82].

7

Mixtures of Liquid Crystals and Polymers

Binary mixtures of flexible polymers and low-molecular-weight liquid crystalline molecules play an important role in electro-optical devices such as light shutters and displays. Because the miscibility and phase separation directly govern material performance, the phase behavior and phase separation kinetics of these mixtures have been of both fundamental and practical interest. In the following, we introduce a free energy framework to describe the phase behavior of polymer-liquid crystal mixtures.

7.1 Free energy of mixing

Let us consider a binary mixture composed of rod-like molecules and flexible polymer chains. The rod-like molecules are characterized by a length L and a diameter D. The polymer chains have a contour length bn_p, where b is the length of a unit segment and n_p denotes the number of segments in a chain. We denote by N_p and N_r the total number of polymers and rods, respectively. In this model, the polymers are assumed to be sufficiently flexible so that they do not contribute to the nematic ordering of the system.

Extending the second virial approximation (Eq. (4.10)) to a binary mixture consisting of flexible polymers and rod-like molecules, the Helmholtz free energy can be expressed as follows [83, 84]:

$$\begin{aligned}\beta F(N_p, N_r) &= \sum_{i=p,r} (N_i \beta \mu_i^\circ - N_i + N_i \ln \rho_i) + v_{pp} N_p \rho_p + 2v_{pr} N_p \rho_r + v_{rr} N_r \rho_r \\ &+ N_r \int f_d(\theta) \ln[4\pi f_d(\theta)] d\mathbf{\Omega} \\ &+ \frac{1}{2} N_r \rho_r \iint A_2(\theta, \theta') f_d(\theta) f_d(\theta') d\mathbf{\Omega} d\mathbf{\Omega}', \end{aligned} \tag{7.1}$$

where μ_i° denotes the standard chemical potential of the pure component i ($i = p, r$), and $\rho_i = N_i/V$ represents the number density of species i. The parameter v_{ij} corresponds to the average excluded volume interaction between molecules i and j. The function $f_d(\theta)$ is the orientational distribution function of the rods, and $A_2(\theta, \theta')$ is the second virial coefficient accounting for the excluded volume and attractive interactions between two rods with orientations θ and θ'.

The free energy given in Eq. (7.1) corresponds to a compressible system, that is, a pseudo-ternary mixture consisting of a flexible polymer, a rod-like molecule, and implicit solvent molecules. To discuss the miscibility and phase behavior of polymer/rod mixtures, it is convenient to introduce the free energy of mixing.

DOI: 10.1201/9781003737711-7

The free energy of mixing is defined as the difference between the free energy of the mixture and the free energies of the pure components in their isotropic states. It can be written as

$$\Delta F = F(N_p, N_r) - F_{iso}(N_p, 0) - F_{iso}(0, N_r), \tag{7.2}$$

where the $F_{iso}(N_p, 0)$ and $F_{iso}(0, N_r)$ denote the reference free energy of the pure polymer and the pure rod-like molecule in the isotropic phase before mixing, respectively. For simplicity, we assume that the mixture is incompressible:

$$\phi_p + \phi_r = 1, \tag{7.3}$$

with the volume fractions $\phi_p = v_p N_p/V$ and $\phi_r = v_r N_r/V$, where v_p and v_r denote the molecular volumes of the polymer and the rod-like molecule, respectively. The molecular volume of the polymer can be expressed as $v_p = b^3 n_p$. The molecular volume of the rod-like molecule is given by $v_r = (\pi/4)D^2 L = b^3 n_r$, where $n_r = L/D$ is the axial ratio of the rod. Here we assume the relation $b^3 \simeq (\pi/4)D^3$ for convenience. Thus, the total volume of the system can be written as

$$V = v_p N_p + v_r N_r = b^3\left(n_p N_p + n_r N_r\right) = b^3 N_t, \tag{7.4}$$

where N_t represents the total number of segments in the system. Then, the volume fractions can be given by

$$\phi_p = n_p N_p/N_t = n_p \rho_p, \tag{7.5}$$

for the polymer segments and

$$\phi_r = n_r N_r/N_t = n_r \rho_r, \tag{7.6}$$

for rod-like molecules.

Using Eq. (7.1), the free energies of the pure components are obtained as

$$\beta F_i(N_p, 0) = \beta\mu_p^\circ N_p - N_p - N_p \ln v_p + v_{pp} N_p/v_p, \tag{7.7}$$

for the pure polymer system, and

$$\beta F_i(0, N_r) = \beta\mu_r^\circ N_r - N_r - N_r \ln v_r + v_{rr} N_r/v_r, \tag{7.8}$$

for the pure rod-like system. Substituting Eqs. (7.7) and (7.8) into Eq. (7.2), we obtain the free energy of mixing as

$$\Delta F = \Delta F_{mix} + F_{nem}, \tag{7.9}$$

where the first term shows the free energy of mixing in the isotropic phase:

$$\begin{aligned} \beta\Delta F_{mix} &= N_t\left(\frac{\phi_p}{n_p}\ln\phi_p + \frac{\phi_r}{n_r}\ln\phi_r + \chi\phi_p\phi_r\right) \\ &= N_p \ln\phi_p + N_r \ln\phi_r + \chi n_p N_p \phi_r, \end{aligned} \tag{7.10}$$

and the second term is the nematic contribution arising from orientational ordering of the rod-like aggregates:

$$\begin{aligned} \beta F_{nem}/V &= \rho_r \int f_d(\theta) \ln[4\pi f_d(\theta)] d\mathbf{\Omega} \\ &+ \frac{1}{2}\rho_r^2 \iint A_2(\theta, \theta') f_d(\theta) f_d(\theta') d\mathbf{\Omega} d\mathbf{\Omega}'. \end{aligned} \tag{7.11}$$

The interaction parameter χ in Eq. (7.10) is defined as

$$\chi = b^3\left(\frac{2v_{pr}}{v_p v_r} - \frac{v_{pp}}{v_p^2} - \frac{v_{rr}}{v_r^2}\right). \tag{7.12}$$

Thus, the free energy ΔF_{mix} (Eq. (7.10)) in the isotropic phase consists of two distinct contributions:

$$\Delta F_{mix} = \Delta E_{mix} - T\Delta S_{mix}, \tag{7.13}$$

where the first term is given by the energy term $\beta\Delta E_{mix}/N_t = \chi\phi_p(1-\phi_p)$, coming from the isotropic interactions between unlike molecules. The second term shows the entropy of mixing ΔS_{mix} of the polymer and the rod-like molecule:

$$\Delta S_{mix}/(k_B N_t) = -\left(\frac{\phi_p}{n_p}\ln\phi_p + \frac{\phi_r}{n_r}\ln\phi_r\right), \tag{7.14}$$

where the first (second) term in Eq. (7.14) corresponds to the translational entropy of molecules consisting of n_p (n_r) segments. The mixing entropy $|\Delta S_{mix}/(k_B N_t)|$ decreases with an increasing molecular weight. The parameter χ (Eq. (7.12)) is well known as the Flory-Huggins interaction parameter between unlike molecules in the isotropic phase. The average excluded volumes v_{rr} between rod-like molecules in Eq. (7.12) is given by $v_{rr} = (\pi/4)L^2D$, while the average excluded volumes v_{pr} and v_{pp} depend on both the polymer chain conformation and temperature. Therefore, the χ parameter is defined as

$$\chi = \frac{z}{k_BT}\left(\epsilon_{pr} - \frac{1}{2}(\epsilon_{pp} + \epsilon_{rr})\right), \tag{7.15}$$

where z is the coordination number of the lattice, ϵ_{pr} is the interaction energy between a polymer segment and a solvent (or rod-like) molecule, and ϵ_{pp} and ϵ_{rr} are the corresponding interaction energies of polymer-polymer and solvent-solvent pairs. A positive χ indicates unfavorable polymer-solvent interactions, leading to phase separation at sufficiently high χ. Conversely, a small or negative χ corresponds to good mixing. Experimentally, χ often depends on temperature and can be expressed as $\chi = A + B/T$, consisting of an entropic contribution (A) and an energetic contribution (B) in the isotropic state. Here we assume that the χ parameter depends only on temperature in the form:

$$\chi = U_0/(k_BT), \tag{7.16}$$

where U_0 is the interaction energy between unlike molecules. Under this assumption, the mixing free energy ΔF_{mix} reduces to the familiar 'Flory-Huggins lattice theory' for polymer blends and solutions of rod-like particles [85,86]. We have extended Onsager's virial theory to binary mixtures under the incompressibility condition (Eq. (7.3)). As a result, the free energy of isotropic mixing results in the Flory-Huggins theory for polymer blends, while nematic ordering is still treated within the second-virial approximation. This formulation thus allows the theory to be applied to more concentrated solutions than the original Onsager theory.

7.1.1 Hybrid model of nematic ordering

For the intermolecular interaction A_2 in Eq. (7.11), we adopt a hybrid model that accounts for both the excluded volume interactions (Eq. (4.23)) and the orientation-dependent attractive interactions between rod-like molecules (Eq. (4.59)). Accordingly, the function A_2 in Eq. (7.11) can be expressed as

$$A_2(\boldsymbol{\Omega}, \boldsymbol{\Omega}') = 2L^2D|\sin\theta_{12}| - v_{rr}\nu_M SP_2(\cos\theta), \tag{7.17}$$

where the first term represents the excluded volume repulsion between rods, and the second term accounts for the orientation-dependent attractive interaction proportional to the nematic order parameter S. Substituting Eq. (7.17) into Eq. (7.11) with the Legendre polynomial expansion (Eq. (4.26)) for $|\sin\theta_{12}|$ in Eq. (7.17), the free energy of the nematic phase is given by

$$\beta F_{nem}/N_t = \frac{\phi_r}{n_r}\int f_d(\theta)\ln[4\pi f_d(\theta)]d\mathbf{\Omega} - \frac{1}{2}(\nu_M + \frac{5}{4})\phi_r^2 S^2. \tag{7.18}$$

The orientational distribution function $f_d(\theta)$ of the rod-like molecules is determined by minimizing the free energy in Eq. (7.18) with respect to $f_d(\theta)$: $(\delta F_{nem}/\delta f_d(\theta))_T = 0$, as

$$f_d(\theta) = \frac{1}{Z_0}\exp\Big[\Gamma_0 S P_2(\cos\theta)\Big], \tag{7.19}$$

where

$$\Gamma_0 = (\nu_M + \frac{5}{4})n_r\phi_r, \tag{7.20}$$

with $\nu_M = U_M/(k_B T)$ and 5/4 arising from the attractive and excluded volume interactions between rod-like molecules, respectively. Substituting Eq. (7.19) into the normalization condition (Eq. (4.11)) yields $Z_0 = 4\pi I_0(\Gamma_0 S)$, where the function I_0 is defined in Eq. (4.32). The orientational order parameter S (Eq. (4.21)) of the rods is then determined by the self-consistent equation

$$S = \frac{I_1(\Gamma_0 S)}{I_0(\Gamma_0 S)}. \tag{7.21}$$

Finally, substituting Eq. (7.19) into Eq. (7.18), we obtain the nematic free energy per segment as a function of S:

$$\beta F_{nem}/N_t = \frac{1}{2}\left(\nu_M + \frac{5}{4}\right)\phi_r^2 S^2 - \frac{\phi_r}{n_r}\ln I_0(\Gamma_0 S), \tag{7.22}$$

which has the same form as the nematic free energy, Eq. (4.38).

The chemical potential of the polymer is given by

$$\begin{aligned}
\beta\Delta\mu_p &= \beta(\mu_p - \mu_p^\circ) \\
&= \beta\Big(\frac{\partial\Delta F}{\partial N_p}\Big)_{N_r,T} \\
&= n_p\Big[f_{LC} + (1-\phi_p)(\frac{\partial f_{LC}}{\partial\phi_p})\Big] \\
&= \ln\phi_p + \Big(1 - \frac{n_p}{n_r}\Big)(1-\phi_p) + n_p\chi(1-\phi_p)^2 \\
&\quad + \frac{n_p}{2}\Big(\nu_M + \frac{5}{4}\Big)S^2(1-\phi_p)^2,
\end{aligned} \tag{7.23}$$

and that of the rod is given by

$$\begin{aligned}
\beta\Delta\mu_r &= \beta(\mu_r - \mu_r^\circ) \\
&= \beta\Big(\frac{\partial\Delta F}{\partial N_r}\Big)_{N_p,T} \\
&= n_r\Big[f_{LC} - \phi_p(\frac{\partial f_{LC}}{\partial\phi_p})\Big] \\
&= \ln(1-\phi_p) + \Big(1 - \frac{n_r}{n_p}\Big)\phi_p + n_r\chi\phi_p^2 \\
&\quad + \frac{n_r}{2}\Big(\nu_M + \frac{5}{4}\Big)S^2(1-\phi_p)^2 - \ln I_0(\Gamma_0 S),
\end{aligned} \tag{7.24}$$

where we define the dimensionless free energy per unit lattice: $f_{LC} = \beta\Delta F/N_t$.

7.2 Nematic-isotropic phase transition

As discussed in Fig. 4.5, the nematic-isotropic phase transition (NIT) temperature T_{NI} is given by $\Gamma_0^{NI} = 4.55$. Using Eq. (7.20), we obtain

$$T_{NI} = \frac{(U_M/k_B)n_r(1-\phi_p)}{4.55 - 1.25n_r(1-\phi_p)}, \tag{7.25}$$

as a function of ϕ_p.

Figure 7.1 shows the NIT curve on the temperature-concentration plane for various values of n_r. The high-temperature side of each curve shows the isotropic phase, while the low-temperature side corresponds to the nematic phase. At the NIT temperature, the orientational order parameter jumps from zero to 0.43, as shown in Fig. 4.5. As the polymer concentration increases, the NIT temperature decreases. For a pure liquid crystal molecule, $\phi_p = 0$, the NIT temperature is given by

$$T_{NI,r} = \frac{(U_M/k_B)n_r}{4.55 - 1.25n_r}. \tag{7.26}$$

The intermolecular excluded volume interaction becomes dominant for rod-like molecules with an axial ratio $n_r > 3.64$, and $T_{NI,r}$ diverges to the high-temperature side. On the other hand, for low-molecular-weight liquid crystals with $n_r < 3.64$, the attractive intermolecular interaction (ν_M) becomes dominant, leading to the NIT curve strongly depending on temperature. Using Eqs. (7.25) and (7.26), the NIT temperature is given by

$$\frac{T_{NI}}{T_{NI,r}} = \frac{(4.55 - 1.25n_r)(1-\phi_p)}{4.55 - 1.25n_r(1-\phi_p)}, \tag{7.27}$$

where the temperature is normalised by the NIT temperature $T_{NI,r}$ of a pure liquid crystal molecule.

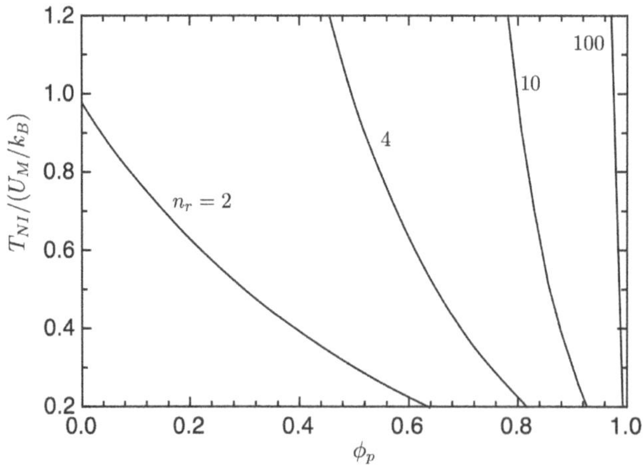

FIGURE 7.1
NIT curve on the temperature-concentration plane for various values of n_r. The left (right) side of each curve shows the nematic (isotropic) phase.

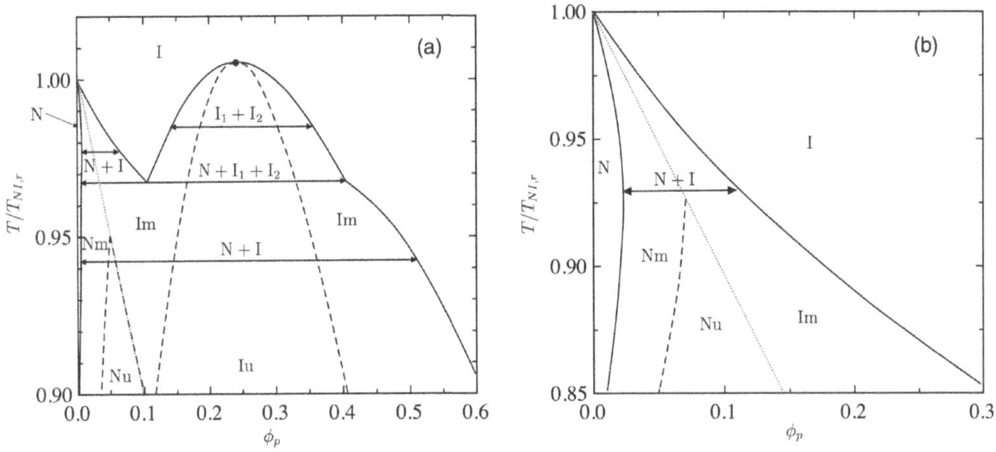

FIGURE 7.2
Typical phase diagrams of polymer/liquid crystal mixtures for $\nu_M/\chi = 3.1$ (a) and $\nu_M/\chi = 5$ (b). The solid curve denotes the binodal line and the dotted line indicates the NIT temperature derived in Eq. (7.27).

7.3 Phase separation in polymer-dispersed liquid crystals

We introduce the phase behavior of a flexible polymer and a low-molecular-weight liquid crystal molecule (nematogen) with $n_r < 3.64$. The conditions of two-phase equilibria are given by

$$\Delta\mu_p(\phi_p') = \Delta\mu_p(\phi_p''), \tag{7.28}$$

$$\Delta\mu_r(\phi_p') = \Delta\mu_r(\phi_p''), \tag{7.29}$$

where ϕ_p' and ϕ_p'' show the volume fraction of polymers in lower and higher concentrations in the coexisting two phases, respectively. The above coupled equations correspond to the common tangent construction, discussed in Fig. 4.8.

Figure 7.2 presents representative phase diagrams calculated for two parameter sets: (a) $\nu_M/\chi = 3.1$, $n_p = 20$, $n_r = 2$, and (b) $\nu_M/\chi = 5$, $n_p = 10$, $n_r = 2$. The temperature is normalised by the NIT temperature $T_{NI,r}$ of the pure liquid crystal. The solid curve denotes the binodal line, while the dotted line indicates the NIT temperature derived in Eq. (7.27). The dashed curve corresponds to the spinodal line defined by $(\partial^2 f_{LC}/\partial\phi_p^2)_T = 0$. The closed circle marks the critical point (CP). Below the spinodal curve, two unstable regions emerge: the isotropic unstable (Iu) and nematic unstable (Nu) regions. These unstable regions are characterized by

$$\left(\frac{\partial^2 f_{\mathrm{LC}}}{\partial\phi_p^2}\right)_T < 0, \tag{7.30}$$

where fluctuations in concentration and orientational order drive spinodal decomposition during the phase separation process [87–89]. Between the binodal and spinodal curves, two metastable regions appear: isotropic metastable (Im) and nematic metastable (Nm). These metastable and unstable regions are separated by the condition $(\partial^2 f_{LC}/\partial\phi_p^2)_T = 0$, which corresponds to the inflection point of the free energy [84, 90–92]. In Fig. 7.2(a), a triple point (TP) appears, where a nematic phase and two isotropic phases can coexist simultaneously (N+I_1+I_2). Above the triple point, a biphasic region consisting of a nematic and an isotropic phase (N+I) is observed at lower concentrations. In addition, an isotropic-isotropic (I_1+I_2) phase separation with an upper critical solution temperature (UCST) occurs, similar to that reported in poly(acrylic acid) (PAA)-n-alkane blends [93], and polystyrene (PS)-7CB blends [94]. Below the TP temperature, broad biphasic regions of nematic and isotropic phases (N+I) appear. In this regime, the nematic phase is composed almost entirely of liquid crystalline molecules, whereas the coexisting isotropic phase contains both liquid crystals and flexible polymers. When the nematic interaction parameter ν_M/χ is increased, the I_1+I_2 phase separation with a UCST disappears, leaving only the N+I phase separation as shown in Fig. 7.2(b). Such phase diagrams have been experimentally observed for various polymer-nematogen mixtures [11,95–97]. When the system is thermally quenched from a stable isotropic phase into an unstable region of the coexistence curve, the interplay between concentration and orientation fluctuations can lead to various morphologies [98–101]. For a detailed review of the spinodal decompositions, see Ref. [44].

Figure 7.3 schematically illustrates the concentration dependence of the free energy corresponding to the phase diagram in Fig. 7.2(a). The free energy is plotted as a function of the polymer concentration ϕ_p at four characteristic temperature ranges: (a) $T_c > T > T_{NI,r}$, (b) $T_{NI,r} > T > T_{TP}$, (c) $T = T_{TP}$, and (d) $T < T_{TP}$. Here, T_c denotes the UCST and

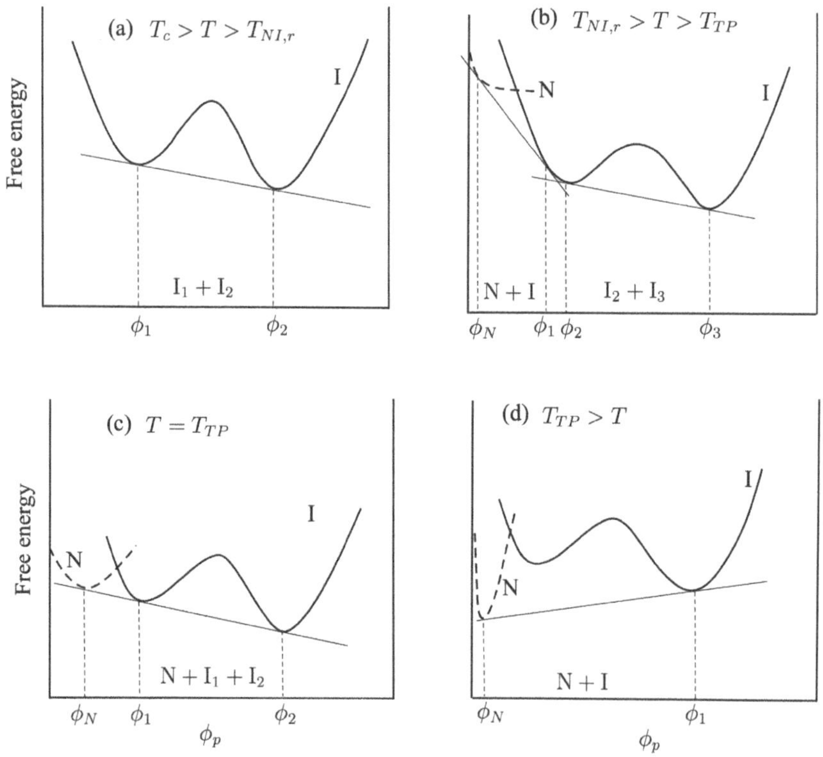

FIGURE 7.3
Schematically illustrated the free energy of Fig. 7.2(a) plotted against the concentration of polymer for a temperature of $T_c > T > T_{NI,r}$ (a), $T_{NI,r} > T > T_{TP}$ (b), $T = T_{TP}$ (c), and $T_{TP} > T$ (d). The straight line corresponds to the common tangent method, which determines coexistence compositions.

T_{TP} represents the triple point temperature. The solid curve represents the isotropic free energy $\beta\Delta F_{mix}/N_t$ (Eq. (7.10)), while the broken curve denotes the nematic free energy $\beta\Delta F_{nem}/N_t$ (Eq. (7.22)). The straight line corresponds to the common tangent construction, which determines coexistence compositions (see also Fig. 4.8). At high temperatures ($T_c > T > T_{NI,r}$), as shown in Fig. 7.3(a), the free energy of the isotropic phase exhibits two minima at $\phi_p = \phi_1$ and $\phi_p = \phi_2$. These two isotropic phases (I_1 and I_2) coexist because the unfavorable polymer-solvent interactions, expressed by the Flory-Huggins parameter χ, dominate in the intermediate composition region. Larger values of χ correspond to poorer miscibility between unlike molecules. Upon decreasing temperature below $T_{NI,r}$ (Fig. 7.3(b)), the nematic free energy branch appears at low polymer concentrations. In this case, two common tangent lines can be drawn: one between the nematic and isotropic phases, and the other between the two isotropic phases. The nematic free energy branch merges into the isotropic free energy at the NIT concentration ϕ_{NI}. At $T = T_{TP}$ (Fig. 7.3(c)), the two isotropic phases become simultaneously stable. This corresponds to the triple point, where a single common tangent touches the free energy curve at three distinct concentrations, ϕ_N, ϕ_1, and ϕ_2. At lower temperatures ($T < T_{TP}$) (Fig. 7.3(d)), the attractive nematic interaction ν_M between liquid crystalline molecules becomes dominant, and the nematic free energy is further stabilized. As a result, the I_1+I_2 phase separation disappears, leaving only a broad biphasic region consisting of nematic and isotropic phases (N+I), in which microdroplets of the N phase are dispersed within the polymer matrix (I). Such structures are referred to as polymer-dispersed liquid crystal (PDLC) [11].

Using the Flory-Huggins theory, the critical point (closed circle) in the isotropic phase with $S = 0$ is obtained from the conditions $(\partial\Delta\mu_r/\partial\phi_p)_T = 0$ and $(\partial^2\Delta\mu_r/\partial\phi_p^2)_T = 0$. From these, the critical concentration ϕ_p^c is given by

$$\phi_p^c = \frac{\sqrt{n_r}}{\sqrt{n_p} + \sqrt{n_r}}, \tag{7.31}$$

and the UCST:

$$\frac{1}{\chi_c} = \frac{k_B T_c}{U_0} = \frac{2n_p n_r}{(\sqrt{n_p} + \sqrt{n_r})^2}. \tag{7.32}$$

As shown in Fig. 7.2, two generic types of phase diagrams can occur, depending on whether I_1+I_2 phase separation appears. The criterion for distinguishing these cases is the relative magnitude of the NIT temperature $T_{NI,r}$ and the UCST: T_c. We introduce the nematic interaction parameter, $\alpha = \nu_M/\chi$, which, from Eqs. (7.26) and (7.32), can be expressed as

$$\alpha = \frac{U_M/(k_B T)}{U_0/(k_B T)} = \frac{2n_p(4.55 - 1.25n_r)}{(\sqrt{n_p} + \sqrt{n_r})^2}\left(\frac{T_{NI,r}}{T_c}\right). \tag{7.33}$$

For high molecular weight polymers ($n_p \gg 1$), it is given by

$$\alpha = \alpha^*\left(\frac{T_{NI,r}}{T_c}\right), \tag{7.34}$$

where $\alpha^* = 2(4.55 - 1.25n_r)$. Thus, when $\alpha < \alpha^*$ (i.e., $T_{NI,r} < T_c$), the phase diagram exhibits the I_1+I_2 phase separation with a UCST (Fig.7.2(a)). Conversely, when $\alpha > \alpha^*$ (i.e., $T_{NI,r} > T_c$), the I_1+I_2 phase separation disappears, and only N+I coexistence remains (Fig.7.2(b)). For example, when $n_r = 2$, we obtain $\alpha^* = 4.1$. In this case, the I_1+I_2 phase separation does not appear for $\alpha = 5$, even for very large n_p. Such behavior has been reported in PS/EBBA mixtures [95]. The mean field theory introduced in this section provides a quantitative description of the observed phase diagrams for blends of flexible polymers with nematogens.

Flexible polymers such as polystyrene (PS) hardly dissolve in nematic liquid crystalline molecules (e.g., EBBA) below the NIT temperature, and a broad biphasic region (N+I) between the nematic and isotropic phases appears. In this case, the free energy is minimized by phase separation rather than by nematic ordering of the polymer chains. This phenomenon is classified as nematic order-induced phase separation. On the other hand, the stiffness of polymer chains plays a crucial role in determining phase behaviors [102, 103]. Liquid crystalline polymers possess anisotropic interactions with nematic molecules, and their blends exhibit a narrower N+I biphasic region, as observed experimentally [104] (we will discuss in Section 7.5). Here, the free energy is minimized by the partial orientational ordering of polymer chains dissolved in the nematic solvent, rather than by macroscopic phase separation. This phenomenon is referred to as polymer-induced stabilization of nematic ordering. Thus, the stiffness of polymer chains strongly influences both nematic ordering and the width of the N+I biphasic region.

7.4 Phase separation in solutions of rigid rod-like polymers

We here consider binary mixtures of a rigid rod-like polymer and a solvent molecule. When $n_r \gg 1$ and $n_p = 1$, the free energy given in Eq. (7.9) can describe the phase behavior of a rod-like molecular solution [84]. Figure 7.4 shows the phase diagram of $\nu_M/\chi = 5$ and the axial ratio of the rod-like molecules is varied with $n_r = 10$ (a) and $n_r = 100$ (b). The vertical axis represents temperature, normalized by the UCST (T_c) for $n_r = 100$. The solid lines indicate coexistence curves, while the black circle denotes the critical point (CP). The dashed lines passing through the CP correspond to spinodal curves. Within the chimney-type phase separation (I+N), the dotted line represents the NIT curve.

When $n_r = 10$, a chimney-shaped I+N phase separation between the isotropic (I) and nematic (N) phases appears. At higher temperatures, the concentrations along the coexistence curve are nearly identical to those predicted by Eqs. (4.55) and (4.56). At lower temperatures, however, attractive interactions slightly shift the nematic phase boundary to lower concentrations. The region enclosed by the coexistence curve corresponds to the metastable region. This chimney-shaped I+N phase separation, similar to that shown in Figs. 1.5 and 1.6, originates from excluded volume interactions between rod-like molecules, as discussed in Chapter 4. As the axial ratio increases, the coexistence curve shifts toward lower concentrations.

For rod-like molecular solutions with a large axial ratio ($n_r = 100$), a three-phase equilibrium (I+N_1+N_2), involving the isotropic phase and two distinct nematic phases, appears on the low-temperature side of the chimney-shaped I+N coexistence region. Below the TP temperature, a broad I+N phase separation is observed. As the attractive interaction (ν_M) between rod-like molecules increases, the concentration gap between coexisting phases becomes larger, leading to the I+N phase separation as shown in Fig. 1.5. Above the TP temperature, the coexistence of two nematic phases (N_1+N_2) with the UCST arises from the interaction parameter (χ) between the polymer and solvent molecules. This type of phase separation is reminiscent of polymer solutions with a UCST, although in this case the CP lies entirely within the nematic phase. At high temperatures, the chimney-shaped I+N phase separation is driven by excluded volume effects (entropy), whereas at lower temperatures, strong intermolecular interactions (χ) dominate, producing phase separation with large concentration differences. Such phase diagrams have been observed experimentally in solutions of rigid rod-like polymers such as poly(γ-benzyl-L-glutamate) (PBLG) [9] and in DNA solutions [10], and can also be interpreted within the Flory lattice model for nematic phases in solutions of rod-like particles [86, 105].

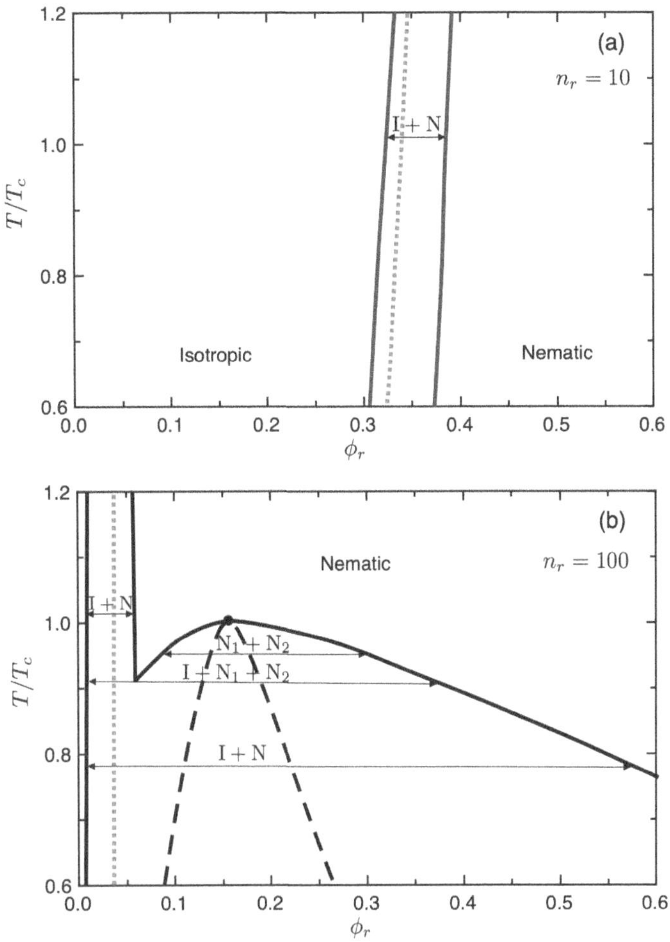

FIGURE 7.4
Phase diagrams on the temperature-concentration plane of rod-like molecular solutions for the axial ratio $n_r = 10$ (a) and 100 (b) with $n_p = 1$ and $\nu_M/\chi = 5$.

7.5 Mixtures of rod-like polymers and liquid crystals

We now extend the mean field framework to binary mixtures of a low-molecular-weight liquid crystal (LC) and a rigid, rod-like polymer with a large axial ratio. Representative examples of such rod-like polymers include carbon nanotubes (CNTs), Kevlar fibers, tobacco mosaic virus, and liquid crystalline polymers. Particular emphasis is placed on the stability and phase behavior of nematic phases in these mixtures.

Consider a binary mixture of N_1 rigid-rod-like polymers of the length L_1 and the diameter D_1, and N_2 LC molecules of the length L_2 and the diameter D_2: $L_1 \gg L_2$. The volume of the rod-like particle and that of the LC are given by $v_1 = (\pi/4)D_1^2 L_1$ and $v_2 = (\pi/4)D_2^2 L_2$, respectively. Let $\rho_1(\mathbf{u})$ and $\rho_2(\mathbf{u})$ be the number density of the rod-like particles and the LCs with an orientation $\mathbf{u}$ (or its solid angle $\mathbf{\Omega}$), respectively. By introducing the orientational distribution function $f_i(\mathbf{u})$ for species $i(=1,2)$, the number densities can be expressed as

$$\rho_i(\mathbf{u}) = \rho_i f_i(\mathbf{u}), \tag{7.35}$$

where $\rho_i = N_i/V$ is the average number density of species i. The free energy ΔF of the binary mixture can be given by

$$\Delta F = \Delta F_{mix} + F_{nem}, \tag{7.36}$$

where the first term represents the free energy of mixing and the second term denotes the nematic contribution. Using the Flory-Huggins theory [85], the free energy of mixing is expressed as (see Eq. (7.10))

$$\beta \Delta F_{mix}/N_t = \frac{\phi_1}{n_1}\ln\phi_1 + \frac{\phi_2}{n_2}\ln\phi_2 + \chi\phi_1\phi_2, \tag{7.37}$$

where $\phi_1 = v_1 N_1/V$ and $\phi_2 = v_2 N_2/V$ are the volume fractions of the rod-like polymer and the LC molecule, respectively. Introducing the axial ratios $n_1 = L_1/D_1$ for the rod and $n_2 = L_2/D_2$ for the LC molecule, the molecular volumes can be expressed as $v_1 = b^3 n_1$ and $v_2 = b^3 n_2$, where we assume a common diameter $D = (D_1 = D_2)$ and define $b^3 \simeq (\pi/4)D^3$. The total volume of the system is then given by

$$V = b^3(n_1 N_1 + n_2 N_2) = b^3 N_t, \tag{7.38}$$

where N_t denotes the total number of segments in the system.

The second term in Eq. (7.36) represents the nematic free energy. Extending Eq. (7.11) to binary mixtures, it is expressed as

$$\begin{aligned}\beta F_{nem}/V &= \sum_{i=1,2} \rho_i \int f_i(\mathbf{u}) \ln[4\pi f_i(\mathbf{u})] d\mathbf{\Omega} \\ &+\frac{1}{2}\rho_1^2 \iint A_{11}(\mathbf{u},\mathbf{u}') f_1(\mathbf{u}) f_1(\mathbf{u}') d\mathbf{\Omega} d\mathbf{\Omega}' \\ &+\frac{1}{2}\rho_2^2 \iint A_{22}(\mathbf{u},\mathbf{u}') f_2(\mathbf{u}) f_2(\mathbf{u}') d\mathbf{\Omega} d\mathbf{\Omega}' \\ &+\rho_1\rho_2 \iint A_{12}(\mathbf{u},\mathbf{u}') f_1(\mathbf{u}) f_2(\mathbf{u}') d\mathbf{\Omega} d\mathbf{\Omega}',\end{aligned} \tag{7.39}$$

where $\beta = 1/(k_B T)$. The first term corresponds to the orientational entropy of rods and LC molecules. The subsequent three terms represent the pairwise interactions: rod-rod (A_{11}), LC-LC (A_{22}), and rod-LC (A_{12}) pair.

For the intermolecular interactions in Eq. (7.39), we adopt the following models. For short LC molecules, we employ the attractive Maier-Saupe model (see Eq. (4.60)) [43]:

$$A_{22} = -v_{22}\nu_{22} P_2(\cos\theta) P_2(\cos\theta'), \tag{7.40}$$

where $\nu_{22}(= U_{22}/(k_B T) > 0)$, U_{22} denotes the anisotropic attractive interaction energy between LC molecules, and $v_{22} = (\pi/4)L_2^2 D$ is the average excluded volume in the isotropic phase. For long rigid rods, we include both excluded volume and attractive interactions (Eq. (7.17)) [106]:

$$A_{11} = 2L_1^2 D|\mathbf{u}\times\mathbf{u}'| - v_{11}\nu_{11} P_2(\cos\theta) P_2(\cos\theta'), \tag{7.41}$$

where the first term represents the excluded volume interaction originally introduced by the Onsager model, $\nu_{11} = U_{11}/(k_B T) > 0$ characterizes the Maier-Saupe attractive interaction between rods, and $v_{11} = (\pi/4)L_1^2 D$ is the average excluded volume in the isotropic phase. The interaction between a rod and an LC molecule is modeled as

$$A_{12} = -v_{12}\nu_{12}P_2(\cos\theta)P_2(\cos\theta'), \tag{7.42}$$

where $\nu_{12} = U_{12}/(k_B T)$ may take positive or negative values, depending on whether the interaction favors or disfavors alignment between rods and LCs. Since the LC molecules are short, their excluded volume effect is assumed to be negligible. The volume $v_{12} = (\pi/4)L_1 L_2 D$ is the average excluded volume between a rod and a short LC molecule in an isotropic orientation. To facilitate analytical calculations, we expand the angle-dependent term $|\mathbf{u}\times\mathbf{u}'| = \sin\theta_{12}$ in terms of Legendre polynomials (see Eq. (4.26)):

$$|\sin\theta_{12}| \simeq \frac{\pi}{4} - \frac{5\pi}{32}P_2(\cos\theta)P_2(\cos\theta'), \tag{7.43}$$

where θ_{12} is the angle between the local orientation $\mathbf{u}$ and $\mathbf{u}'$.

The orientational order parameter S_i of the molecule $i(=1,2)$ is given by

$$S_i = \int P_2(\cos\theta)f_i(\theta)d\mathbf{\Omega}, \tag{7.44}$$

where $P_2(\cos\theta) = 3(\cos^2\theta - 1/3)/2$. Substituting Eqs. (7.40), (7.41), and (7.42) into Eq. (7.39), the nematic contribution to the free energy per unit volume is given by

$$\begin{aligned} b^3\beta F_{nem}/V &= \sum_{i=1,2}\frac{\phi_i}{n_i}\int f_i(\theta)\ln[4\pi f_i(\theta)]d\mathbf{\Omega} \\ &\quad -\frac{1}{2}(\nu_{11}+5/4)\phi_1^2 S_1^2 - \frac{1}{2}\nu_{22}\phi_2^2 S_2^2 - \nu_{12}\phi_1\phi_2 S_1 S_2, \end{aligned} \tag{7.45}$$

where the first term represents the loss of orientational entropy due to nematic ordering. The terms ν_{11} and $5/4$ arise from the attractive and excluded volume interactions between rods, respectively. When $\nu_{12} > 0$, the LC molecules preferentially align parallel to the rods.

In a thermal equilibrium state, the distribution function $f_1(\theta)$ of a rod is determined by minimizing the free energy (7.45) with respect to this function: $(\delta F/\delta f_1(\theta))_{\{f_2(\theta)\}} = 0$. We then obtain

$$f_1(\theta) = \frac{1}{Z_1}\exp\Big[\Gamma_1 P_2(\cos\theta)\Big], \tag{7.46}$$

where

$$\Gamma_1 = n_1\big((\nu_{11}+5/4)\phi_1 S_1 + \nu_{12}\phi_2 S_2\big). \tag{7.47}$$

The constant Z_1 is determined by the normalization condition:

$$\int f_1(\theta)d\mathbf{\Omega} = 1, \tag{7.48}$$

as

$$Z_1 = 4\pi I_0(\Gamma_1), \tag{7.49}$$

where the function $I_0(\Gamma_1)$ is defined by

$$I_m(\Gamma_1) = \int_0^1 [P_2(x)]^m \exp[\Gamma_1 P_2(x)]dx, \tag{7.50}$$

m = 0,1,2. Substitution of Eq. (7.46) into Eq. (7.44) yields the self-consistent equation for the orientational order parameter S_1:

$$S_1 = \frac{I_1(\Gamma_1)}{I_0(\Gamma_1)}. \tag{7.51}$$

Similarly, for the LC molecules, the distribution function is obtained as

$$f_2(\theta) = \frac{1}{Z_2} \exp\Big[\Gamma_2 P_2(\cos\theta)\Big], \tag{7.52}$$

with

$$\Gamma_2 = n_2\left(\nu_{22}\phi_2 S_2 + \nu_{12}\phi_1 S_1\right). \tag{7.53}$$

The normalization constant is

$$Z_2 = 4\pi I_0(\Gamma_2), \tag{7.54}$$

and the self-consistent equation for the order parameter S_2 is

$$S_2 = \frac{I_1(\Gamma_2)}{I_0(\Gamma_2)}. \tag{7.55}$$

By numerically solving the coupled equations (7.51) and (7.55), the nematic order parameters can be obtained as functions of temperature and concentration.

Substituting Eqs. (7.46) and (7.52) into Eq. (7.45), the free energy F_{nem} of the nematic phase is given by

$$\begin{aligned} \beta F_{nem}/N_t &= \frac{1}{2}(\nu_{11}+5/4)\phi_1^2 S_1^2 + \frac{1}{2}\nu_{22}\phi_2^2 S_2^2 + \nu_{12}\phi_1\phi_2 S_1 S_2 \\ &\quad -\frac{\phi_1}{n_1}\ln I_0(\Gamma_1) - \frac{\phi_2}{n_2}\ln I_0(\Gamma_2). \end{aligned} \tag{7.56}$$

The total free energy (7.36) is given by the sum of Eqs. (7.37) and (7.56). Using $f_{LC} = \beta\Delta F/N_t$, the chemical potential of the rod is given by

$$\begin{aligned} \beta\Delta\mu_1 &= \beta(\partial\Delta F/\partial N_1)_{N_2} \\ &= n_1\Big[f_{LC} + (1-\phi_1)\Big(\frac{\partial f_{LC}}{\partial\phi_1}\Big)\Big] \\ &= n_1\Big[\frac{1}{n_1}\ln\phi_1 + \Big(\frac{1}{n_1} - \frac{1}{n_2}\Big)(1-\phi_1) + \chi(1-\phi_1)^2 \\ &\quad +\frac{1}{2}(\nu_{11}+5/4)S_1^2\phi_1^2 + \frac{1}{2}\nu_{22}S_2^2\phi_2^2 \\ &\quad +\nu_{12}S_1S_2\phi_1\phi_2 - \frac{1}{n_1}\ln I_0(\Gamma_1)\Big], \end{aligned} \tag{7.57}$$

and that of the LC is given by

$$\begin{aligned} \beta\Delta\mu_2 &= \beta(\partial\Delta F/\partial N_2)_{N_1} \\ &= n_2\Big[f_{LC} - \phi_1\Big(\frac{\partial f_{LC}}{\partial\phi_1}\Big)\Big] \\ &= n_2\Big[\frac{1}{n_2}\ln(1-\phi_1) + \Big(\frac{1}{n_2} - \frac{1}{n_1}\Big)\phi_1 + \chi\phi_1^2 \\ &\quad +\frac{1}{2}(\nu_{11}+5/4)S_1^2\phi_1^2 + \frac{1}{2}\nu_{22}S_2^2\phi_2^2 \\ &\quad +\nu_{12}S_1S_2\phi_1\phi_2 - \frac{1}{n_2}\ln I_0(\Gamma_2)\Big]. \end{aligned} \tag{7.58}$$

7.5.1 Nematic-isotropic transitions of binary rods

The coexistence curve (binodal) is obtained by solving the two-phase equilibrium conditions (Eqs. (7.28) and (7.29)), which are equivalent to the common tangent construction, where the equilibrium compositions correspond to the two points on the free energy curve sharing the same tangent line. The spinodal line, which separates metastable from unstable regions, is determined by the inflexion condition of the free energy: $(\partial^2 f_{LC}/\partial\phi_1^2)_T = 0$:

$$\frac{1}{n_1\phi_1} + \frac{1}{n_2\phi_2} - 2\chi - (\nu_{11} + 5/4)S_1^2 + 2\nu_{12}S_1S_2 - \nu_{22}S_2^2 = 0. \tag{7.59}$$

In the isotropic limit ($S_1 = S_2 = 0$), Eq. (7.59) reduces to the classical spinodal condition for isotropic polymer blends of the Flory-Huggins theory. When $\phi_2 = 1$ for a pure LC, from Eq. (7.53), the NIT occurs at

$$n_2\nu_{22} = n_2U_{22}/(k_BT^{\circ}_{NI,2}) = 4.55, \tag{7.60}$$

where $T^{\circ}_{NI,2}$ shows the NIT temperature of the pure LC. This relation yields the reduced temperature,

$$T/T^{\circ}_{NI,2} = 4.55/(n_2\nu_{22}). \tag{7.61}$$

To analyse the phase behavior of LC/rod mixtures in a more general form, it is convenient to introduce the following dimensionless parameters: the relative strength of the anisotropic attractive interaction between LC molecules and the isotropic repulsion parameter,

$$\alpha_{22} = \nu_{22}/\chi, \tag{7.62}$$

the relative anisotropic interaction strength between rods,

$$c_1 = \nu_{11}/\nu_{22}, \tag{7.63}$$

and the relative anisotropic interaction strength between rods and LCs,

$$c_{12} = \nu_{12}/\nu_{22}. \tag{7.64}$$

When $c_{12} = 0$, i.e., $\nu_{12} = 0$, from Eq. (7.47), the NIT of the rods occurs when

$$n_1(\nu_{11} + 5/4)\phi_1 = 4.55, \tag{7.65}$$

so that the NIT temperature of the rods is given by

$$T_{NI,1}/T^{\circ}_{NI,2} = \frac{4.55c_1n_1\phi_1}{n_2(4.55 - 1.25n_1\phi_1)}. \tag{7.66}$$

Similarly, from Eq. (7.53), when $c_{12} = 0$, or $\nu_{12} = 0$, the NIT of the LC component takes place at

$$n_2\nu_{22}\phi_2 = 4.55, \tag{7.67}$$

and the NIT temperature of the LCs is obtained as

$$T_{NI,2}/T^{\circ}_{NI,2} = 1 - \phi_1. \tag{7.68}$$

Figure 7.5 shows the NIT curves, Eqs. (7.66) and (7.68), plotted on the temperature ($T/T^{\circ}_{NI,2}$)-concentration (ϕ_1) plane for $c_{12} = 0$ and $c_1 > 0$. The vertical broken line corresponds to $c_1 = 0$, where the attractive interaction between rods can be neglected. In this case, the NIT occurs solely due to the excluded volume interaction at

$$\phi_1^* = 3.64/n_1. \tag{7.69}$$

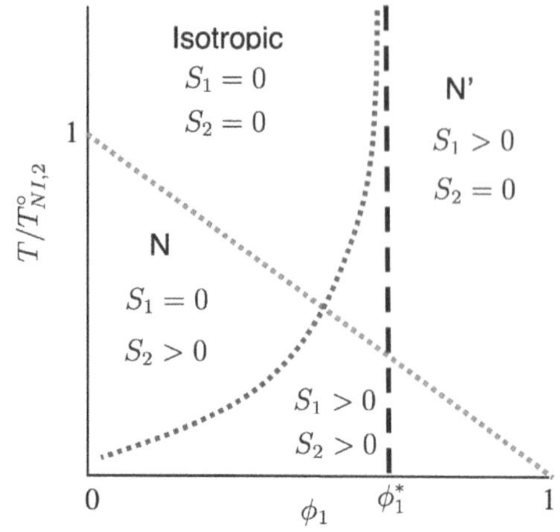

FIGURE 7.5
Nematic-isotropic phase transition curves (Eqs. (7.66) and (7.68)) plotted against the volume fraction of rods for $c_1 > 0$ with $c_{12} = 0$.

For $\phi_1 > \phi_1^*$, we have a nematic phase. When $c_1 > 0$, at higher temperatures, increasing the concentration of rods drives the isotropic phase into the rod-dominated nematic phase (N'), characterized by $S_1 > 0$ and $S_2 = 0$, through the excluded-volume mechanism. As the temperature decreases, the NIT concentration of rods decreases because the attractive interaction (c_1) between rods becomes stronger. On the other hand, below the NIT line of LCs (Eq. (7.68)), the system exhibits the LC-dominated nematic phase (N), where the LC molecules order ($S_2 > 0$) while rods remain isotropic ($S_1 = 0$). At sufficiently low temperatures, below both NIT lines, both components order simultaneously, resulting in a nematic phase with $S_1 > 0$ and $S_2 > 0$.

Figure 7.6(a) shows the phase diagram for $n_1 = 10$, $n_2 = 1$, $\alpha_{22} = 10$, and $c_1 = 0$. The solid line represents the binodal curve, while the dotted lines indicate the first-order NIT lines for the LC and the rod, and the dashed curve shows the spinodal line. At high temperatures ($T/T^\circ_{NI,2} > 1$), increasing the rod concentration induces the NIT of the rods, leading to a chimney-type I+N' phase separation. This behavior has been described by the Onsager theory (Fig. 7.4). At lower temperatures, below the NIT line of the LCs (Eq. (7.68)), the system exhibits an LC-dominated nematic (N) phase, in which the LC molecules are ordered ($S_2 > 0$), while the rods remain isotropic ($S_1 = 0$). The phase diagram thus features a chimney-type I + N' coexistence region and a triple point (N + I + N'), where isotropic, LC-nematic, and rod-nematic phases coexist. Below this triple point, a broad nematic-nematic (N+N') coexistence region appears. The N phase at low rod concentrations ($\phi_1 \simeq 0$) corresponds to nearly pure LC molecules, whereas the N' phase is dominated by rod ordering. Near $T/T^\circ_{NI,2} < 1$, an additional N+I coexistence region is also observed. Within the binodal, metastable and unstable regions can be identified, namely, the nematic metastable (Nm), isotropic metastable (Im), and nematic unstable (Nu) regions. The spinodal boundary shifts to lower temperatures with increasing α_{22}, since larger α_{22} corresponds to a smaller χ [106].

Figure 7.6(b) shows the orientational order parameters of the rods (S_1) and the LCs (S_2) as a function of the rod concentration ϕ_1 at several reduced temperatures $T/T^\circ_{NI,2} = 1.05, 0.96, 0.88$, and 0.82, corresponding to the phase diagram in Fig. 7.6(a). At high temperature ($T/T^\circ_{NI,2} = 1.05$), both order parameters vanish ($S_1 = S_2 = 0$) at low $\phi_1 < 0.38$, indicating the isotropic phase. With increasing ϕ_1, rods undergo nematic ordering,

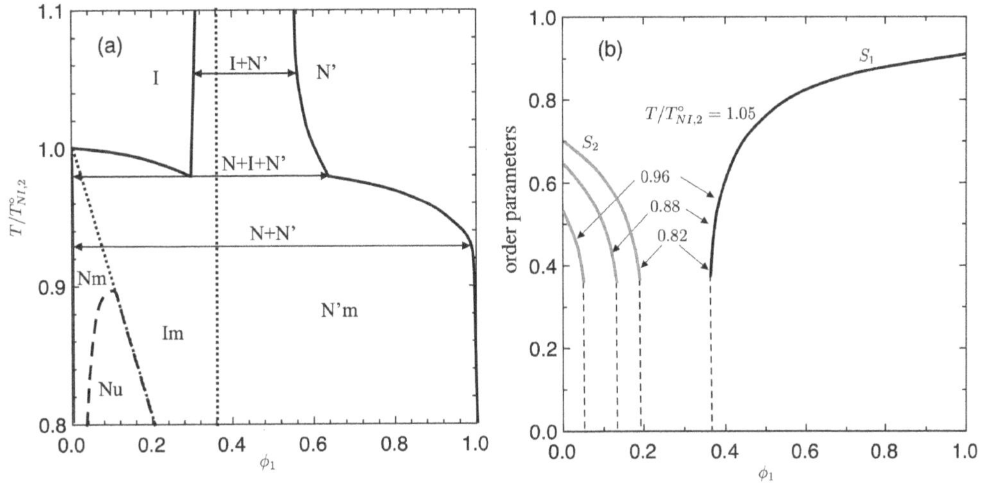

FIGURE 7.6
(a) Phase diagram on the temperature-concentration plane for $c_1 = c_{12} = 0$. The solid curve shows the binodal. The dotted lines show the first-order NIT line of an LC (Eq. (7.68)) and of a rod (Eq. (7.69)), respectively. (b) Orientational order parameter S_1 of the rod and that S_2 of the LC plotted against the concentration ϕ_1 for $T/T^\circ_{NI,2} = 1.05, 0.96, 0.88$ and 0.82 in Fig. 7.6(a). Reprinted with permission from Ref. [106].

giving rise to the N' phase with $S_1 > 0$ and $S_2 = 0$. At intermediate temperatures (e.g., $T/T^\circ_{NI,2} = 0.96$), both order parameters remain zero ($S_1 = S_2 = 0$) in the Im region. At $T/T^\circ_{NI,2} = 0.82$, an LC-dominated nematic (N) phase with $S_1 = 0$ and $S_2 > 0$ appears for $\phi_1 < 0.2$, corresponding to the Nm and Nu regions, whereas a rod-dominated nematic (N') phase with $S_1 > 0$ and $S_2 = 0$ emerges for $\phi_1 > 0.38$.

7.5.2 Induced-nematic phase

As the coupling constant c_{12} increases, the NIT curves for rods and LCs shift toward higher temperatures and lower rod concentrations. Since c_{12} represents the anisotropic interaction between rods and LCs, a positive value of c_{12} promotes their parallel alignment. With sufficiently strong coupling, the two separate NIT curves shown in Fig. 7.5 merge into a single transition line, reflecting the cooperative orientational ordering of rods and LCs.

Figure 7.7 shows the phase diagrams on the temperature-concentration plane for $c_{12} = 0.3$ (a) and $c_{12} = 0.4$ (b). In Fig. 7.7(a), below the NIT line (dotted line), a nematic N_0 phase appears, where both rods and LCs are aligned parallel to each other ($S_1 > 0$ and $S_2 > 0$). The width of the chimney-type I+N'_0 coexistence region decreases with decreasing temperature, and a triple point (N_0+I+N'_0) appears, where nematic N_0, isotropic (I), and nematic N'_0 phases coexist. Below the triple point, nematic-nematic (N_0+N'_0) phase separation occurs, where two nematic phases with different concentrations coexist. As the coupling parameter c_{12} increases, the binodal line of the N_0 phase shifts to higher concentrations, while that of the N'_0 phase shifts to lower concentrations, and the two binodal curves merge into one.

In Fig. 7.7(b), the binodal curve is split into two parts: one corresponds to the isotropic-nematic (I+N_0) phase separation with a lower critical solution temperature (LCST) at $T/T^\circ_{NI,2} = 1$, and the other corresponds to the nematic-nematic (N_0+N'_0) phase separation with an UCST [107]. Between the NIT temperature $T^\circ_{NI,2}$ and the UCST, a stable nematic N_0 phase appears. In this N_0 phase ($S_1 > 0$ and $S_2 > 0$), the order parameter S_1 of the

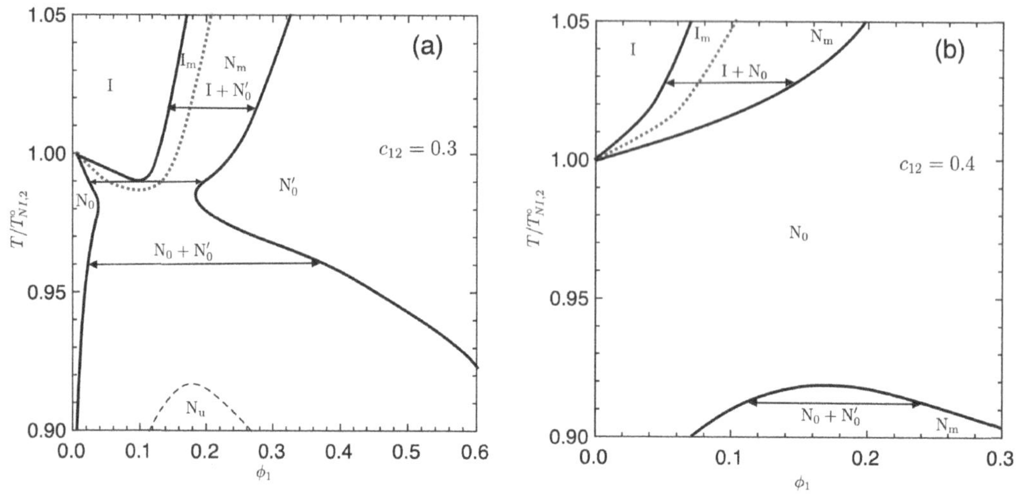

FIGURE 7.7
Phase diagram on the temperature-concentration plane for c_{12}=0.3 (a) and 0.4 (b). The solid curve shows the binodal, and the dotted line shows the first-order NIT. Reprinted with permission from Ref. [106].

rods is larger than that of the LCs. The attractive interaction c_{12} between unlike molecules promotes the nematic ordering of the LCs, stabilized by the alignment of long rods, and leads to an induced-nematic phase (N_0) between the LCST and UCST. Such LCST-type phase diagrams have been reported experimentally, for instance, in mixtures of a main-chain nematic polyester poly[oxy(chloro-1,4-phenylene)oxycarbonyl][(trifluoromethyl)-1,4-phenylene]carbonyl (PTFC) with a nematic liquid crystal *p*-azoxyanisole (PAAd14) [108]. The present theory qualitatively captures this behavior. A positive $c_{12} = \nu_{12}/\nu_{22}$ enhances both order parameters S_1 and S_2 through the self-consistent Eqs. (7.51) and (7.55), thereby lowering the free energy of the mixed nematic state (Eq. (7.56)). At high temperatures, the excluded volume interaction between rods dominates, resulting in the I+N_0 phase separation. Upon cooling, however, the attractive coupling term c_{12} reduces the biphasic region and stabilizes the homogeneous N_0 phase, thus giving rise to the LCST-type phase diagram.

7.6 Semiflexible polymers mixed with liquid crystalline molecules

For rod-like polymers that are semiflexible, such as main-chain or side-chain liquid crystalline polymers (LCPs), the excluded volume contribution is much smaller than for rigid rods. Semiflexible chains can bend and partially interpenetrate, reducing the effective volume excluded to other molecules. Accordingly, the excluded volume term in Eq. (7.41) can be neglected, and the interaction term in Eq. (7.65) is modified as

$$\nu_{11} + 5/4 \longrightarrow \nu_{11}, \tag{7.70}$$

where the excluded volume contribution (5/4) is omitted. We further assume that ν_{12} is proportional to the square root of the product of ν_{11} and ν_{22},

$$\nu_{12} = c\sqrt{\nu_{11}\nu_{22}}, \tag{7.71}$$

where c is a proportionality constant that characterizes the relative strength of the cross-interaction between dissimilar mesogens in comparison with those between identical mesogens.

In binary mixtures of short LC or semiflexible LCPs with low-molecular-weight LCs, the phase diagrams do not exhibit the chimney-type phase separation characteristic of long rigid rods (see Fig. 7.6). Instead, the nematic ordering is primarily driven by the anisotropic attractive interactions between molecules. As a result, the phase behavior displays I+I, I+N, and N+N coexistence regions. The phase behavior of binary mixtures of LCs has been summarized by de Gennes [1] and Brochard [109].

Figure 7.8 shows the temperature-composition phase diagrams of side-chain liquid crystal epoxy polymer/LC mixtures. These phase diagrams depend strongly on the NIT temperatures of both components and, in particular, on the relative strength of the cross-nematic interactions c between them: (a) $c = 0.8$, (b) $c = 0.97$, (c) $c = 0.94$, (d) $c = 0.96$. The calculated phase diagrams for various values of the cross-interaction parameter c are compared

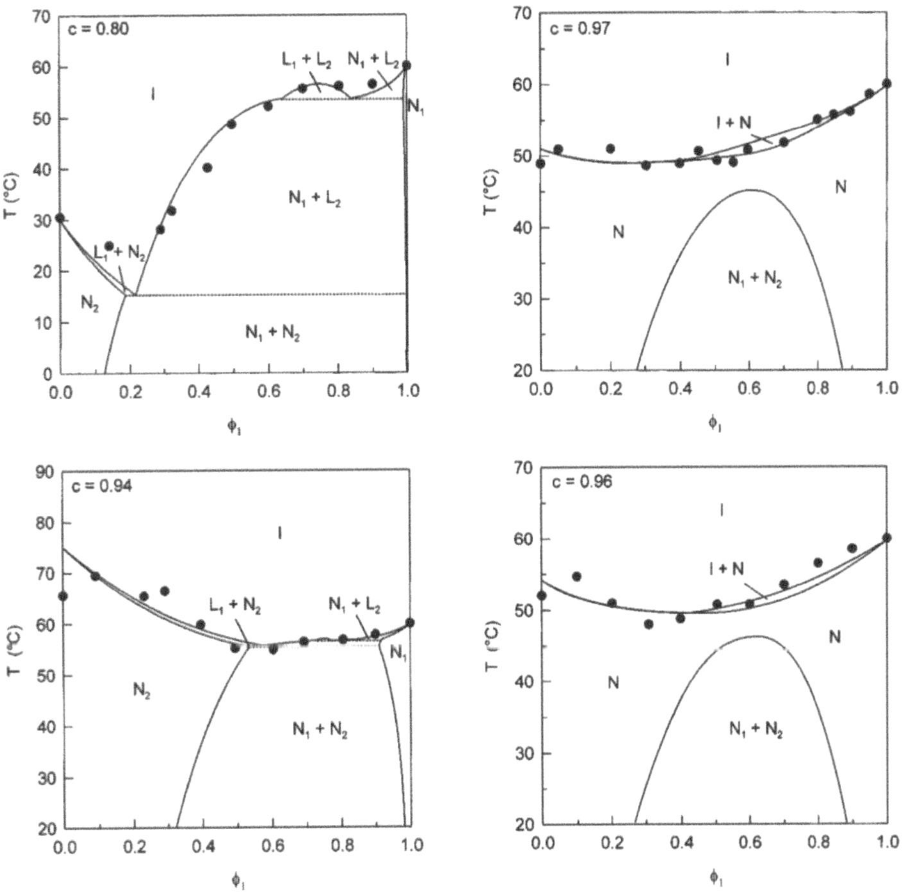

FIGURE 7.8
Temperature versus composition phase diagrams of a side-chain LCP/LC mixture for various values of c: (a) EGDE-C6/E7 ($c = 0.80$), (b) EGDE-C7/E7 ($c = 0.97$), (c) EGDE-C8/E7 ($c = 0.94$), and (d) EGDE-C9/E7 ($c = 0.96$), in comparison with the theoretical phase diagrams calculated (solid curves). Reprinted with permission from Ref. [15].

with the experimental ones for the EGDE-C6/E7 (a), EGDE-C7/E7 (b), EGDE-C8/E7 (c), and EGDE-C9/E7 (d) systems, where ϕ_1 is the volume fraction of the LC. The calculated phase diagram reveals the existence of two distinct nematic regions: one at high E7 content (N_1) and the other at high EGDE-C6 concentration (N_2). The two dotted lines indicate triple points, which correspond to the coexistence of three phases, either two liquids plus one nematic, or one liquid plus two nematic phases. Narrow nematic, isotropic coexistence regions are observed, labeled $N_1 + L_2$ and $L_1 + N_2$. Below the second triple-point line, a two-nematic coexistence region ($N_1 + N_2$) appears. As the temperature decreases, the system transitions from an isotropic-nematic coexistence region (I+N) to a single-phase nematic region (N). Upon further cooling, the $N_1 + N_2$ phase separation emerges, terminating at a critical point with a UCST within the nematic phase, driven by strong nematic interactions. The calculated diagrams successfully reproduce the isotropic and nematic coexistence regions (I+I, I+N, N+N) without exhibiting the chimney-type phase separation typically observed in rigid rod-like molecular solutions. Therefore, the emergence of chimney-type phase separation in rod/LC mixtures is strongly dependent on the polymer stiffness: rigid rods with large axial ratios exhibit excluded volume-driven nematic ordering and chimney-type biphasic regions, whereas semiflexible LCPs display phase coexistence primarily governed by anisotropic attractive interactions.

Flory (1956) proposed a lattice theory for semiflexible polymer chains in isotropic solutions, in which successive bonds along a chain can adopt either bent or straight conformations [102]. Extension of this model to describe nematic ordering can be achieved by combining it with the Maier-Saupe theory [103]. Mean field approaches have also been applied to nanoparticles dispersed in LCs [110–112]. In such hybrid systems, the physical or chemical properties of the colloids or nanoparticles are coupled with the orientational order of the LCs, leading to the formation of a self-assembled nanostructure [113] and a transparent nematic phase [114]. These systems offer enhanced functionality and potential applications in displays, tunable optical devices, sensors, and drug delivery, as reviewed in Refs. [115–117].

8

Volume Phase Transitions of Liquid Crystalline Gels

Liquid crystalline gels (LC gels) have attracted considerable attention in fields such as artificial muscles, electronic materials, and biotechnology. Incorporating liquid crystal structures, such as nematic or smectic phases, etc., into polymer gels enhances their responsiveness and enables the development of various novel functionalities. The coupling between the orientation of liquid crystalline molecules and the deformation of the polymer network gives rise to characteristic phenomena in liquid crystal elastomers, including soft elasticity [19, 118]. Meanwhile, polymer gels swollen with solvent molecules are known to undergo volume phase transitions in response to changes in temperature or pH. In this chapter, we present a mean field theoretical framework for the volume phase transitions of liquid crystalline polymer gels. In Section 8.1, we introduce the volume phase transitions of isotropic gels, while Sections 8.2 and 8.3 discuss nematic gels and cholesteric gels, respectively.

8.1 Statistical theory of rubber elasticity

Figure 8.1 is a schematic illustration of a polymer gel swollen by absorbing solvent molecules. Although real gels have more complex structures, we do not consider such details here. Polymer chains of equal length are crosslinked to form a three-dimensional network. The swelling and shrinking of the gel occur as solvent molecules move in and out of the network.

The method for deriving the elastic force of a rubber closely corresponds to the statistical mechanical treatment of an ideal gas. Therefore, let us first briefly review how the pressure of an ideal gas is derived [85]. Let us calculate the probability W that N_J gas molecules, initially distributed throughout a container of volume V_0, spontaneously occupy a subvolume V of the container. The probability that all molecules are found within this subvolume is given by

$$W = (V/V_0)^{N_J}. \tag{8.1}$$

For the compression of the gas from V_0 to V, the associated entropy change is given by

$$\Delta S_v = k_B \ln W = k_B N_J \ln(V/V_0). \tag{8.2}$$

Using the thermodynamic relation for the Helmholtz free energy F, the pressure is given by

$$P = -(\partial F/\partial V)_T = -(\partial E/\partial V)_T + T(\partial S_v/\partial V)_T, \tag{8.3}$$

and for an ideal gas $(\partial E/\partial V)_T = 0$, leading to

$$P = T(\partial S_v/\partial V)_T = k_B T N_J/V, \tag{8.4}$$

which is the familiar ideal gas law.

DOI: 10.1201/9781003737711-8

The elastic free energy F_{el} of a polymer gel arises from two contributions: the change in elastic entropy ΔS_e, which is associated with the deformation of polymer chains, and the entropy change due to confinement ΔS_v, as discussed in Eq. (8.2). Therefore, the total elastic free energy can be written as

$$F_{el} = -T(\Delta S_e + \Delta S_v). \tag{8.5}$$

Consider a single polymer chain connecting two crosslinking points. Let b denote the length of a single polymer segment and n the total number of segments in the chain. The probability that a polymer chain of contour length $L_c(= bn)$ has an end-to-end vector $\mathbf{R}_e = (R_x, R_y, R_z)$ is approximated by a Gaussian distribution:

$$p(\mathbf{R}_e) = \left(\frac{3}{2\pi nb^2}\right)^{3/2} \exp\left(-\frac{3\mathbf{R}_e^2}{2nb^2}\right). \tag{8.6}$$

Therefore, the configurational entropy of a single polymer chain is given by

$$S_e(\mathbf{R}_e) = k_B \ln p(\mathbf{R}_e) + \text{const.} = -\frac{3k_B\mathbf{R}_e^2}{2nb^2} + \text{const.} \tag{8.7}$$

When a polymer chain is stretched from its thermal equilibrium state, $\mathbf{R}_e^2$ increases and its entropy decreases. The elastic free energy of a single polymer chain is therefore given by $F_{el,1} = -TS_e(\mathbf{R}_e)$. If one end of the polymer is subjected to an external force $\mathbf{f}_e$, the resulting tension is expressed as

$$\mathbf{f}_e = -\frac{\partial F_{el,1}(\mathbf{R}_e)}{\partial \mathbf{R}_e} = -K_p\mathbf{R}_e, \tag{8.8}$$

where $K_p = 3k_BT/(nb^2)$ is the effective spring constant. This relation is formally equivalent to Hooke's law, describing the linear elastic response of a stretched spring.

Let $\mathbf{R}_0 = (R_{0x}, R_{0y}, R_{0z})$ denote the end-to-end vector of an ideal polymer chain at equilibrium. The change in the configurational entropy per chain due to a deformation from this equilibrium configuration is given by

$$\begin{aligned} \Delta S_1(\mathbf{R}_e) &= S_e(\mathbf{R}_e) - S_e(\mathbf{R}_0) = -\frac{3k_B(\mathbf{R}_e^2 - \mathbf{R}_0^2)}{2nb^2} \\ &= -\frac{3k_B}{2nb^2}\Big((\alpha_x^2-1)R_{0x}^2 + (\alpha_y^2-1)R_{0y}^2 + (\alpha_z^2-1)R_{0z}^2\Big), \end{aligned} \tag{8.9}$$

where

$$\alpha_i = R_i/R_{0i}, \tag{8.10}$$

denotes the deformation ratio (expansion factor) along each coordinate axis. In an isotropic phase, the distribution of segments on the ideal chain is spherical symmetry. In terms of the bond vectors $\mathbf{u}_\alpha$ of length b, the end-to-end distance is given by

$$\mathbf{R}_0 = \sum_{\alpha=1}^{n} \mathbf{u}_\alpha. \tag{8.11}$$

Since the bond vectors $\mathbf{u}_\alpha$ of the ideal chain are uncorrelated in their direction, we obtain the average $\langle u_\alpha u_\beta\rangle = \delta_{\alpha\beta}b^2$ and the mean-square end-to-end distance

$$\langle \mathbf{R}_0^2\rangle = \sum_{\alpha,\beta} u_\alpha u_\beta = b^2 n, \tag{8.12}$$

where $\delta_{\alpha\beta}$ is the Kronecker delta function (Eq. (2.31)). Therefore, the mean-square end-to-end distance along each Cartesian axis can be expressed, by the equipartition theorem, as

$$\langle R_{0x}^2\rangle = \langle R_{0y}^2\rangle = \langle R_{0z}^2\rangle = \frac{1}{3}\langle R_0^2\rangle = \frac{1}{3}b^2 n = \frac{1}{3}bL_c. \tag{8.13}$$

We then obtain the change in entropy for a single polymer chain under macroscopic deformation:

$$\Delta S_1(\mathbf{R}_e) = -\frac{1}{2}k_B(\alpha_x^2 + \alpha_y^2 + \alpha_z^2 - 3). \tag{8.14}$$

The total configurational entropy change of a gel consisting of N_g polymer chains is then given by

$$\Delta S_e = N_g \Delta S_1 = -\frac{N_g k_B}{2}(\alpha_x^2 + \alpha_y^2 + \alpha_z^2 - 3), \tag{8.15}$$

where it is assumed that the macroscopic deformation of the gel is proportional to the microscopic deformation of the polymer chains between crosslinking points. This assumption is known as affine deformation.

When polymer chains are crosslinked, the probability of simultaneously finding N_J crosslinking points within the volume V occupied by the gel is given by $(V/\Omega)^{N_J}$, where Ω is the total volume of the system. Consequently, the confinement entropy of the gel is expressed as

$$S_v(V) = k_B \ln(V/\Omega)^{N_J} + const. \tag{8.16}$$

This term accounts for the reduction in entropy due to the spatial confinement of crosslinking points within the gel (Eq. (8.2)). Accordingly, when the initial gel volume $V_0(= N_g R_{0x} R_{0y} R_{0z})$ changes to $V(= N_g R_x R_y R_z)$ due to solvent absorption, the resulting change in confinement entropy is given by

$$\begin{aligned} \Delta S_v &= S_v(V) - S_v(V_0) = k_B N_J \ln(V/V_0) \\ &= k_B N_J \ln(\alpha_x \alpha_y \alpha_z). \end{aligned} \tag{8.17}$$

If J denotes the functionality (i.e., the number of polymer chains connected to each crosslinking point), the total number of polymer chains N_g formed by N_J crosslinking points in the gel can be approximated as

$$N_g \simeq J N_J / 2. \tag{8.18}$$

Thus, the elastic free energy of the polymer gel is given by the sum of Eqs. (8.15) and (8.17):

$$\begin{aligned} \beta F_{el} &= \frac{1}{2}N_g(\alpha_x^2 + \alpha_y^2 + \alpha_z^2 - 3) - N_J \ln(\alpha_x \alpha_y \alpha_z) \\ &= N_g\left(\frac{1}{2}(\alpha_x^2 + \alpha_y^2 + \alpha_z^2 - 3) - \frac{2}{J}\ln(\alpha_x \alpha_y \alpha_z)\right). \end{aligned} \tag{8.19}$$

Figure 8.1 illustrates the case of a network with functionality $J \simeq 4$. In the limiting case of $J = 2$, the gel network reduces to a linear chain without crosslinking.

Let N_0 be the number of solvent molecules contained within the polymer gel (see Fig. 8.1). Then, the gel volume can be expressed as $V = b^3 N_t$, where $N_t = N_0 + nN_g$ is the total number of lattice sites of the gel. Accordingly, the volume fraction of polymer chains in the gel is

$$\phi = \frac{b^3 n N_g}{V} = \frac{nN_g}{N_t}. \tag{8.20}$$

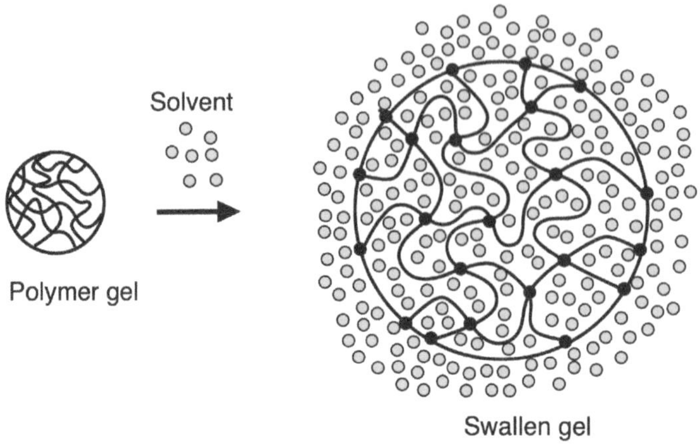

FIGURE 8.1
A polymer gel absorbs solvent molecules from the surrounding medium, leading to an expansion of its internal polymer network and an increase in volume. Solvent remains present both inside and outside the gel, indicating a state approaching equilibrium.

8.1.1 Isotropic deformations

We consider an isotropic deformation of the gel and take $\alpha_x = \alpha_y = \alpha_z = \alpha_e$. The ratio of the root-mean-square end-to-end distance $R = \sqrt{\langle \mathbf{R}_e^2 \rangle}$ of a swollen (or contracted) polymer chain to that of the corresponding ideal chain $R_0 = \sqrt{\langle R_0^2 \rangle} = b\sqrt{n}$ is defined as the expansion factor:

$$\alpha_e = \frac{R}{R_0} = \frac{R}{b\sqrt{n}}. \tag{8.21}$$

Since the volume occupied by a single polymer chain is approximately R^3, the total volume of the gel can be estimated as $V \simeq R^3 N_g$. Then, the volume fraction of polymer chains in the gel can be expressed using the expansion factor α_e (Eq. (8.21)) as

$$\phi = \frac{b^3 n N_g}{R^3 N_g} = \frac{1}{\alpha_e^3 \sqrt{n}}. \tag{8.22}$$

Let V_0 be the initial volume of the gel, and let $\phi_i (= b^3 n N_g / V_0)$ denote the corresponding initial polymer volume fraction. Then, the degree of swelling of the gel after mixing with the solvent molecules is given by

$$Q_e = \frac{V}{V_0} = \frac{\phi_i}{\phi} = \sqrt{n}\phi_i \alpha_e^3 = \alpha_e^3. \tag{8.23}$$

The polymer chains forming the gel before mixing with solvent molecules are assumed to be in a melt state. In this regime, the high concentration of polymer segments effectively screens the excluded volume interactions between chains. As a result, the intrachain excluded volume effect is canceled on average, and the chains can be treated as ideal Gaussian chains. This justifies the assumption that the end-to-end distance of a chain scales as $R_0 \sim b\sqrt{n}$, leading to the initial polymer volume fraction $\phi_i = b^3 n / R_0^3 = 1/\sqrt{n}$. Substituting this into Eq. (8.23) gives the simplified relation $Q_e = \alpha_e^3$. Then, Eq. (8.19) can be expressed in terms of Q_e as

$$\beta F_{el} = N_g\left(\frac{3}{2}(Q_e^{2/3} - 1) - \frac{2}{J}\ln Q_e\right). \tag{8.24}$$

8.1.2 Volume phase transitions of an isotropic gel

The total free energy of a polymer gel is composed of the elastic contribution from the network and the mixing contribution due to the solvent and is expressed as

$$F = F_{el} + \Delta F_{mix}, \tag{8.25}$$

where F_{el} is the elastic free energy of the polymer network (Eq. (8.24)) and ΔF_{mix} accounts for the free energy change due to mixing between polymer chains and solvent molecules. Based on the Flory-Huggins theory for polymer solutions (Eq. (7.10)), the mixing free energy of a gel can be expressed as

$$\begin{aligned} \beta\Delta F_{mix} &= N_t\Big((1-\phi)\ln(1-\phi) + \chi\phi(1-\phi)\Big) \\ &= nN_g\Big(\big(\frac{Q_e}{\phi_i} - 1\big)\ln\big(1 - \frac{\phi_i}{Q_e}\big) + \chi\big(1 - \frac{\phi_i}{Q_e}\big)\Big), \end{aligned} \tag{8.26}$$

where the polymer volume fraction ϕ is expressed using the expansion factor as $\phi = \phi_i/Q_e$ (Eq. 8.23). Since the number of polymer segments forming the gel is considered very large, the translational entropy of the polymer network, $(\phi/(nN_g))\ln\phi$, is negligible. The second term, $\chi\phi(1-\phi)$, represents the enthalpic contribution due to interactions between polymer and solvent molecules, where $\chi = U_0/(k_B T)$ is the Flory-Huggins interaction parameter between polymer and solvent [85].

The total free energy of the gel, given by Eq. (8.25), is obtained by summing the elastic free energy (Eq. (8.24)) and the mixing free energy (Eq. (8.26)):

$$\begin{aligned} \beta F/N_g &= \frac{3}{2}(Q_e^{2/3} - 1) - \frac{2}{J}\ln Q_e \\ &+ n\Big(\big(\frac{Q_e}{\phi_i} - 1\big)\ln\big(1 - \frac{\phi_i}{Q_e}\big) + \chi\big(1 - \frac{\phi_i}{Q_e}\big)\Big). \end{aligned} \tag{8.27}$$

At thermal equilibrium, the chemical potential μ_0 of solvent molecules inside the gel equals that μ_0° of the solvent outside the gel: $\mu_0(Q_e) = \mu_0^\circ$. Consequently, the equilibrium state of the gel is determined by the condition of osmotic pressure $\Pi = 0$:

$$\begin{aligned} \Pi &= -(\mu_0(Q_e) - \mu_0^\circ) = -\left(\frac{\partial F}{\partial N_0}\right)_{N_g,T} \\ &= -b^3\left(\frac{\partial F}{\partial V}\right)_{N_g,T} = -\frac{b^3}{V_0}\left(\frac{\partial F}{\partial Q_e}\right)_{N_g,T} = 0, \end{aligned} \tag{8.28}$$

where we have used the relation $V = b^3(N_0 + nN_g)$ and $Q_e = V/V_0$. This condition, $\Pi = 0$, defines the equilibrium degree of swelling Q_e of the gel. Using Eq. (8.27), the osmotic pressure Π is obtained as

$$\beta\Pi = -\left[\frac{\phi_i}{n}\big(Q_e^{-1/3} - \frac{2}{JQ_e}\big) + \ln\big(1 - \frac{\phi_i}{Q_e}\big) + \frac{\phi_i}{Q_e} + \chi\big(\frac{\phi_i}{Q_e}\big)^2\right] = 0, \tag{8.29}$$

where the coefficient of the first term, $\phi_i/n = b^3 N_g/V_0$, represents the number density of polymer chains forming the gel. To eliminate the explicit dependence on the segment number

n, we introduce a parameter s_g that characterizes the number density of crosslinking points in the gel. The coefficient is then rewritten as

$$\frac{\phi_i}{n} \to s_g \phi_i^3. \tag{8.30}$$

In realistic polymer gels, the network structure formed by crosslinking is often far from ideal. Structural heterogeneities, such as inhomogeneous crosslink density, dangling chains, entanglements, and loop formations, lead to deviations from the ideal affine deformation model. These microscopic features can significantly influence the macroscopic swelling behavior and elastic properties of the gel. To account for such deviations in a coarse-grained framework, we introduce the phenomenological parameter s_g, representing the effective contribution of the crosslink density to the elastic response of the gel.

The parameter s_g characterizes deviations from the ideal polymer network structure. When $s_g = 1$, the system corresponds to an ideal network composed of Gaussian chains, where the chain number density is given by $\phi_i/n \sim n^{-3/2}$. For example, assuming a typical segment number $n = 10^4$, we obtain $\phi_i/n = 10^{-6}$. In contrast, when $s_g > 1$, the network is more densely crosslinked or contains additional topological constraints such as entanglements or loop formations, leading to a higher effective chain density. For instance, if we take $s_g = 3000$ and $\phi_i = 0.01$, then the effective chain number density becomes $\phi_i/n \to s_g\phi_i^3 = 3 \times 10^{-3}$, which is three orders of magnitude greater than that of the ideal case. This demonstrates that the parameter s_g enables the model to describe polymer gels with significantly enhanced chain densities and more complex internal structures than those predicted by the ideal Gaussian chain model.

Figure 8.2 shows the osmotic pressure Π of a polymer gel as a function of the degree of swelling Q_e on a semi-logarithmic scale, under the conditions $s_g = 3000$ and $\phi_i = 0.01$. The Flory-Huggins parameter χ, which reflects the temperature dependence of solvent quality (with $\chi \propto 1/T$), is varied to represent thermal effects. As χ increases, corresponding to poorer solvent conditions (or lower temperatures), the equilibrium swelling degree Q_e decreases due to stronger polymer-polymer attraction. The thermodynamic stability of the

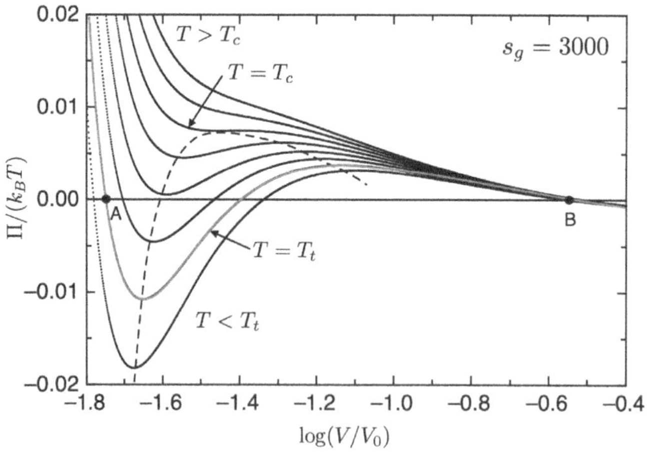

FIGURE 8.2
Osmotic pressure Π of a polymer gel plotted as a function of the degree of swelling $Q_e(= V/V_0)$ on a semi-logarithmic scale for $s_g = 3000$.

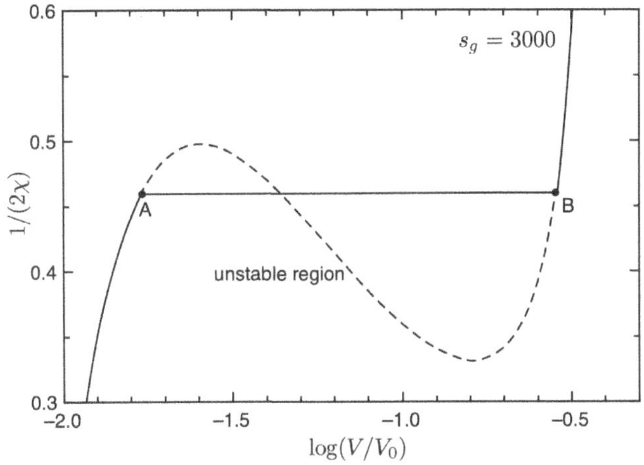

FIGURE 8.3
Volume phase transition of the isotropic gel.

polymer gel is ensured by the condition (thermodynamic inequality)

$$\left(\frac{\partial \Pi}{\partial Q_e}\right)_T < 0, \tag{8.31}$$

which guarantees that the osmotic pressure Π decreases with increasing the swelling degree Q_e under isothermal conditions. This condition is analogous in structure to the thermodynamic stability criterion for an ideal gas, $(\partial P/\partial V)_T < 0$. The dashed line in Fig. 8.2 indicates the spinodal point, defined by the condition $(\partial \Pi/\partial Q_e)_T = 0$. At high temperatures ($T > T_c$), the swelling degree Q_e at which the osmotic pressure $\Pi = 0$ lies to the right of point B, and the gel is in a thermodynamically stable swollen state. As the temperature decreases, a critical temperature $T = T_c$ appears, below which the gel exhibits behavior reminiscent of a van der Waals loop: the osmotic pressure curve develops both a maximum and a minimum. Within the region between these extrema, Π increases with increasing V, indicating thermodynamic instability. The stable gel volumes are determined by the Maxwell equal-area construction. At a specific temperature $T = T_t$, the areas above and below the pressure curve $\Pi = 0$ become equal, resulting in a discontinuous jump in gel volume between points A and B. This phenomenon is referred to as the first-order volume phase transitions of the gel. Upon further cooling ($T < T_t$), the condensed state of the gel (to the left of point A) becomes thermodynamically stable.

Figure 8.3 shows the equilibrium swelling curve of the gel for $s_g = 3000$. The vertical axis represents the temperature $1/(2\chi) \propto T$, and the horizontal axis represents the gel swelling degree, $\log Q_e$. Points A and B correspond to points A and B in Fig. 8.2. At $1/(2\chi) \simeq 0.45$, a volume phase transition occurs, and the region between points A and B corresponds to unstable or metastable states. At high temperatures, the mixing entropy dominates, allowing the polymer gel to absorb solvent molecules and swell. At low temperatures, the repulsive interaction between polymer and solvent molecules, characterized by the Flory-Huggins parameter χ, becomes dominant, leading to solvent expulsion and a collapsed gel state. This behavior is analogous to UCST-type phase separation, in which polymer chains condense at low temperatures. The swelling and shrinking of the gel thus follow the same thermodynamic principle as a gas-liquid phase transition. At large polymer chain number densities s_g, the gel exhibits a discontinuous volume phase transition. As s_g decreases, the width of the volume phase transition decreases, and at $s_g \simeq 100$ the transition becomes continuous.

Such volume phase transitions of gels were first reported in the late 1970s by Tanaka and co-workers [119,120]. For ionic gels, the gel volume can change by as much as three orders of magnitude in response to temperature variation [121]. In water-soluble polymer gels, an inverse volume phase transition is observed, where the gel swells at lower temperatures due to the significant role of hydrogen bonding between polymer and solvent molecules [122,123].

8.1.3 Coil-globule transitions of a single polymer chain

Here, we introduce the coil-globule transition of a single flexible polymer chain. When $J = 2$ and $N_g = 1$ in Eq. (8.24), the elastic free energy of a single polymer chain is given by

$$\beta F_{el} = \frac{3}{2}(\alpha_e^2 - 1) - 3\ln\alpha_e, \tag{8.32}$$

as a function of the expansion factor α_e. The mixing free energy of a single chain dissolved in solvent molecules can be written, following Eq. (8.26), as

$$\beta\Delta F_{mix} = \frac{R^3}{b^3}\Big((1-\phi)\ln(1-\phi) + \chi\phi(1-\phi)\Big), \tag{8.33}$$

where the relation $b^3 N_t = R^3$ has been used for a system consisting of a single polymer chain dispersed in solvent. The total free energy is given by $F = F_{el} + \Delta F_{mix}$.

When $\phi \ll 1$, the first term in Eq. (8.33) can be expanded in powers of ϕ as

$$\beta\Delta F_{mix} = \alpha_e^3 n^{3/2}\Big((\chi - 1)\phi + (\frac{1}{2} - \chi)\phi^2 + \frac{B_3}{2}\phi^3 + \cdots\Big), \tag{8.34}$$

where the coefficient of the quadratic term (the second virial coefficient),

$$B_2 = \tfrac{1}{2} - \chi, \tag{8.35}$$

represents the contribution from two-body interactions (excluded volume effect), and the B_3 term corresponds to three-body interactions (the third virial coefficient) that prevent excessive collapse of the chain.

In a thermal equilibrium state, the expansion factor α_e can be determined by the condition

$$\left(\frac{\partial F}{\partial \alpha_e}\right)_T = 0. \tag{8.36}$$

Substituting the polymer volume fraction ϕ from Eq. (8.22), we obtain the following relation for α_e:

$$\alpha_e^5 - \alpha_e^3 - \frac{B_3}{\alpha_e^3} = B_2\sqrt{n}, \tag{8.37}$$

which describes the balance between the elastic free energy of the polymer chain and the effective interactions between monomers.

To understand how these interactions, B_2 and B_3, influence the conformational size of the polymer, we now turn to the scaling behavior of the expansion factor α_e and, consequently, of the end-to-end distance R under different solvent conditions.

- Θ solvent ($B_2 = 0$): At the Θ condition, the excluded-volume effect vanishes, i.e., $B_2 = 0$. For $n \to \infty$, α_e approaches a constant of order unity. Therefore, under the Θ conditions, the chain behaves as an ideal Gaussian chain, with size

$$R \sim b n^{1/2}. \tag{8.38}$$

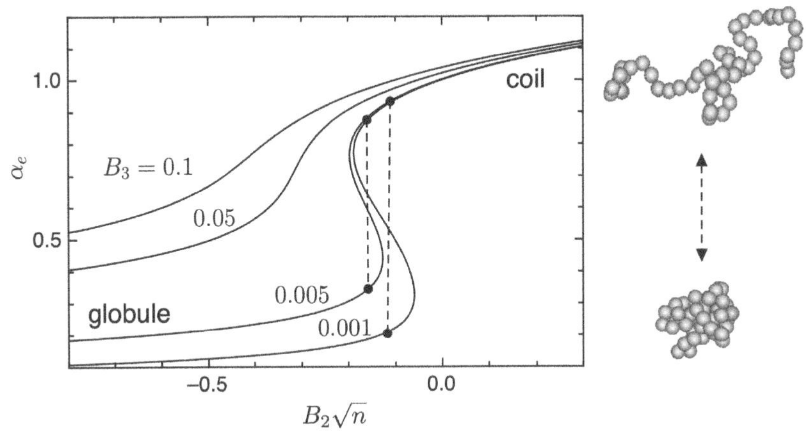

FIGURE 8.4
Expansion factor α_e plotted against $B_2\sqrt{n}$ for various values of B_3 in Eq. (8.37).

- Good solvent ($B_2 > 0$): For large n, the right-hand side $B_2\sqrt{n}$ dominates: $\alpha_e^5 \sim B_2\sqrt{n}$. The chain dimension is then
$$R = \alpha_e R_0 \sim bn^{3/5}, \tag{8.39}$$
reproducing the Flory scaling law for a polymer chain in good solvent conditions.
- Poor solvent ($B_2 < 0$): When $B_2 < 0$, two-body interactions favor chain collapse. The collapse is counterbalanced by the three-body interaction term B_3/α_e^3. In the large n limit, the balance condition is $B_3/\alpha_e^3 \sim B_2\sqrt{n}$. Accordingly, the chain dimension scales as
$$R \sim bn^{1/3}, \tag{8.40}$$
corresponding to the globule state in poor solvent.

The coil-globule transition can be represented in a phase diagram as a function of temperature (or, equivalently, the interaction parameter χ) and the polymerization degree n. At high temperatures ($\chi < 1/2$), the excluded volume interactions dominate ($B_2 > 0$), and the chain is in a swollen conformation with $R \sim bn^{3/5}$. At low temperatures ($\chi > 1/2$), attractive interactions dominate ($B_2 < 0$), and the chain collapses into a compact globule of size $R \sim bn^{1/3}$.

Figure 8.4 shows the expansion factor α_e as a function of $B_2\sqrt{n}$ for various values of B_3, calculated from Eq. (8.37). For larger B_3, the expansion factor α_e decreases continuously with decreasing temperature. In contrast, for smaller B_3, the value of α_e exhibits a discontinuous change, indicating a first-order coil-globule transition. The broken lines denote the coexistence region associated with this first-order transition. Such coil-globule transitions have been observed in many polymer solutions, including polystyrene chains dispersed in cyclohexane [124, 125].

It should be noted that the scaling laws introduced above are valid for flexible polymer chains, where the persistence length is much smaller than the overall chain dimension. As the chain stiffness increases, the conformational statistics deviate from those of Gaussian chains, and the scaling relations no longer hold.

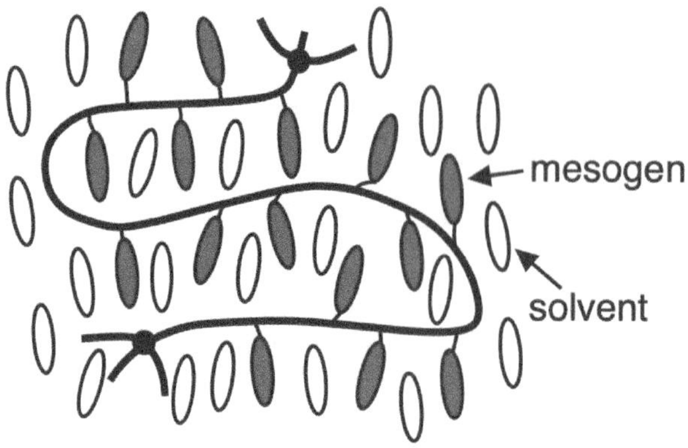

FIGURE 8.5
Schematic illustration of a side-chain liquid crystalline gel immersed in liquid crystal solvent molecules. The closed circles show crosslinking points.

8.2 Nematic gels

Nematic liquid crystalline (LC) gels are a class of soft materials that couple the orientational order of nematic liquid crystals with the elastic properties of polymer networks. These hybrid systems exhibit a wide range of phase behavior, among which the volume phase transition is particularly important. In response to external stimuli such as temperature, solvent composition, or electric fields, nematic gels can undergo abrupt yet reversible volume changes. In the following, we present a free energy framework for nematic gels and discuss the mechanisms underlying their volume phase transitions.

8.2.1 Elastic free energy of nematic gels

Consider a liquid crystalline polymer gel swollen by low-molecular-weight liquid crystal solvent molecules. Figure 8.5 schematically illustrates a side-chain liquid crystalline gel immersed in such solvent molecules. We focus on a "subchain" between two crosslinks of the polymer network, where n denotes the number of segments in the subchain. Each repeating unit (monomer) consists of a rigid liquid crystal side-chain moiety (mesogen) with an axial ratio n_m, and a flexible backbone spacer composed of n_s segments. The total number of segments in a subchain between cross-links is therefore $n = (n_m + n_s)t$, where t is the number of repeating units. Let N_g and N_0 denote the number of subchains and solvent molecules within the gel, respectively. The total volume fraction of polymer in the gel is then expressed as

$$\phi = \frac{b^3 n N_g}{V} = \frac{n N_g}{N_t}, \tag{8.41}$$

where b^3 is the volume of a unit segment and $V = b^3 N_t$ is the total volume of the gel, including solvent molecules. Here, $N_t(= n_0 N_0 + n N_g)$ represents the total number of unit segments (or unit cells), with n_0 being the axial ratio of the low-molecular-weight liquid crystal solvent molecules. The average volume per subchain is given by $R^3 = V/N_g$. The volume fraction of mesogenic units is expressed as

$$\phi_m = \frac{b^3 n_m t N_g}{V} = (1-p)\phi, \tag{8.42}$$

where $p = n_s/(n_s + n_m)$ denotes the fraction of spacer segments in the chain.

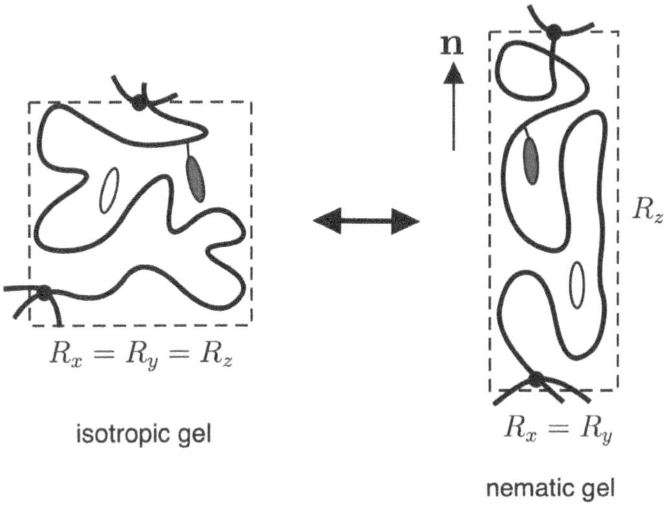

FIGURE 8.6
Schematic illustration of an isotropic and nematic gel. The depiction of liquid crystalline molecules is omitted for simplicity.

Nematic gels differ fundamentally from isotropic gels because the orientational order of mesogens imposes an intrinsic anisotropy on the polymer network. When the system enters the nematic phase, the alignment of mesogenic units along the director **n** tends to stretch the polymer network in that direction. As a result, the gel elongates along the z-axis (parallel to **n**) while contracting in the perpendicular x- and y-directions (see Fig. 8.6). This anisotropic deformation is a direct manifestation of the coupling between orientational order and network elasticity [126–129]. Under this assumption, the expansion factors in Eq. (8.19) satisfy $\alpha_x = \alpha_y$, and the elastic free energy is given by

$$\beta F_{el} = \frac{N_g}{2}\left(2\alpha_x^2 + \alpha_z^2 - 3 - \frac{4}{J}\ln(\alpha_x^2 \alpha_z)\right). \tag{8.43}$$

The expansion factors are defined as

$$\alpha_x = \alpha_y = R_x/R_{0x}, \tag{8.44}$$

$$\alpha_z = R_z/R_{0z}, \tag{8.45}$$

where R_{0x}^2 and R_{0z}^2 denote the spontaneous mean square end-to-end distances of the polymer chain in the x- and z-directions, respectively. In the isotropic phase, the polymer chain has no preferred direction. Its spontaneous size corresponds to that of an ideal Gaussian chain: $R_0 = b\sqrt{n}$, and $R_{0i}^2 = R_0^2/3$, where the mean-square end-to-end distance is equally distributed among the three spatial directions. In the nematic phase, however, the orientational order of mesogens induces anisotropy in the chain conformation. To describe this effect, we adopt a freely-jointed rod model. In this model, each bond of length b makes an angle θ with the z-axis (the nematic director). Its projection along the z-direction is $b\cos\theta$, while the projection in the perpendicular directions is $b\sin\theta$.

We then obtain

$$R_{0x}^2 = R_{0y}^2 = \frac{1}{2}\langle b\sin\theta\rangle^2 n = \frac{1}{3} l_\perp L_c, \tag{8.46}$$

$$R_{0z}^2 = \langle b\cos\theta\rangle^2 n = \frac{1}{3} l_{\parallel} L_c, \tag{8.47}$$

where $L_c = nb$ is the contour length of the chain. The $l_{\perp} = (3/2)b\langle \sin^2\theta\rangle$ and $l_{\parallel} = 3b\langle \cos^2\theta\rangle$ represent the effective step length perpendicular and parallel to the nematic director $\mathbf{n}$, respectively. When the mesogens are randomly oriented (an isotropic phase), the orientational averages are $\langle \sin^2\theta\rangle = 2/3$ and $\langle \cos^2\theta\rangle = 1/3$, we have $l_{\parallel} = l_{\perp} = b$. Using the orientational order parameter of the mesogens $S_m = (3/2)\langle \cos^2\theta\rangle - 1/2$ of the gel, we have

$$l_{\parallel} = b(1 + 2S_m), \tag{8.48}$$

$$l_{\perp} = b(1 - S_m). \tag{8.49}$$

We define the deformation ratio, relative to the ideal chain, in the direction i ($i = x, y, z$) as

$$\kappa_i = \frac{R_i}{R_0}, \tag{8.50}$$

then the volume fraction of the polymer in the gel is given by

$$\phi = \frac{b^3 n}{R_x R_y R_z} = \frac{1}{\kappa_x^2 \kappa_z \sqrt{n}}. \tag{8.51}$$

When there is no volume change before and after deformation, the volume fraction of the gel is $\phi = 1$. Using κ_z, Eqs. (8.44) and (8.45) can be rewritten as

$$\alpha_x = (1 - S_m)^{-1/2}(\phi\kappa_z\sqrt{n})^{-1/2}, \tag{8.52}$$

$$\alpha_z = (1 + 2S_m)^{-1/2}\kappa_z, \tag{8.53}$$

where all numerical coefficients, such as the factor 1/3, have been omitted for simplicity. Substituting Eqs. (8.52) and (8.53) into Eq. (8.43), the elastic free energy can be expressed as

$$\frac{\beta F_{el}}{N_t} = \frac{\phi}{2n}\left[\frac{2}{\phi\sqrt{n}(1 - S_m)\kappa_z} + \frac{\kappa_z^2}{1 + 2S_m} - 3 + \frac{4}{J}\ln\left(\phi\sqrt{nA(S_m)}\right)\right], \tag{8.54}$$

where we define

$$A(S_m) = (1 + 2S_m)(1 - S_m)^2, \tag{8.55}$$

and the function A corresponds to $\sqrt{A} = R_{0x}^2 R_{0z}/R_0^3$. Thus, the elastic free energy F_{el} explicitly depends on both the orientational order parameter S_m and the deformation ratio κ_z. This coupling indicates that the elastic response of the LC gel is not solely determined by network deformation, but is strongly influenced by the degree of orientational ordering of the mesogens.

The equilibrium value of the elongation κ_z is determined from $(\partial F_{el}/\partial \kappa_z)_{\phi,S} = 0$, yielding

$$\kappa_z = \left(\frac{1 + 2S_m}{(1 - S_m)\phi\sqrt{n}}\right)^{1/3}. \tag{8.56}$$

The elongation κ_z increases with increasing orientational order parameter S_m. Substituting Eq. (8.56) into Eq. (8.54), the elastic free energy becomes

$$\beta F_{el}/N_t = \frac{3}{2n}\left(\frac{\phi}{nA(S_m)}\right)^{1/3} + \frac{2}{nJ}\phi\ln(\phi\sqrt{nA(S_m)}), \tag{8.57}$$

which is expressed as a function of ϕ and S_m. When $S_m = 0$, or equivalently $A = 1$, Eq. (8.57) reduces to the elastic free energy for isotropic deformation, Eq. (8.24).

8.2.2 Equilibrium swelling of nematic gels

The free energy of the liquid crystalline gel is expressed as

$$F = F_{el} + \Delta F_{mix} + F_{nem}, \tag{8.58}$$

where the first term, F_{el} (Eq. (8.57)), represents the elastic free energy arising from the deformation of the gel. The second term, ΔF_{mix}, corresponds to the free energy associated with mixing the gel with solvent molecules. According to Flory-Huggins theory for polymer blends, the free energy of mixing is given by

$$\beta \Delta F_{mix}/N_t = \frac{(1-\phi)}{n_0} \ln(1-\phi) + \chi p\phi(1-p\phi), \tag{8.59}$$

where χ is the Flory-Huggins interaction parameter between the spacer segments and liquid crystalline molecules. Here, we assume that the liquid crystal solvent acts as a good solvent for the mesogenic units: $\chi \simeq 0$.

The third term in Eq. (8.58) represents the free energy associated with nematic ordering. Based on Maier-Saupe theory for nematic phases, we consider three coupling contributions arising from anisotropic interactions. Let ν_{mm} denote the orientation-dependent interaction between mesogens, ν_{m0} the interaction between mesogens and solvent molecules, and ν_{00} the interaction between solvent molecules. The nematic free energy is then expressed as

$$\begin{aligned}
\beta F_{nem}/N_t \quad = \quad & \frac{\phi_m}{n_m} \int f_m(\theta) \ln(4\pi f_m(\theta)) d\mathbf{\Omega} \\
& + \frac{\phi_0}{n_0} \int f_0(\theta) \ln(4\pi f_0(\theta)) d\mathbf{\Omega} \\
& - \frac{1}{2}\nu_{mm}\phi_m^2 S_m^2 - \nu_{m0}\phi_m(1-\phi)S_m S_0 - \frac{1}{2}\nu_{00}(1-\phi)^2 S_0^2,
\end{aligned} \tag{8.60}$$

where $d\mathbf{\Omega} = 2\pi \sin\theta d\theta$. Here, $f_m(\theta)$ and $f_0(\theta)$ represent the orientational distribution functions of the mesogens and solvent molecules, respectively. The orientational order parameter of the mesogens is defined as

$$S_m = \int P_2(\cos\theta) f_m(\theta) d\mathbf{\Omega}, \tag{8.61}$$

and that of the liquid crystal solvent molecules is defined as

$$S_0 = \int P_2(\cos\theta) f_0(\theta) d\mathbf{\Omega}, \tag{8.62}$$

where $P_2(\cos\theta) = (3/2)\cos^2\theta - 1/2$ is the second-order Legendre polynomial.

The orientational distribution functions $f_m(\theta)$ for the mesogens and $f_0(\theta)$ for the solvent molecules are determined by minimizing the free energy (Eq. (8.58)) with respect to these functions:

$$(\delta F/\delta f_m(\theta))_{f_0} = 0, \tag{8.63}$$

$$(\delta F/\delta f_0(\theta))_{f_m} = 0, \tag{8.64}$$

subject to the normalization conditions:

$$\int f_i(\theta) d\mathbf{\Omega} = 1, \tag{8.65}$$

where $i = m, 0$. For example, the required derivatives of F are given by

$$\beta\left(\frac{\delta F_{\rm el}}{\delta f_m(\theta)}\right)_{f_0} = \frac{3}{nA(S_m)}\left[\left(\frac{\phi}{nA(S_m)}\right)^{1/3} - \frac{2}{J}\phi\right](1-S_m)S_m \int P_2(\cos\theta)\, d\mathbf{\Omega}, \quad (8.66)$$

$$\beta\left(\frac{\delta F_{\rm nem}}{\delta f_m(\theta)}\right)_{f_0} = \frac{\phi_m}{n_m}\int\left(\ln(4\pi f_m(\theta)) + \frac{1}{4\pi}\right) d\mathbf{\Omega}$$
$$- \left(\nu_{mm}\phi_m^2 S_m + \nu_{m0}\phi_m(1-\phi)S_0\right)\int P_2(\cos\theta)\, d\mathbf{\Omega}. \quad (8.67)$$

We then obtain the orientational distribution function for the mesogen:

$$f_m(\theta) = \frac{1}{Z_m}\exp\left(\Gamma_m P_2(\cos\theta)\right), \quad (8.68)$$

$$\Gamma_m = n_m\left(\nu_{mm}\phi_m S_m + \nu_{m0}(1-\phi)S_0 - B(\phi, S_m)\right), \quad (8.69)$$

$$B(\phi, S_m) = \frac{3}{nA(S_m)\phi_m}\left[\left(\frac{\phi}{nA(S_m)}\right)^{1/3} - \frac{2}{J}\phi\right]S_m(1-S_m), \quad (8.70)$$

and that for the LC solvent molecule:

$$f_0(\theta) = \frac{1}{Z_0}\exp\left(\Gamma_0 P_2(\cos\theta)\right), \quad (8.71)$$

$$\Gamma_0 = n_0\left(\nu_{m0}\phi_m S_m + \nu_{00}(1-\phi)S_0\right). \quad (8.72)$$

The constants Z_i $(i = m, 0)$ are determined by the normalization condition, Eq. (8.65), as

$$Z_i = 4\pi I_0(\Gamma_i), \quad (8.73)$$

where the function $I_0(\Gamma_i)$ is defined as

$$I_q(\Gamma_i) = \int_0^1 (P_2(x))^q \exp\left(\Gamma_i P_2(x)\right) dx, \quad (8.74)$$

with $q = 0, 1$. Using these distribution functions, the orientational order parameters are determined by the following self-consistent equations:

$$S_m = I_1(\Gamma_m)/I_0(\Gamma_m), \quad (8.75)$$

for the mesogens, and

$$S_0 = I_1(\Gamma_0)/I_0(\Gamma_0), \quad (8.76)$$

for the LC solvent molecules. Substituting Eqs.(8.68) and (8.71) into (8.60), the nematic free energy is given by

$$\beta F_{nem}/N_t = \frac{1}{2}\nu_{mm}\phi_m^2 S_m^2 + \nu_{m0}\phi_m(1-\phi)S_m S_0$$
$$+\frac{1}{2}\nu_{00}(1-\phi)^2 S_0^2 - \frac{\phi_m}{n_m}\ln I_0(\Gamma_m)$$
$$-\frac{1-\phi}{n_0}\ln I_0(\Gamma_0) - \phi_m S_m B(\phi, S_m). \quad (8.77)$$

The chemical potential μ_0 of the LC solvent molecules within the gel is given by

$$\begin{aligned}
\beta(\mu_0 - \mu_0^{iso})/n_0 &= \frac{\beta}{n_0}\left(\frac{\partial F}{\partial N_0}\right)_{N_g} \\
&= f_{LC} - \phi\left(\frac{\partial f_{LC}}{\partial \phi}\right)_{S_i} \\
&= \frac{1}{n}\left[\left(\frac{\phi}{nA(S_m)}\right)^{1/3} - \frac{2}{J}\phi\right] + \frac{1}{n_0}\ln(1-\phi). \\
&+ \frac{\phi}{n_0} + \chi p^2\phi^2 + \frac{1}{2}\nu_{mm}S_m^2\phi_m^2 + \nu_{m0}S_mS_0\phi_m(1-\phi) \\
&+ \frac{1}{2}\nu_{00}S_0^2(1-\phi)^2 - \frac{1}{n_0}\ln I_0(\Gamma_0),
\end{aligned} \tag{8.78}$$

where $f_{LC} = \beta F/N_t$ and μ_0^{iso} is the standard chemical potential of the solvent molecules in the isotropic phase.

The chemical potential of the pure LC solvent molecules outside the gel, μ_0°, is obtained by setting $\phi = 0$ in Eq. (8.78), yielding as following:

$$\beta(\mu_0^{\circ} - \mu_0^{iso})/n_0 = \frac{1}{2}\nu_{00}S_b^2 - \frac{1}{n_0}\ln I_0(\Gamma_b), \tag{8.79}$$

where

$$\Gamma_b = n_0\nu_{00}S_b, \tag{8.80}$$

and S_b is the orientational order parameter of the LC solvent molecules outside the gel, given by

$$S_b = I_1(\Gamma_b)/I_0(\Gamma_b). \tag{8.81}$$

The NIT of the LC solvent molecules outside the gel occurs at $n_0\nu_{00} = 4.55$ (see Fig. 4.4). Using the definition $\nu_{00} = U_{00}/(k_BT)$, the NIT temperature of the LC solvent molecules is given by

$$T_{NI,b} = \frac{n_0U_{00}}{4.55k_B}. \tag{8.82}$$

The equilibrium concentration ϕ (i.e., the volume fraction) of the gel is determined by the condition that the osmotic pressure vanishes:

$$\Pi = \mu_0^{\circ}(S_b) - \mu_0(\phi, S_m, S_0) = 0. \tag{8.83}$$

The degree of swelling, defined as $Q_e = V/V_0$, is inversely proportional to the volume fraction ϕ (Eq. (8.23)). When the osmotic pressure Π is plotted as a function of ϕ, the appearance of van der Waals loops indicates the occurrence of a phase transition. In such cases, the thermodynamic stability conditions are determined by Maxwell equal-area construction. The equilibrium swelling of the gel corresponds to the condition $\Pi = 0$, while the region where $(\partial\Pi/\partial\phi)_T < 0$ represents an unstable spinodal region, as illustrated in Fig. 8.2.

Figure 8.7 shows the osmotic pressure Π (a) and the orientational order parameters (b) as a function of the polymer volume fraction ϕ for $T/T_{NI,b} = 0.9$. The parameters used are $n_m = 4$, $n_s = 2$, $t = 20$, $n_0 = 2$, $J = 4$, with all interaction parameters set equal: $\nu_M = \nu_{mm} = \nu_{m0} = \nu_{00}$, and $\nu_M/\chi = 6$. There are three points, A, B, and C, where the osmotic pressure satisfies $\Pi = 0$. The NIT of the gel occurs at the volume fraction corresponding to point B. At $T/T_{NI,b} = 0.9$, the area (A) under the curve at point A is larger than the area (B) at point C: area (A) > area (B). This indicates that the stable volume fraction of the gel corresponds to point A, which lies in the nematic gel with $S_m > 0$,

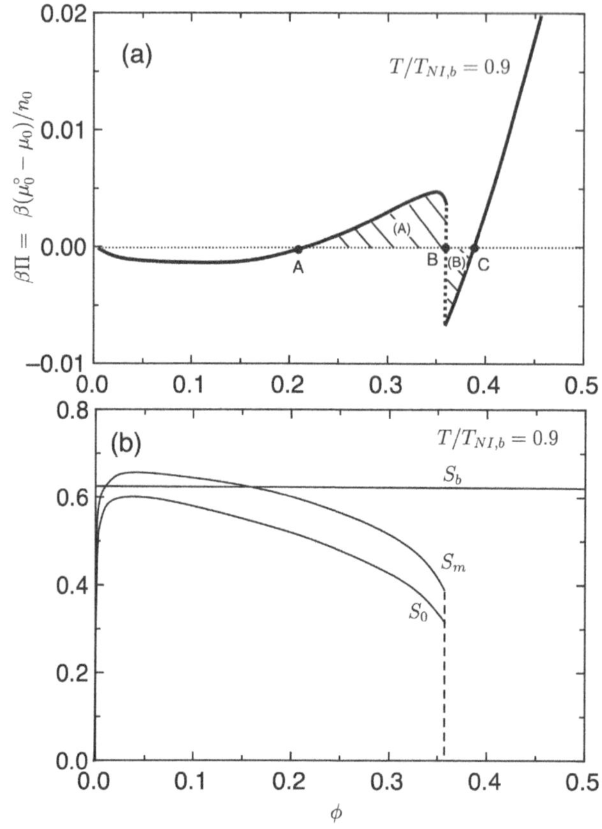

FIGURE 8.7
Osmotic pressure Π (a) and the orientational order parameters (b) plotted against ϕ for $T/T_{NI,b} = 0.9$. The stable volume fraction of the gel is determined by Maxwell's equal-area construction.

$S_0 > 0$, and $S_b > 0$. As the temperature increases, the area (A) decreases while the area (B) increases. At the NIT temperature of the gel, $T_{NI,g}$, the two areas become equal, area (A) = area (B), and the nematic gel transforms into the isotropic gel. For $T_{NI,g} < T < T_{NI,b}$, point C, corresponding to the isotropic gel with $S_m = S_0 = 0$ and $S_b > 0$, represents the thermodynamically stable state.

8.2.3 Volume phase transition of nematic gels

This section introduces volume phase transitions in systems consisting of a liquid crystalline gel swollen by either an isotropic solvent or LC solvents. By comparing the chemical potential of solvent molecules inside the gel (Eq. (8.78)) with that outside the gel (Eq. (8.79)), we can analyze four distinct cases for satisfying $\Pi = (\mu_0^\circ - \mu_0) = 0$. Figure 8.8 schematically depicts these volume phase transitions for nematic gels, with temperature shown on the vertical axis and the gel volume fraction on the horizontal axis [128]. This representation provides a clear view of how the gel's volume responds to temperature and nematic ordering.

(a) Isotropic gel immersed in LC solvents: When $S_m = S_0 = 0$ ($I_0 = 1$) in Eq. (8.78), the system corresponds to an isotropic gel mixed with liquid crystal solvent molecules (Figure

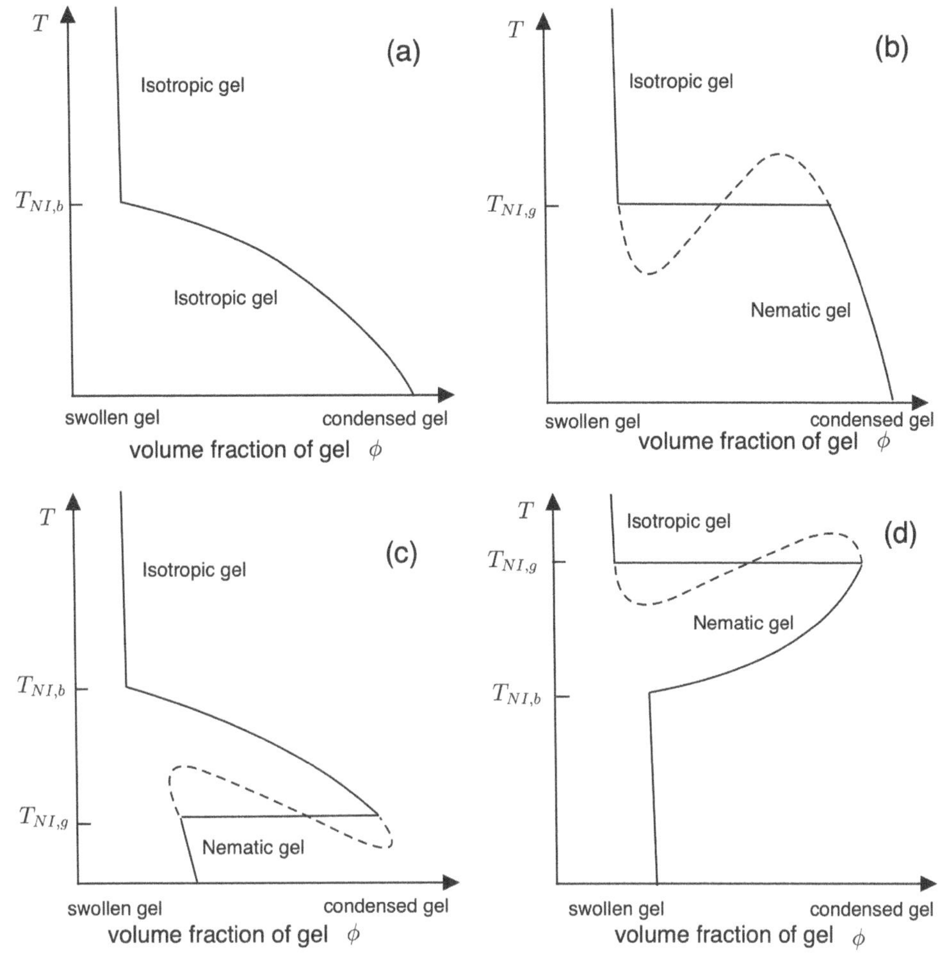

FIGURE 8.8
Schematic illustration of volume phase transitions in nematic gels: (a) isotropic gel mixed with LC solvent, (b) LC gel mixed with isotropic solvent, (c) LC gel mixed with LC solvent ($T_{NI,b} > T_{NI,g}$), (d) LC gel mixed with LC solvent ($T_{NI,b} < T_{NI,g}$).

8.8(a)). Here, the NIT temperature $T_{NI,b}$ exists only for the LC solvent molecules outside the gel (Eq. (8.82)). For $T > T_{NI,b}$, the isotropic gel is swollen by the solvent. As the temperature decreases to $T_{NI,b}$, the bulk isotropic phase transitions to the nematic phase, causing the gel to contract and its volume fraction to increase continuously, because the osmotic pressure has no van der Waals loops, such as Fig. 8.7. For $T < T_{NI,b}$, the isotropic gel expels the nematic solvent molecules. In mixtures of polystyrene and low-molecular-weight liquid crystals such as EBBA (4-Ethyl-4'-butylbiphenyl), phase separation into nematic and isotropic phases occurs below the NIT temperature of EBBA. To preserve the orientational order of the nematic phase, polymer chains are excluded from the nematic domains, leading to phase separation [130].

(b) Nematic gel mixed with isotropic solvents: When $S_b = 0$ in Eq. (8.79), the system corresponds to an LC gel mixed with isotropic solvent molecules (see Fig. 8.8(b)). In this case, the NIT temperature $T_{NI,g}$ of the gel exists only. As the temperature decreases, the gel undergoes a discontinuous volume contraction at the NIT temperature $T_{NI,g}$ of the gel, resulting in a first-order volume phase transition. The dashed curve indicates the unstable region associated with van der Waals loops, as shown in Fig. 8.7(a).

(c) Nematic gel mixed with nematic solvents. (see Fig. 8.8(c)). In this case, two NIT temperatures exist: $T_{NI,b}$ for the bulk solvent and $T_{NI,g}$ for the gel. When $T_{NI,b} > T_{NI,g}$:

- For $T > T_{NI,b}$: Both the gel and the solvent molecules are isotropic.
- At $T = T_{NI,b}$: The bulk solvent undergoes a transition to the nematic phase, leading to a continuous increase in the gel's volume fraction, like Fig. 8.8(a).
- At $T = T_{NI,g}$: The gel undergoes a discontinuous volume phase transition, resulting in a swollen nematic gel.

Thus, the swelling behavior of the gel follows the curve shown in Fig. 8.8(c). The osmotic pressure in the regime $T < T_{NI,b}$ is shown in Fig. 8.7.

(d) Nematic gel mixed with nematic solvents for $T_{NI,b} < T_{NI,g}$ (see Fig. 8.8(d)).

- For $T > T_{NI,g}$: Both the gel and the solvent molecules are isotropic.
- At $T = T_{NI,g}$: The gel undergoes a discontinuous volume phase transition, forming a condensed nematic gel.
- For $T_{NI,b} < T < T_{NI,g}$: The mesogens and LC solvents inside the gel exhibit nematic order ($S_m, S_0 > 0$), while the solvent molecules outside the gel remain isotropic.
- At $T = T_{NI,b}$: The bulk solvent undergoes a nematic transition, during which the gel's volume fraction ϕ changes continuously.

Figure 8.9 presents numerical results of the phase diagram in the temperature-concentration ($T/T_{NI,b} - \phi$) plane (a) and the corresponding order parameters (b), for the case $T_{NI,b} < T_{NI,g}$ (Fig. 8.8(d)). For $T > T_{NI,g}$, both the gel and the LC solvent are in the isotropic state. Upon cooling, the isotropic gel undergoes a first-order volume phase transition at $T_{NI,g}$, where it discontinuously condenses into the nematic phase. The dashed curve indicates the unstable region of the gel. As shown in Fig. 8.9(b), at $T_{NI,g}$, both the orientational order parameter of the mesogens in the gel (S_m) and that of the LC solvent inside the gel (S_0) exhibit discontinuous jumps. Upon further cooling, the nematic gel reswells, and the swelling curve displays a kink at the NIT temperature $T_{NI,b}$, where the order parameter S_b of the LC solvent outside the gel exhibits discontinuous jumps. Such reentrant volume phase transitions have been experimentally observed by Urayama et al. (see Fig. 1.8), and the present theoretical predictions are consistent with these experiments [18]. Note that the degree of swelling is related to the polymer volume fraction by $Q_e = V/V_0 \propto 1/\phi$.

When the NIT temperature of the bulk solvent is higher than that of the gel ($T_{NI,b} > T_{NI,g}$, case (c)), the gel first responds smoothly to the nematic ordering of the solvent. Subsequently, it undergoes its own discontinuous volume change at $T_{NI,g}$. In contrast, when the NIT temperature of the gel is higher than that of the bulk solvent ($T_{NI,b} < T_{NI,g}$, case (d)), the gel contracts discontinuously into the nematic state at $T_{NI,g}$ while the bulk solvent remains isotropic, followed by a continuous change at $T_{NI,b}$. Thus, the relative ordering of $T_{NI,g}$ and $T_{NI,b}$ determines whether the gel exhibits swelling or contraction during the coupled NIT. Furthermore, as the axial ratio n_m of the mesogen decreases, the phase diagram changes from case (d) to case (c) [128].

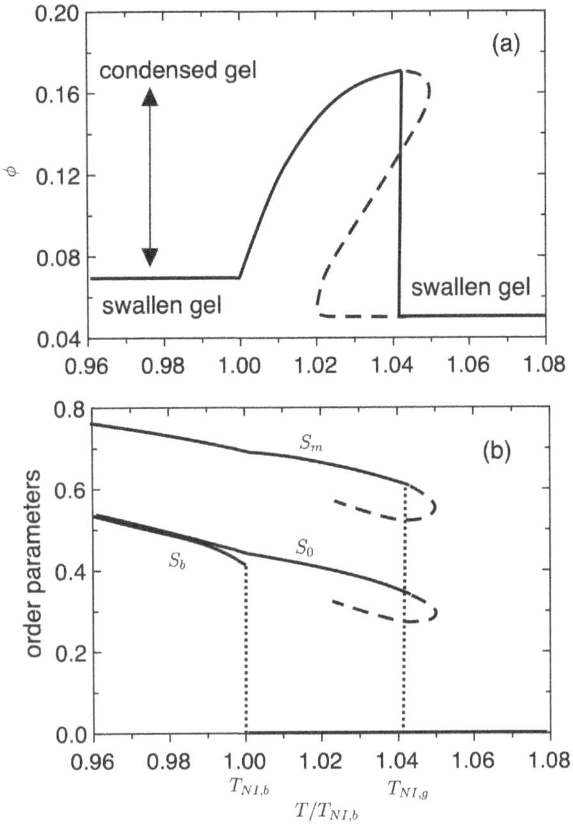

FIGURE 8.9
Phase diagram on the temperature ($T/T_{NI,b}$)-concentration (ϕ) plane (a) and the order parameters (b), corresponding to Fig. 8.8(d): $T_{NI,b} < T_{NI,g}$.

The orientational ordering of the gel itself drives the discontinuous, first-order volume phase transitions of nematic gels. In contrast, the discontinuous volume phase transition of an isotropic gel, as discussed in Fig. 8.3, is induced by the intermolecular interactions between the gel and solvent molecules (i.e., the χ parameter). In nematic gels, the interplay between the orientational order of liquid crystals and the elastic properties of the polymer network leads to discontinuous volume phase transitions. For further details, the reader is referred to the review by Urayama [131].

8.3 Cholesteric gels

Cholesteric gels are soft materials that combine the anisotropic properties of cholesteric liquid crystals with the elastic characteristics of polymer networks. These gels exhibit a helical molecular arrangement and can respond to external stimuli, such as temperature, light, or mechanical stress, by altering their optical and mechanical properties. Due to their unique ability to reflect selective wavelengths of light, cholesteric gels are of interest for applications in sensors, actuators, and tunable photonic devices.

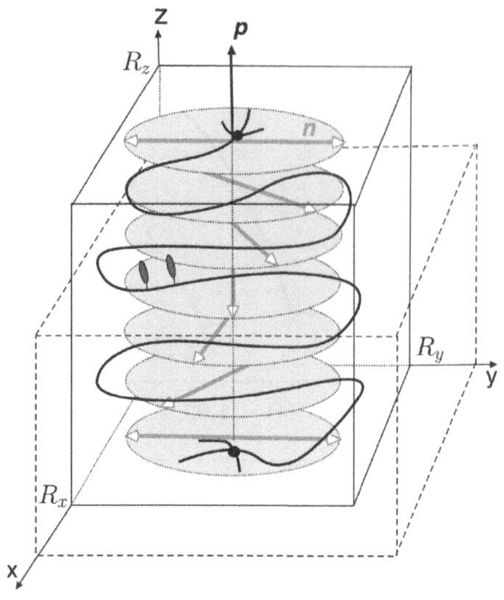

FIGURE 8.10
Schematic representation of a cholesteric liquid crystalline polymer (or subchain) between two crosslinks in a cholesteric elastomer (or gel). The director **n** (indicated by arrows) rotates helically along the pitch axis **p**, which is aligned with the z-axis. The solid box illustrates a prolate conformation of the subchain with spontaneous elongation along the z-axis (R_z) and uniform contraction in the perpendicular x and y directions ($R_x = R_y$). The dashed box represents an oblate shape, corresponding to spontaneous compression along the z-axis (parallel to the pitch axis **p**).

8.3.1 Elastic free energy of cholesteric gels

In this section, we introduce the fundamental aspects of cholesteric elastomers and gels. We first describe the deformation of cholesteric elastomers (without solvent), and then the swelling behavior of cholesteric gels immersed in an isotropic solvent. To capture their unique structural features, we derive the elastic free energy and the twist free energy (discussed in Chapter 5), which govern the equilibrium shape of the system. On this basis, we analyze how the anisotropic deformation, orientational order parameter, cholesteric pitch, and gel volume vary with temperature under thermal equilibrium.

Consider a cholesteric liquid-crystalline gel immersed in an isotropic solvent. We focus on a liquid crystalline subchain connecting two crosslinks in the polymer network, where n denotes the number of segments in the subchain, as shown in Fig. 8.10. Each repeating unit (monomer) consists of a rigid side-chain mesogenic moiety with an axial ratio n_m, and a flexible backbone spacer comprising n_s segments. The total number of segments in the subchain is therefore $n = (n_m + n_s)t$, where t is the number of repeating units. For the mesogen, with length L and diameter D, the axial ratio is $n_m = L/D$. Its volume is given by $v_m = (\pi/4)D^2L$, while the volume of the flexible spacer is $v_s = b^3 n_s$, where b^3 denotes the volume of a single spacer segment. This coarse-grained model, within the mean-field approximation, may be applied to both side-chain and main-chain liquid crystalline gels.

Let N_g denote the number of subchains and N_0 the number of solvent molecules contained within the gel. The polymer volume fraction ϕ is given by

$$\phi = b^3 n N_g / V = n N_g / N_t, \tag{8.84}$$

where $V = b^3 N_t$ is the volume of the gel, and $N_t(= nN_g + N_0)$ is the total number of segments inside the gel. In this model, each solvent molecule is assumed to occupy the same volume b^3 as a single polymer segment. The volume fraction of mesogenic units is given by

$$\phi_m = x_m \phi, \tag{8.85}$$

where $x_m = n_m/(n_m + n_s)$ represents the fraction of mesogens within the repeating unit. The volume occupied by a single subchain within the gel is given by

$$R_x R_y R_z = V/N_g, \tag{8.86}$$

where R_i denotes the chain length along the i-axis in the cholesteric polymer network.

We consider uniaxial deformations along the z-axis, which is parallel to the cholesteric pitch axis $\mathbf{p}$. Under this assumption, the deformation is symmetric in the transverse directions, so that $R_x = R_y$ (see Fig. 8.10). We define the expansion factors of the cholesteric gel as

$$\alpha_i = R_i/R_{0i}, \tag{8.87}$$

where R_{0i} is the spontaneous mean-square end-to-end distance (or radius) of the subchain along the i-axis ($i = x, y, z$). Using the effective step length tensor l_{ii}, which characterizes the anisotropy of a Gaussian chain, R_{0i}^2 can be expressed as

$$R_{0i}^2 = \frac{1}{3} l_{ii} L_c = \frac{1}{3} R_0^2 \left(\frac{l_{ii}}{b}\right), \tag{8.88}$$

where $R_0 = b\sqrt{n}$ is the end-to-end distance of an isotropic ideal chain. When $l_{ii} = b$, the chain behaves as an isotropic Gaussian chain. In a general coordinate frame, the effective step length tensor l_{ij}, which characterizes the anisotropic flexibility of the subchain in the presence of a director field $\mathbf{n}$, is given by [19]

$$l_{ij} = l_\perp \delta_{ij} + (l_\parallel - l_\perp) n_i n_j, \tag{8.89}$$

where n_i is the i-th component of the director $\mathbf{n}$, and $l_\parallel$ is the step length parallel to the director, $l_\perp$ is the step length perpendicular to the director, δ_{ij} is the Kronecker delta. This tensor captures the anisotropy in chain statistics induced by the orientational ordering of mesogenic units in the network. In the isotropic limit, $l_\parallel = l_\perp$, the tensor reduces to the isotropic form $l_{ij} = l_\perp \delta_{ij}$. When the director lies along the z-axis: $\mathbf{n} = (0, 0, 1)$, the tensor components become $l_{zz} = l_\parallel$ and $l_{xx} = l_{yy} = l_\perp$.

In the cholesteric phase, the director $\mathbf{n}$ varies along the helical axis $\mathbf{p}$ as shown in Fig. 8.11. Taking the helical axis to lie along the z-axis, the director at position z is given by

$$\mathbf{n} = (\cos\omega, \sin\omega, 0), \tag{8.90}$$

where $\omega = qz$ is the angle between the director and the x-axis at position z. The cholesteric pitch p is defined as the spatial period of the helical structure: $p = 2\pi/|q|$, with q being the pitch wavenumber. A right-handed helix corresponds to $q > 0$, while a left-handed helix corresponds to $q < 0$. Figure 8.11 illustrates the spatial rotation of the director $\mathbf{n}$ around the helical axis $\mathbf{p}$, forming a helical texture characteristic of the cholesteric phase.

Substituting Eq. (8.90) into Eq. (8.89), the diagonal components of the effective step length tensor are obtained as

$$l_{xx} = l_\perp + (l_\parallel - l_\perp)\langle\cos^2\omega\rangle, \tag{8.91}$$

$$l_{yy} = l_\perp + (l_\parallel - l_\perp)\langle\sin^2\omega\rangle, \tag{8.92}$$

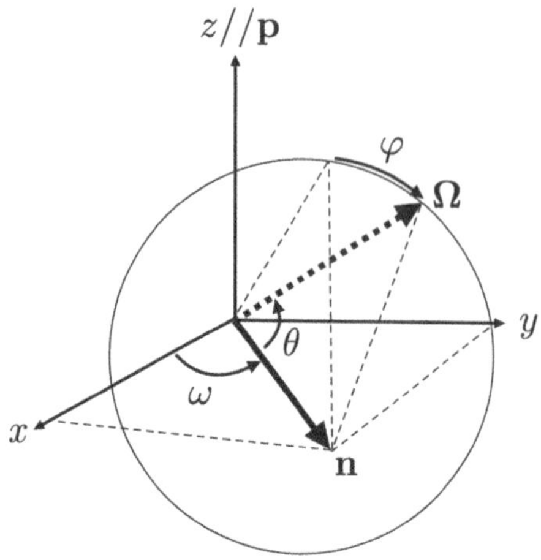

FIGURE 8.11
Schematic representation of the director $\mathbf{n}$ in the x-y plane at a given position z, using the same coordinate system as in Fig. 8.10. The cholesteric pitch axis $\mathbf{p}$ is aligned with the z-axis. A mesogenic unit on a subchain is represented by a bond vector $\boldsymbol{\Omega}$, which forms an angle θ with the local director $\mathbf{n}$. The angle φ denotes the azimuthal rotation about the director $\mathbf{n}$, characterizing the orientation of the mesogen relative to the local director field.

$$l_{zz} = l_{\perp}. \tag{8.93}$$

For the uniaxial deformation of the cholesteric gel, the average over the angular variable ω, which characterizes the rotation of the director $\mathbf{n}$ on the plane perpendicular to the pitch axis z, is given by

$$\langle \cos^2 \omega \rangle = \langle \sin^2 \omega \rangle = \frac{1}{2}. \tag{8.94}$$

Substituting Eqs. (8.94), (8.48), and (8.49) into (8.91)–(8.93), we obtain

$$l_{xx} = l_{yy} = b(1 + \frac{S_m}{2}), \tag{8.95}$$

$$l_{zz} = b(1 - S_m). \tag{8.96}$$

As the orientational order parameter S_m increases, the effective step length in the x-y plane increases spontaneously, while it decreases along the z-direction (i.e., parallel to the pitch axis).

The volume fraction ϕ of the gel is expressed as

$$\phi = \frac{b^3 n}{R_z R_x^2} = \frac{1}{\sqrt{n}\kappa_z \kappa_x^2}, \tag{8.97}$$

where R_x, R_y, and R_z are the principal dimensions of the deformed network. The deformation ratio κ_i are defined relative to the size R_0 of an isotropic ideal chain:

$$\kappa_z = R_z / R_0, \tag{8.98}$$

$$\kappa_x = R_x / R_0 = R_y / R_0. \tag{8.99}$$

Solving Eq. (8.97) for κ_x^2 yields

$$\kappa_x^2 = \frac{1}{\sqrt{n}\phi\kappa_z}. \tag{8.100}$$

Using the deformation ratio κ_z, the expansion factor α_i in Eq. (8.87) can be expressed as functions of the orientational order parameter S_m. The in-plane components are given by

$$\alpha_x = \alpha_y = (1 + S_m/2)^{-1/2}(\sqrt{n}\phi\kappa_z)^{-1/2}, \tag{8.101}$$

while the strain along the pitch axis (taken to be the z-axis) is

$$\alpha_z = (1 - S_m)^{-1/2}\kappa_z. \tag{8.102}$$

Substituting Eqs. (8.101) and (8.102) into (8.19), we finally obtain the elastic free energy of cholesteric elastomers as

$$\begin{aligned} \beta F_{el}/N_t &= \frac{\phi}{n}\left[\frac{1}{2}(2\alpha_x^2 + \alpha_z^2 - 3) - \frac{2}{J}\ln\alpha_x^2\alpha_z\right] \\ &= \frac{\phi}{2n}\left[\frac{\kappa_z^2}{1 - S_m} + \frac{2}{\phi\sqrt{n}(1 + S_m/2)\kappa_z} - 3 + \frac{4}{J}\ln\left(\phi\sqrt{nA(S_m)}\right)\right], \end{aligned} \tag{8.103}$$

where $A(S_m) = (1 + S_m/2)^2(1 - S_m)$. The elastic free energy couples the volume fraction ϕ of the gel, the orientational order S_m, and the uniaxial deformation κ_z along the pitch axis.

8.3.2 Equilibrium swelling of cholesteric gels

The free energy of a cholesteric gel is given by

$$F = F_{el} + \Delta F_{mix} + F_{ch}, \tag{8.104}$$

where F_{el} is the elastic free energy (Eq. (8.103)) associated with the deformation of the cholesteric elastomer network. The second term ΔF_{mix} represents the isotropic mixing contribution of the gel and solvent molecules, which governs the osmotic balance between the polymer network and the surrounding solvent. According to the Flory-Huggins theory for polymer solutions, the free energy of mixing is given by [44, 85]

$$\beta\Delta F_{mix}/N_t = (1 - \phi)\ln(1 - \phi) + \chi\phi(1 - \phi), \tag{8.105}$$

where χ is the Flory-Huggins interaction parameter between the gel and the solvent molecules (see Eq. (8.26)).

The last term F_{ch} in Eq. (8.104) represents the free energy of the cholesteric (twist) phase. The configuration of the constituent mesogens within the subchains is characterized by their position vector $\mathbf{r}$ and an orientation unit vector $\boldsymbol{\Omega}$, which is specified by the polar angle θ and azimuthal angle φ with respect to a fixed coordinate frame. The orientation can equivalently be described by the solid angle element $d\boldsymbol{\Omega} = \sin\theta d\theta d\varphi$, as illustrated in Fig. 8.11.

Let $f_d(\mathbf{n}(\mathbf{r}) \cdot \boldsymbol{\Omega})$ denote the orientational distribution function of the mesogenic units, where $\mathbf{n}(\mathbf{r})$ is the local director at position $\mathbf{r}$, and $\boldsymbol{\Omega}$ is a unit vector representing the molecular orientation. The distribution function depends only on the relative angle θ between the local director $\mathbf{n}(\mathbf{r})$ (as defined in Eq. (8.90)) and the orientation vector $\boldsymbol{\Omega}$. Specifically, this dependence enters through their scalar product: $\cos\theta = \mathbf{n}(\mathbf{r}) \cdot \boldsymbol{\Omega}$, which reflects the rotational symmetry around the director axis.

Using the distribution function, the orientational order parameter of the mesogens is given by

$$S_m = \int P_2(\cos\theta) f_d(\mathbf{n}(\mathbf{r}) \cdot \boldsymbol{\Omega}) d\boldsymbol{\Omega}, \tag{8.106}$$

where $P_2(x) = (3/2)(x^2 - 1/3)$.

The distortion free energy $F_{\rm ch}$, associated with the spatial variation of the director in a cholesteric phase, can be written as the sum of the orientational entropy (Eq. (5.60)) and the intermolecular interaction contribution (Eq. (5.61)):

$$\begin{aligned}\beta F_{ch}/N_t &= \frac{\phi_m}{n_m}\int f_d(\mathbf{n}(\mathbf{r})\cdot\mathbf{\Omega})\ln[4\pi f_d(\mathbf{n}(\mathbf{r})\cdot\mathbf{\Omega})]d\mathbf{\Omega}\\ &\quad -\frac{1}{2}\nu_M\phi_m^2 S_m^2(1+g(Q)),\end{aligned} \tag{8.107}$$

where the first term represents the entropic cost of restricting mesogen orientations, and the second term captures the nematic mean field interaction energy modified by twist. Here, the excluded volume term (5/4) in Eqs. (5.60) and (5.61) is neglected, since the mesogenic units are relatively short. The function $g(Q)$ encodes the effect of the helical modulation of the director field and is defined as

$$g(Q) = Q_0 Q - \frac{1}{2}Q^2, \tag{8.108}$$

where $Q = qb = 2\pi b/p$ is the dimensionless wavenumber, and

$$Q_0 = 2\pi b/p_0 = c_L/\nu_M, \tag{8.109}$$

is a chirality parameter (see Eq. (5.63)), where p_0 is the pitch length of the cholesteric gel before deformations. The quadratic form of $g(Q)$ implies that the twist free energy is minimized at a finite equilibrium pitch p, which results from a balance between the chiral interaction (linear in Q) favoring twist and the elastic cost (quadratic in Q) penalizing distortions. The chirality parameter Q_0, or c_L quantifies the strength of the molecular twist.

The pitch wavenumber Q is influenced by the deformation κ_z along the pitch (or z-) axis in a cholesteric gel. Assuming affine deformation of the subchains, the pitch stretches proportionally to the network deformation: $p = p_0\kappa_z$. Since $p \propto Q^{-1}$, the dimensionless wavenumber becomes a function of the deformation ratio:

$$Q = \frac{2\pi b}{p} = \frac{Q_0}{\kappa_z}. \tag{8.110}$$

Thus, stretching the gel along the pitch axis ($\kappa_z > 1$) reduces the wavenumber Q, whereas compression ($\kappa_z < 1$) increases the value of Q, reflecting the coupling between the network elasticity and the cholesteric structure.

The orientational distribution function $f_d(\mathbf{n}(\mathbf{r})\cdot\mathbf{\Omega})$ of the mesogens is determined by the free energy (Eq. (8.104)) with respect to this function:

$$\left(\frac{\delta F}{\delta f_d(x)}\right)_{\phi,\kappa_z} = 0, \tag{8.111}$$

under the normalization condition

$$\int f_d(\mathbf{n}(\mathbf{r})\cdot\mathbf{\Omega})d\mathbf{\Omega} = 1. \tag{8.112}$$

We then obtain the orientational distribution function of the mesogens:

$$f_d(x) = \frac{1}{Z_0}\exp\left[\eta P_2(x)\right], \tag{8.113}$$

where we define

$$\eta = n_m\left[\nu_M\phi_m S_m(1+g(Q)) - B(S_m,\phi)\right], \tag{8.114}$$

with

$$B(S_m,\phi) = \frac{1}{2nx_m}\left[\frac{\kappa_z^2}{(1-S_m)^2} - \frac{1}{\sqrt{n}\phi(1+S_m/2)^2\kappa_z} - \frac{6S_m}{J(1-S_m)(1+S_m/2)}\right]. \tag{8.115}$$

The parameter η represents the effective field experienced by the mesogens due to the combined effects of nematic interactions, twist deformation, and network elasticity, while $B(S_m,\phi)$ accounts for the elastic contribution from the polymer network and its coupling to the orientational order. The normalization constant Z_0 is determined by the normalization condition as $Z_0 = 4\pi I_0(S_m)$ and the function $I_k(S_m)$ is defined as

$$I_k(S_m) = \int_0^1 [P_2(x)]^k \exp\left[\eta P_2(x)\right] dx, \tag{8.116}$$

where $k = 0, 1, \ldots$. Substituting Eq. (8.113) into (8.106), the orientational order parameter S_m can be determined self-consistently as

$$S_m = I_1(S_m)/I_0(S_m). \tag{8.117}$$

Using the distribution function (8.113), the cholesteric free energy (8.107) can be expressed as

$$\beta F_{ch}/N_t = \frac{1}{2}\nu_M \phi_m^2 S_m^2(1+g(Q)) - \frac{\phi_m}{n_m}\ln I_0(S_m) - \phi_m S_m B(S_m,\phi). \tag{8.118}$$

The total free energy $F(\phi, S_m, \kappa_z)$ of the cholesteric gel is then given by the sum of Eqs. (8.103), (8.105), and (8.118). Clearly, in the isotropic phase where $S_m = 0$, the cholesteric contribution (8.118) vanishes.

The equilibrium deformation of the cholesteric gel (or elastomer) at thermal equilibrium is determined by minimizing the dimensionless free energy $f_{LC} = \beta F/N_t$. This condition is expressed as

$$\left(\frac{\partial f_{LC}}{\partial \kappa_z}\right)_{S_m,\phi} = 0, \tag{8.119}$$

or equivalently, $(\partial f_{LC}/\partial Q)_{S_m,\phi} = 0$, where the partial derivative is taken at fixed orientational order parameter S_m and volume fraction ϕ. This condition ensures that the system adopts the deformation state that minimizes the total free energy while maintaining constant orientational and density conditions. We then obtain the equilibrium condition for the deformation κ_z:

$$\frac{\phi}{n}\left[\frac{\kappa_z}{1-S_m} - \frac{1}{\sqrt{n}\phi(1+S_m/2)\kappa_z^2}\right] + \frac{1}{2}\nu_M\phi_m^2 S_m^2\left(1-\frac{1}{\kappa_z}\right)\left(\frac{Q_0}{\kappa_z}\right)^2 = 0, \tag{8.120}$$

where the first two terms arise from the elastic free energy, and the last term represents the twist distortion energy associated with the cholesteric ordering. The coefficient of the last term is the dimensionless twist elastic constant: $\tilde{K}_2 = (1/2)\nu_M\phi_m^2 S_m^2$ (see Eq. (5.100)). For small values of Q_0, the last term of Eq. (8.120) can be neglected and the deformation ratio κ_z is approximately given by

$$\kappa_z \simeq \left(\frac{1-S_m}{\sqrt{n}\phi(1+S_m/2)}\right)^{1/3}. \tag{8.121}$$

In this regime, κ_z decreases with increasing orientational order parameter S_m, meaning that the cholesteric elastomer (or gel) tends to compress along the pitch axis. In contrast, for large Q_0, the twist term dominates Eq. (8.120), and the elastomer tends to adopt $\kappa_z \sim 1$(corresponding to $Q \sim Q_0$), preserving the intrinsic cholesteric pitch.

8.3.3 Phase transitions of cholesteric gels

The chemical potential of the solvent molecules within the cholesteric gel is expressed as

$$\begin{aligned}\beta(\mu_0 - \mu_0^\circ) &= \beta(\partial F/\partial N_0)_{N_g} \\ &= f_{LC} - \phi(\partial f_{LC}/\partial \phi),\end{aligned} \tag{8.122}$$

where μ_0 is the chemical potential of a solvent molecule inside the gel and μ_0° is the chemical potential of the pure solvent outside the gel. As shown in Eq. (8.97), the deformation κ_z depends on the gel volume fraction ϕ. Therefore, the derivative of the elastic free energy with respect to ϕ can be expanded as

$$(\frac{df_{LC}}{d\phi}) = (\frac{\partial f_{LC}}{\partial \phi})_{S_m,\kappa_z} + (\frac{\partial f_{LC}}{\partial \kappa_z})_{S_m,\phi}(\frac{\partial \kappa_z}{\partial \phi}), \tag{8.123}$$

where the second term vanishes due to the equilibrium condition for κ_z, Eq. (8.119). Consequently, only the explicit dependence of f_{LC} on ϕ contributes to the chemical potential. We thus obtain the following expression for the chemical potential:

$$\begin{aligned}\beta(\mu_0 - \mu_0^\circ) &= \frac{1}{n\sqrt{n}(1+S_m/2)\kappa_z} + (1 - \frac{2}{nJ})\phi + \ln(1-\phi) + \chi\phi^2 \\ &+\frac{1}{2}\nu_M \phi_m^2 S_m^2 (1 + g(Q)).\end{aligned} \tag{8.124}$$

where the term $2/(nJ)$ can be neglected because $2/(nJ) \ll 1$. The equilibrium swelling (or the volume fraction ϕ) of a gel immersed in solvent molecules is determined by the balance of chemical potentials (i.e., osmotic pressure) between the solvent molecules inside and outside the gel:

$$\beta\Pi = \beta(\mu_0^\circ - \mu_0) = 0. \tag{8.125}$$

If the osmotic pressure Π as a function of ϕ exhibits a van der Waals loop, the equilibrium volume fraction is obtained using Maxwell's equal-area construction, such as Fig. 8.7. The region where $(\partial\Pi/\partial\phi)_T < 0$ corresponds to a thermodynamically unstable spinodal region.

In the following, we show numerical results for the equilibrium values of the gel volume fraction ϕ, the deformation ratios κ_i $(i = x, y, z)$, and the orientational order parameter S_m as functions of the reduced temperature $\tau = 1/\nu_M$.

8.3.3.1 Deformation of cholesteric elastomers

We now turn to the case of a cholesteric elastomer melt in the absence of solvent molecules. In this situation, the elastomer preserves a constant volume, $V = R_0^3 N_g$, as given in Eq. (8.86). Accordingly, the volume fraction of the elastomer melt is fixed at $\phi = b^3 n/R_0^3 = 1/\sqrt{n}$. The orientational order parameter S_m and the deformation ratio κ_z of the elastomer can then be determined as functions of the reduced temperature τ, by simultaneously solving Eqs. (8.117) and (8.120).

Figure 8.12(a) shows the deformation ratios κ_i as a function of the reduced temperature T/T_{CI}, where T_{CI} denotes the cholesteric-isotropic transition (CIT) temperature. The parameters are chosen as $n = 100$, $n_m = 2$, $n_s = 2$, and $Q_0 = 0.45$. Figure 8.12(b)

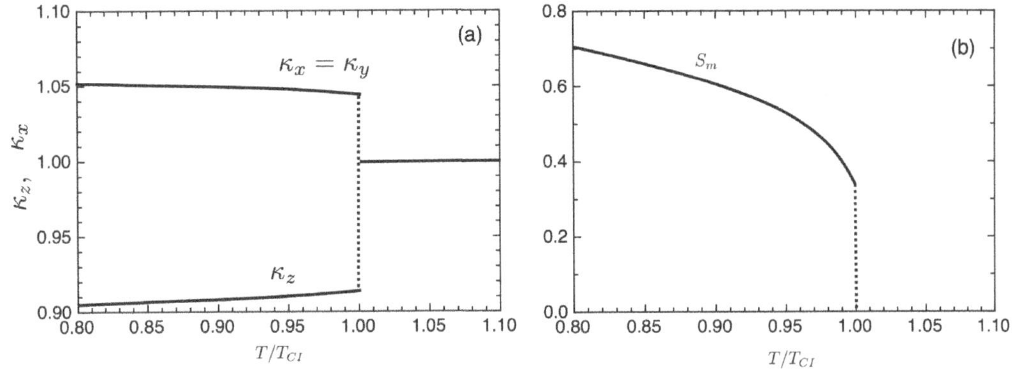

FIGURE 8.12
(a) The deformation ratios κ_i are plotted as a function of the normalized temperature T/T_{CI}, where T_{CI} denotes the CIT temperature. (b) The orientational order parameter S_m is shown as a function of temperature. The cholesteric elastomer exhibits spontaneous compression along the z-axis, which is parallel to the pitch axis, and simultaneous equal elongation in the x and y directions.

presents the orientational order parameter S_m as a function of temperature. As the temperature decreases, a first-order CIT takes place, accompanied by discontinuous changes in the deformation ratios κ_i at $T = T_{CI}$. The cholesteric elastomer undergoes spontaneous compression ($\kappa_z < 1$) along the z-axis, parallel to the pitch axis, while it elongates equally in the transverse directions ($\kappa_x = \kappa_y > 1$). Consequently, $\kappa_z < \kappa_x$, and the elastomer adopts an oblate shape. The theoretical predictions are in qualitative agreement with the experimental observation (Fig. 1.9). However, in contrast to the discontinuous change predicted by the mean field theory [132], the experimental data exhibit a continuous deformation [17, 133], as shown in Fig. 1.9.

Figure 8.13 shows the deformation ratio κ_z as a function of the orientational order parameter S_m for different values of the chirality parameter Q_0. Larger values of S_m correspond to lower temperatures. The curve for $Q_0 = 0.45$ corresponds to the results in Fig. 8.12, and the dotted line indicates the CIT. Note that the pitch length p is proportional to κ_z according to Eq. (8.110). For small Q_0, κ_z decreases with increasing S_m (i.e., decreasing temperature), as described by Eq. (8.121), where the elastic contribution dominates. In contrast, for larger Q_0, κ_z remains nearly constant. In this regime, the twist term (the last term in Eq. (8.120)) becomes dominant, yielding $\kappa_z \sim 1$ (or equivalently, $Q \sim Q_0$). Thus, the shape of the cholesteric elastomer is governed by the competition between elastic free energy and twist distortion. While the elastic contribution favors an elongation of the elastomer in the plane perpendicular to the pitch axis (z-axis), the twist distortion tends to maintain $\kappa_z \sim 1$, thereby preserving the original cholesteric pitch.

8.3.3.2 Volume phase transitions of cholesteric gels

We now discuss the volume phase transitions and deformation behavior of a cholesteric gel immersed in an isotropic solvent. To derive the equilibrium values of κ_i, S_m, and ϕ, we numerically solve three coupled Eqs. (8.117), (8.120), and (8.125) for $n_m = n_s = 2$, $n = 100$, $Q_0 = 0.05$, and $\chi = 0$. As the temperature decreases, the gel undergoes a transition from an isotropic (swollen) state to a cholesteric (condensed) state at the CIT temperature. At this transition, the gel volume exhibits a discontinuous change, closely resembling the behavior observed in nematic gels, as illustrated in Fig. 8.8(b).

Figure 8.14 shows the deformation ratio κ_z/κ_x plotted as a function of the reduced temperature T/T_{CI} for various values of the chirality parameter Q_0. The shape of the

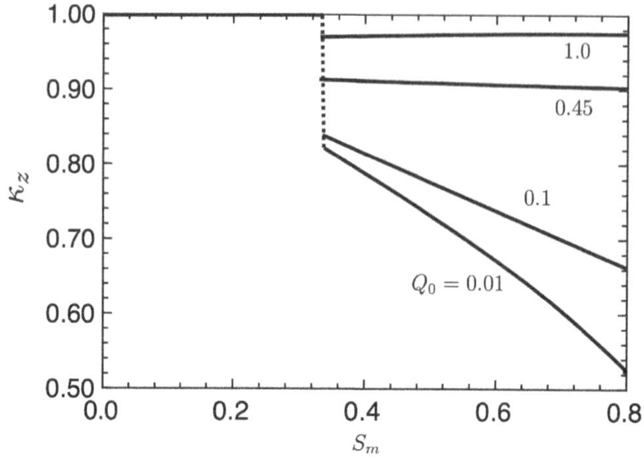

FIGURE 8.13
Deformation ratio κ_z as a function of the orientational order parameter S_m for various values of the chirality parameter Q_0. The results demonstrate that the deformation of the elastomer is strongly affected by Q_0, highlighting that the shape of the cholesteric elastomer is determined by the competition between elastic free energy and twist distortion. Reprinted with permission from Ref. [132].

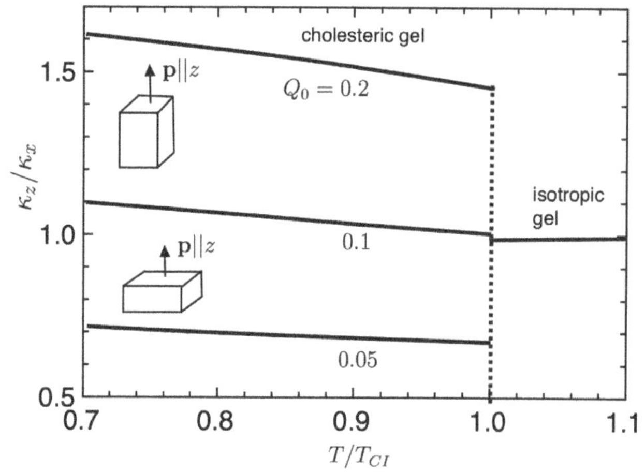

FIGURE 8.14
Deformation ratio κ_z/κ_x plotted as a function of the temperature T/T_{CI} for various values of the chirality parameter Q_0. The shape of the condensed cholesteric gel at low temperatures ($T < T_{CI}$) strongly depends on Q_0. For large values of Q_0, the cholesteric gel adopts a prolate shape, whereas for small Q_0, it exhibits an oblate shape. Reprinted with permission from Ref. [132].

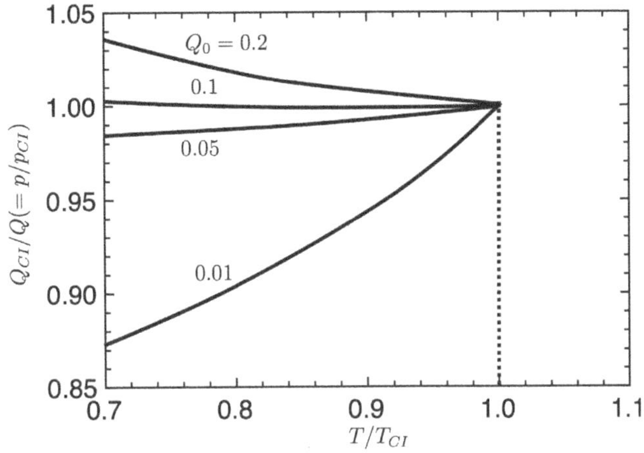

FIGURE 8.15
Cholesteric pitch length $p/p_{CI}(= Q_{CI}/Q)$ plotted against the temperature T/T_{CI} for various values of the chirality parameter Q_0, where the value of $p_{CI}(= 2\pi a/Q_{CI})$ is the pitch length at $T = T_{CI}$. The chirality Q_0 plays a crucial role in determining both the temperature dependence of the pitch and the resulting shape of cholesteric gels. Reprinted with permission from Ref. [132].

condensed cholesteric gel at low temperatures ($T < T_{CI}$) strongly depends on Q_0. For large Q_0, we find $\kappa_z/\kappa_x > 1$, indicating that the gel adopts a prolate shape. In this regime, the twist distortion energy dominates the free energy, suppressing large deformations along the pitch axis and leading the gel to contract within the x-y plane. In contrast, for small Q_0, we obtain $\kappa_z/\kappa_x < 1$, and the gel exhibits an oblate shape. Here, the elastic energy dominates, favouring elongation along the director, which lies in the plane perpendicular to the pitch axis.

Figure 8.15 shows the normalized cholesteric pitch $p/p_{CI}(= Q_{CI}/Q)$ plotted as a function of temperature T/T_{CI} for various values of the chirality parameter Q_0, where $p_{CI}(= 2\pi b/Q_{CI})$ is the pitch length at the CIT temperature $T = T_{CI}$. The pitch p is calculated using Eq. (8.110). The corresponding temperature dependence of the gel deformations is shown in Fig. 8.14. For large values of Q_0, the pitch increases as the temperature decreases. When $Q_0 \sim 0.1$, the pitch remains nearly constant and becomes insensitive to temperature. In contrast, for small values of Q_0, the pitch decreases with decreasing temperature, since the gel becomes more condensed and elongates along the director perpendicular to the pitch axis. The chirality parameter Q_0, Eq. (8.109), thus plays a crucial role in determining both the pitch and the shape of cholesteric gels. The coupling between the elastic energy and twist distortion will also be important in biological cholesteric liquid crystals [134].

9

Self-Assembled Liquid Crystals

Soft matter and biophysical systems often contain monomers that reversibly assemble into spherical, cylindrical, and rod-like aggregates [135]. At sufficiently high concentrations, these aggregates can organize into liquid crystal phases—such as nematic, smectic, or cholesteric—depending on temperature and concentration [136, 137]. The interplay between the rod-like aggregates and liquid crystal ordering presents a rich and active area of research, owing to the complex phase behavior and unique material properties that emerge from their interactions.

The bonding energy considered here is on the order of the thermal energy, so temperature variations establish an equilibrium between bond formation and dissociation, known as thermoreversible association. For example, hydrogen (physical) bonds become increasingly stable at lower temperatures. Under thermal equilibrium, such intermolecular bonding results in polydisperse molecular aggregates. To describe the size distribution of these aggregates, we examine the thermodynamics of the system.

9.1 Size distribution of molecular aggregates

We now discuss the size distribution of molecular aggregates. As illustrated in Fig. 9.1, an aggregate composed of m monomer molecules is referred to as an m-mer ($m = 1, 2, \ldots, \infty$). We consider a thermoreversible association system in which aggregates can form or dissociate as a function of temperature and concentration. Let N_m denote the number of m-mers in the solution, and let v_0 be the volume of a single monomer. Then, the volume fraction of m-mers is given by $\phi_m = v_0 m N_m / V$, where V is the total volume of the system. The total

FIGURE 9.1
Schematic representation of an m-mer formed by the aggregation of m monomer molecules. The system includes polydisperse molecular aggregates.

DOI: 10.1201/9781003737711-9

volume fraction of solute molecules is given by

$$\phi = \sum_{m=1}^{\infty} \phi_m. \tag{9.1}$$

The condition for association equilibrium is that the chemical potential of the solute monomers equals that of the solute molecules forming aggregates. This condition is expressed as

$$m\mu_1 = \mu_m, \tag{9.2}$$

where μ_1 is the chemical potential of a monomer and μ_m is that of an m-mer. Equation (9.2) is referred to as the chemical equilibrium condition. We here assume that the chemical potential of an m-mer in solution is given by

$$\mu_m = \mu_m^\circ + k_B T \ln(\frac{\phi_m}{m}), \tag{9.3}$$

for $m = 1, 2, 3, \cdots$. The first term μ_m° denotes the standard chemical potential associated with the formation of an m-mer from m monomers. The second term represents the contribution from the translational entropy of the micellar (i.e., m-mer) species, corresponding to the translational entropy of an ideal gas.

By substituting Eq. (9.2) into Eq. (9.3), we obtain

$$\phi_m = m \left[\phi_1 \exp\left(\frac{\mu_1^\circ - \mu_m^\circ/m}{k_B T}\right) \right]^m, \tag{9.4}$$

which shows that the volume fraction of m-mers can be expressed in terms of the monomer volume fraction ϕ_1. Equation (9.4) can be interpreted in two limiting cases:

(1) When $\mu_1^\circ - \mu_m^\circ/m < 0$, ϕ_m decreases with increasing m, indicating that most solute molecules remain as monomers.

(2) When $\mu_1^\circ - \mu_m^\circ/m > 0$, the formation of m-mer aggregates is favored.

The functional dependence of μ_m°/m on m determines whether the aggregates attain a finite size or grow indefinitely [135].

9.1.1 Spherical micelles

Spherical micelles (Fig. 9.2(a)) are a fundamental type of self-assembled structure commonly observed in dilute solutions of amphiphilic or stacking molecules. In these aggregates, a finite number of monomer units associate to form a compact spherical structure, minimizing the interfacial energy. When m particles, each with volume v_0, associate to form a spherical aggregate, the volume of the resulting sphere is given by $(4/3)\pi r^3 = m v_0$, where r is the radius of the aggregate. The interfacial energy of such an aggregate is given by $4\pi r^2 \sigma$, where σ N/m is the interfacial tension at the surface of the aggregate. Therefore, the standard chemical potential per molecule in an m-mer is given by

$$\mu_m^\circ/m = \mu_\infty^\circ + 4\pi r^2 \sigma/m, \tag{9.5}$$

where μ_∞° represents the bulk chemical potential per molecule in an infinitely large aggregate, and the second term corresponds to the interfacial energy contribution per molecule. By eliminating the radius r in Eq. (9.5) using the relation $r = (3mv_0/4\pi)^{1/3}$, the standard chemical potential per molecule in an m-mer can be expressed as

$$\frac{\mu_m^\circ}{m} = \mu_\infty^\circ + \frac{\Lambda k_B T}{m^{1/3}}, \tag{9.6}$$

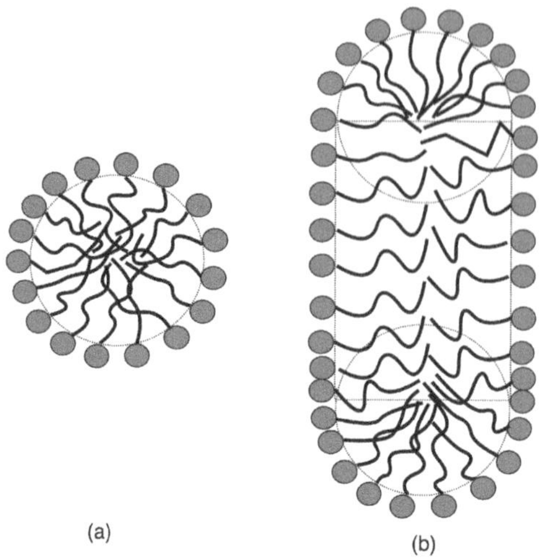

FIGURE 9.2
Schematic diagrams of spherical (a) and cylindrical (b) micelles, composed of amphiphilic molecules that contain both hydrophilic (water-attracting) and hydrophobic (water-repelling) segments within a single molecular structure.

where

$$\Lambda = \frac{4\pi\sigma\left(\frac{3v_0}{4\pi}\right)^{2/3}}{k_B T}, \tag{9.7}$$

is a dimensionless parameter characterizing the aggregation energy. By substituting Eq. (9.6) into Eq. (9.4), the volume fraction of m-mers can be written as

$$\begin{aligned} \phi_m &= m\left[\phi_1 \exp\left(\Lambda(1 - m^{-1/3})\right)\right]^m \\ &\simeq m\left[\phi_1 \exp(\Lambda)\right]^m, \end{aligned} \tag{9.8}$$

where the last approximation is valid for sufficiently large m. When the monomer volume fraction ϕ_1 is sufficiently small such that $\phi_1 \exp(\Lambda) < 1$, the formation of large aggregates is unfavorable, and most solute molecules remain as monomers. As ϕ_1 approaches $\exp(-\Lambda)$ from below, the volume fraction of m-mers, ϕ_m, decreases with increasing m, indicating that small aggregates dominate. In contrast, when $\phi_1 \exp(\Lambda) > 1$, ϕ_m increases with m, signaling the formation of large aggregates in solution. Therefore, there exists a critical monomer concentration that determines whether the solute molecules remain primarily in the monomeric state or associate to form aggregates. This threshold concentration is known as the critical micelle concentration (CMC) and is given by

$$\phi_{\mathrm{CMC}} \simeq \exp(-\Lambda). \tag{9.9}$$

For monomer concentrations above this value ($\phi > \phi_{\mathrm{CMC}}$), micelle formation becomes thermodynamically favorable, and a significant fraction of solute molecules exists as aggregates.

9.1.2 Cylindrical micelles

In addition to spherical micelles, cylindrical micelles (Fig.9.2(b)) are frequently observed in self-assembling systems, especially at higher concentrations or for molecules with specific geometries. In these structures, amphiphilic or stacking molecules associate to form elongated, rod-like aggregates with a nearly constant cross-sectional area, while their length varies with solution conditions and molecular interactions. The free energy change associated with adding a single surfactant molecule to the cylindrical portion of a micelle is essentially independent of the aggregation number m of the cylinder. Consequently, unlike spherical micelles, cylindrical micelles do not possess a well-defined optimal aggregation number and can grow continuously in length without a preferred size. In this case, the total energy of the cylindrical aggregate is dominated not by the cylindrical body, but by the two spherical end-caps at either end of the cylinder, which resemble the hemispheres of a spherical micelle. If the energy of the hemispherical micelle at each end is denoted by $\Delta\epsilon_{\text{endcap}}$, then the energy per molecule associated with the end-cap regions is given by $2\Delta\epsilon_{\text{endcap}}/m = \Lambda k_B T/m$, where Λ is a dimensionless aggregation energy parameter characterizing the end-cap contribution. Therefore, the standard chemical potential per molecule for a cylindrical micelle can be expressed as the sum of the bulk energy of the cylindrical portion, μ_∞°, and the end-cap contribution:

$$\frac{\mu_m^\circ}{m} = \mu_\infty^\circ + \frac{\Lambda k_B T}{m}. \tag{9.10}$$

Substituting Eq. (9.10) into Eq. (9.4), the volume fraction of m-mers is given by

$$\phi_m = m\left[\phi_1 \exp(\Lambda)\right]^m \exp(-\Lambda), \tag{9.11}$$

which describes the size distribution of cylindrical micelles. The CMC is approximately given by the condition $\phi_{\text{CMC}} \exp(\Lambda) \simeq 1$. The total volume fraction ϕ of amphiphilic molecules is then given by the sum over:

$$\phi = \sum_{m=1}^{\infty} \phi_m = \frac{\phi_1}{\left[1 - \phi_1 \exp(\Lambda)\right]^2}, \tag{9.12}$$

where ϕ_1 is the volume fraction of monomers. Introducing the association constant

$$K_a = \exp(\Lambda), \tag{9.13}$$

Equation (9.12) can be inverted to yield the monomer fraction as a function of the total amphiphile concentration ϕ:

$$\phi_1 = \frac{1}{2K_a^2\phi}\left(1 + 2K_a\phi - \sqrt{1 + 4K_a\phi}\right). \tag{9.14}$$

This expression allows the calculation of the monomer concentration ϕ_1 for a given total concentration ϕ of amphiphilic molecules.

Let us now derive an approximate solution using Eq. (9.14). In the high concentration regime above the CMC, where $K_a\phi \gg 1$, the monomer volume fraction ϕ_1 can be approximated up to the order of $(K_a\phi)^{-1/2}$ as

$$\phi_1 \simeq \frac{1}{K_a}\left(1 - \frac{1}{\sqrt{\phi K_a}}\right). \tag{9.15}$$

Substituting Eq. (9.15) into Eq. (9.11) yields an approximate expression for the volume fraction of m-mers as

$$\phi_m \simeq \frac{m}{K_a}\left(1 - \frac{1}{\sqrt{\phi K_a}}\right)^m. \tag{9.16}$$

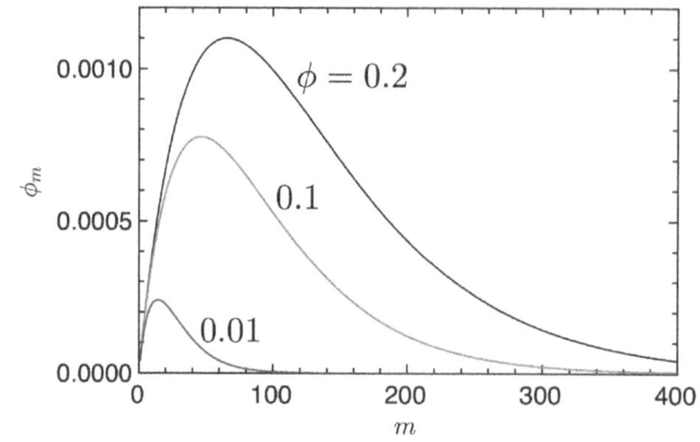

FIGURE 9.3
Volume fraction ϕ_m of aggregates plotted as a function of the aggregation number m for $\phi = 0.01$, 0.1, and 0.2.

Figure 9.3 presents the distribution of the volume fraction ϕ_m (Eq. (9.16)) of m-mers as a function of the aggregation number m. The calculation is carried out for an aggregation energy parameter of $\Lambda = 10$, while varying the total volume fraction as $\phi = 0.01$, 0.1, and 0.2. Compared with spherical micelles, cylindrical micelles display a significantly broader size distribution (i.e., higher polydispersity), which becomes progressively wider as the concentration increases. The aggregation number m_{max} at which ϕ_m attains its maximum is given by

$$m_{\text{max}} = \sqrt{\phi K_a}. \tag{9.17}$$

Micelles in solution continuously undergo association and dissociation. When two cylindrical aggregates merge, the end-caps are eliminated, thereby reducing the energetic cost. Conversely, long cylindrical micelles may split to gain translational (mixing) entropy. The association of two cylindrical micelles lowers the free energy by reducing the number of end-caps. However, from an entropic standpoint, the system disfavors the formation of excessively large micelles, and thus larger aggregates tend to undergo fission to increase entropy. The equilibrium size distribution of micelles is therefore governed by the competition between energetic stabilization from eliminating end-caps and entropic stabilization from producing more, smaller aggregates. The equilibrium distribution will be analyzed in detail using free energy considerations in the following sections.

9.2 Rod-like aggregates

In this section, we introduce a mean field theory to describe the nematic-isotropic phase transition (NIT) and the phase behavior of solutions containing rigid, rod-like aggregates. By combining the Onsager model for rod-like aggregates [138–140] with the Flory-Huggins model for polydisperse associating polymer solutions [141], we derive the free energy of solutions of rod-like aggregates.

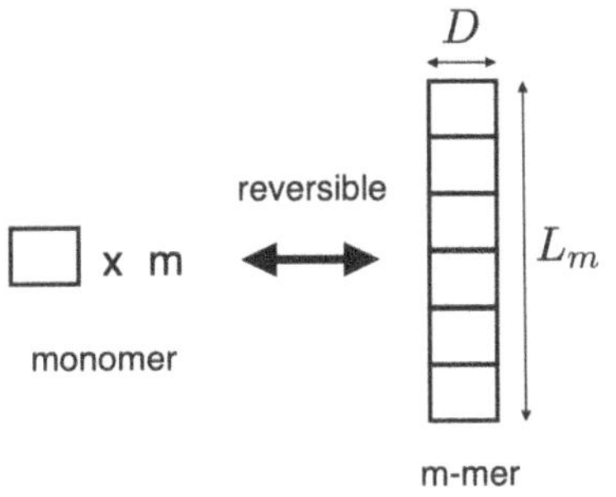

FIGURE 9.4
Schematic representation of a rod-like m-mer formed by the end-to-end aggregation of $m(= 1, 2, 3, \ldots, \infty)$ monomer molecules. These polydisperse anisotropic rod-like aggregates can order into a nematic phase.

9.2.1 Free energy of associating molecules

Consider a binary mixture consisting of solute and solvent molecules. The solute molecules can self-assemble into rod-like aggregates, hereafter referred to as m-mers, each composed of m solute molecules (Fig. 9.4). We focus on the case of thermoreversible association. Let N_m be the number of m-mers ($m = 1, 2, \ldots, \infty$), and N_0 be the number of solvent molecules. The geometry of an m-mer is characterized by its axial ratio $L_m/D = n_r m$, where $L_m(= mL)$ is the length of the m-mer, D is its diameter, and $n_r = L/D$ denotes the axial ratio of a single solute monomer. The volume of the m-mer is given by $v_m = (\pi/4)D^2 L_m = v_0 n_r m$, where $v_0 = (\pi/4)D^3$ is defined as the reference volume corresponding to a single lattice cell. We assume that the volume of a solvent molecule is $v_s = v_0$. The total volume of our system is given by $V = v_0 N_t$, where

$$N_t = N_0 + n_r \sum_{m=1}^{\infty} m N_m, \tag{9.18}$$

is the total number of lattice cells in the system. The volume fraction of m-mers is given by

$$\phi_m = v_0 n_r m N_m / V = n_r m N_m / N_t, \tag{9.19}$$

and the volume fraction of solvent molecules is

$$\phi_0 = v_0 N_0 / V = N_0 / N_t. \tag{9.20}$$

The total volume fraction ϕ of solute molecules, i.e., the combined contribution of all m-mers, is given by

$$\phi = \sum_{m=1}^{\infty} \phi_m = 1 - \phi_0, \tag{9.21}$$

where the last equality follows directly from the definitions of ϕ_m and ϕ_0, since in the Flory-Huggins lattice model every lattice site must be occupied by either a solute or a solvent molecule (see Section 7.1).

The free energy of the system is given by [142]

$$F = F_{ref} + \Delta F_{mix} + F_{nem}, \tag{9.22}$$

where the first term F_{ref} represents the reference free energy of the system, i.e., the intrinsic standard-state free energy of the m-mers and solvent molecules before mixing in an isotropic phase. The second term ΔF_{mix} accounts for the entropic contribution due to mixing between the solvent and the polydisperse rod-like aggregates, and the third term F_{nem} represents the excess free energy arising from nematic ordering of the aggregates. Using the standard chemical potential μ_0° of a pure solvent molecule and the standard chemical potential μ_m° of an m-mer before mixing, the reference free energy is given by

$$F_{ref} = N_0\mu_0^\circ + \sum_{m=1}^{\infty} N_m\mu_m^\circ = N_t\left[\phi_0\mu_0^\circ + \sum_{m=1}^{\infty}\frac{\phi_m}{n_r m}\mu_m^\circ\right]. \tag{9.23}$$

The second term in Eq. (9.22) represents the free energy of mixing the m-mers with solvent molecules. According to the Flory-Huggins theory for polydisperse polymer solutions [85], it is given by

$$\begin{aligned}\beta\Delta F_{mix} &= N_0\ln\phi_0 + \sum_{m=1}^{\infty} N_m\ln\phi_m + \chi\phi N_0 \\ &= N_t\left[\phi_0\ln\phi_0 + \sum_{m=1}^{\infty}\frac{\phi_m}{n_r m}\ln\phi_m + \chi\phi(1-\phi)\right],\end{aligned} \tag{9.24}$$

where $\beta = 1/(k_BT)$ is the inverse thermal energy. The first two terms correspond to the entropy of mixing of the m-mers and solvent molecules, while the last term represents the enthalpic contribution arising from solute-solvent interactions. The dimensionless parameter $\chi = U_0/(k_BT)$ is the Flory-Huggins interaction parameter, where U_0 is the mean interaction energy (primarily dispersion interactions, excluding specific interactions such as hydrogen bonding).

The last term in Eq. (9.22) corresponds to the free energy associated with nematic ordering of the rod-like aggregates. Let $f_m(\theta)$ denote the orientational distribution function of an m-mer. By extending the Onsager model for solutions of rod-like molecules (Eq. (4.10)) to a polydisperse system, the nematic free energy of the polydisperse rod-like aggregates can be written as [143–145]

$$\begin{aligned}\beta F_{nem}/V &= \sum_{m=1}^{\infty}\rho_m\int f_m(\theta)\ln[4\pi f_m(\theta)]d\mathbf{\Omega} \\ &+\frac{1}{2}\sum_{m=1}^{\infty}\sum_{l=1}^{\infty}\rho_m\rho_l\iint f_m(\theta)f_l(\theta')A_2(m,l;\theta,\theta')d\mathbf{\Omega}d\mathbf{\Omega}',\end{aligned} \tag{9.25}$$

where $\rho_m = \phi_m/(n_r m)$ is the number density of the m-mers, and $d\mathbf{\Omega} = 2\pi\sin\theta\, d\theta$ denotes the solid-angle element. The first term in Eq. (9.25) is the entropy change due to the orientational ordering of the m-mers. The second term accounts for the excluded volume interaction between m- and l-mers. The excluded volume between an m-mer of length $L_m(= mL)$ and an l-mer of length $L_l(= lL)$ is given by [146]

$$A_2(m,l;\theta,\theta') \simeq 2L_mL_lD|\sin\theta_{12}|, \tag{9.26}$$

where θ_{12} is the angle between the two rods. We expand $|\sin\theta_{12}|$ in terms of the Legendre polynomial $P_2(x) = (3/2)x^2 - 1/2$ as (see Section 4.1.3)

$$|\sin\theta_{12}| \simeq \frac{\pi}{4} - \frac{5\pi}{32}P_2(\cos\theta)P_2(\cos\theta'). \tag{9.27}$$

Substituting Eq. (9.26) into Eq. (9.25), the orientational distribution function $f_m(\theta)$ of an m-mer is obtained by minimizing the nematic free energy (Eq. (9.25)) with respect to $f_m(\theta)$.

The condition for an extremum is given by

$$\left(\frac{\delta F_{nem}}{\delta f_m(\theta)}\right)_T = 0, \tag{9.28}$$

together with the normalization constraint:

$$\int f_m(\theta) d\mathbf{\Omega} = 1. \tag{9.29}$$

We then obtain

$$f_m(\theta) = \frac{1}{Z_m} \exp\left(\frac{5}{4} n_r m \phi \langle S \rangle P_2(\cos\theta)\right), \tag{9.30}$$

where Z_m is the normalization constant, and

$$\langle S \rangle = \sum_{m=1}^{\infty} S_m \phi_m / \sum_{m=1}^{\infty} \phi_m = \sum_{m=1}^{\infty} S_m \phi_m / \phi, \tag{9.31}$$

denotes the average orientational order parameter over all aggregates. The orientational order parameter of the m-mers is given by

$$S_m = \int P_2(\cos\theta) f_m(\theta) d\mathbf{\Omega}, \tag{9.32}$$

which can be evaluated in terms of the function $I_{[q,m]}$ as

$$S_m = I_{[1,m]} / I_{[0,m]}, \tag{9.33}$$

where we define the integral function:

$$I_{[q,m]} = \int_0^1 \left[P_2(x)\right]^q \exp\left(\Gamma_m \langle S \rangle P_2(x)\right) dx, \tag{9.34}$$

with $q = 0, 1$. The nematic factor Γ_m is defined as

$$\Gamma_m = \frac{5}{4} n_r m \phi, \tag{9.35}$$

for $m = 1, 2, \cdots \infty$. Substituting Eq. (9.30) into Eq. (9.25), the nematic free energy is given by

$$\begin{aligned} \beta F_{nem} &= \frac{5}{8} \langle S \rangle^2 \phi \sum_{m=1}^{\infty} n_r m N_m - \sum_{m=1}^{\infty} N_m \ln I_{[0,m]} \\ &= N_t \left(\frac{5}{8} \langle S \rangle^2 \phi^2 - \sum_{m=1}^{\infty} \frac{\phi_m}{n_r m} \ln I_{[0,m]}\right). \end{aligned} \tag{9.36}$$

The total free energy F is given by the sum of Eqs. (9.23), (9.24), and (9.36). The chemical potential of an m-mer is obtained from the free energy as

$$\begin{aligned} \beta \mu_m &= \beta (\partial F / \partial N_m)_{N_0} \\ &= \beta \mu_m^\circ + \ln \phi_m + n_r m \phi \left(1 - \frac{1}{n_r \langle m \rangle}\right) - n_r m + 1 + \chi n_r m (1 - \phi)^2 \\ &\quad + \frac{5}{8} n_r m \langle S \rangle^2 \phi^2 - \ln I_{[0,m]}, \end{aligned} \tag{9.37}$$

where the average aggregation number is given by

$$\langle m \rangle = \sum_{m=1}^{\infty} mN_m / \sum_{m=1}^{\infty} N_m = \frac{\phi}{n_r} / \sum_{m=1}^{\infty} \frac{\phi_m}{n_r m}. \tag{9.38}$$

Note that

$$\begin{aligned}
\left(\frac{\partial \ln I_{[0,m]}}{\partial N_m}\right)_{N_0} &= \frac{1}{I_{[0,m]}}\left(\frac{\partial I_{[0,m]}}{\partial N_m}\right)_{N_0} \\
&= \frac{1}{I_{[0,m]}}\left(\frac{\partial I_{[0,m]}}{\partial \Gamma_m}\right)\left(\frac{\partial \Gamma_m}{\partial N_m}\right)_{N_0} \\
&= \frac{5}{4} S_m \langle S \rangle (n_r m)^2 \phi_0 / N_t.
\end{aligned} \tag{9.39}$$

Using Eq. (9.37), the chemical potential of the monomer ($m = 1$) is given by

$$\begin{aligned}
\beta\mu_1 &= \beta\mu_1^\circ + \ln \phi_1 + n_r \phi \left(1 - \frac{1}{n_r \langle m \rangle}\right) - n_r + 1 + \chi n_r (1-\phi)^2 \\
&\quad + \frac{5}{8} n_r \langle S \rangle^2 \phi^2 - \ln I_{[0,1]}.
\end{aligned} \tag{9.40}$$

The chemical potential of the solvent molecule is given by

$$\begin{aligned}
\beta\mu_0 &= \beta(\partial F/\partial N_0)_{N_m} \\
&= \beta\mu_0^\circ + \ln(1-\phi) + \phi\left(1 - \frac{1}{n_r \langle m \rangle}\right) + \chi\phi^2 + \frac{5}{8}\langle S \rangle^2 \phi^2,
\end{aligned} \tag{9.41}$$

and the osmotic pressure Π of the solution is given by

$$\beta\Pi = \beta(\mu_0^\circ - \mu_0) = -\ln(1-\phi) - \phi\left(1 - \frac{1}{n_r \langle m \rangle}\right) - \chi\phi^2 - \frac{5}{8}\langle S \rangle^2 \phi^2. \tag{9.42}$$

When $\phi \ll 1$ we can expand as

$$\beta\Pi \simeq \frac{\phi}{n_r \langle m \rangle} + \left(\frac{1}{2} - \chi\right)\phi^2 - \frac{5}{8}\langle S \rangle^2 \phi^2 + \cdots. \tag{9.43}$$

As increasing ϕ, the osmotic pressure increases for $\chi < 1/2$ (good solvent), and discontinuously decreases at the NIT concentration due to a nonzero $\langle S \rangle$, where the osmotic pressure exhibits a jump of $-5\langle S \rangle^2/8$. Further increasing ϕ, the osmotic pressure rises again. It is informative to plot $\beta\Pi/\phi$ against ϕ ($\ll 1$): the intercept corresponds to $1/(n_r\langle m \rangle)$, and the slope represents $(1/2 - \chi)$ in the isotropic phase [85].

9.2.2 Chemical equilibrium conditions

In thermal equilibrium, each m-mer is stabilized through continuous bonding and unbonding processes. The condition of chemical equilibrium for the formation of m-mers (Fig. 9.4) is given by

$$m\mu_1 = \mu_m, (m = 0, 1, \cdots,). \tag{9.44}$$

Substituting Eqs. (9.37) and (9.40) into the equilibrium condition Eq. (9.44), we obtain the volume fraction of the m-mer as

$$\phi_m = \phi_1^m \exp\big(-m\beta f_0 + m - 1\big) D_m, \tag{9.45}$$

where

$$\beta f_0 = \beta(\mu_m^\circ/m - \mu_1^\circ), \tag{9.46}$$

represents the local free energy change associated with the formation of the m-mer from m monomers. We also define the m-function

$$D_m = \frac{I_{[0,m]}}{(I_{[0,1]})^m}, \tag{9.47}$$

which accounts for the orientational contribution to the size distribution. In an isotropic phase, i.e., $\langle S \rangle = 0$, the orientational contribution disappears, and we have $D_m = 1$. In contrast, when $\langle S \rangle \neq 0$, the m-function D_m explicitly depends on both the aggregate size m and the average order parameter $\langle S \rangle$, as defined in Eq. (9.34). The value of D_m increases with increasing m, starting from 1.

The free energy change f_0 associated with the formation of the aggregate consists of energetic and entropic contributions:

$$\beta f_0 = \beta(\Delta h - T\Delta s), \tag{9.48}$$

where Δh is the enthalpic change and Δs is the entropic change. The enthalpy Δh arises from the physical bonding between reactive molecules and scales with the number of associations:

$$\Delta h = -(m-1)\varepsilon_b, \tag{9.49}$$

where $\varepsilon_b(> 0)$ denotes the bonding energy, such as that due to hydrogen bonding. For the entropy, let W_0 denote the number of states available to a monomer molecule. The entropy of the m-mer is then expressed as $s_m = k_B \ln(mW_0)$, and the entropy change upon forming the m-mer is given by

$$\Delta s = s_m - ms_1 = k_B \ln\left(mW_0^{1-m}\right). \tag{9.50}$$

From Eqs. (9.49) and (9.50), the total free energy change for forming the m-mer is

$$m\beta f_0 = -(m-1)\beta\varepsilon_b - \ln\left(mW_0^{1-m}\right). \tag{9.51}$$

Substituting Eq. (9.51) into Eq. (9.45), the volume fraction of the m-mer is given by

$$\phi_m = \frac{m}{K_a}(\phi_1 K_a)^m D_m, \tag{9.52}$$

where the association constant for a single bond formation is defined as

$$K_a = \omega \exp(\beta\varepsilon_b), \tag{9.53}$$

with $\omega = e/W_0$, representing the entropic contribution per bond, and $s_0 = k_B \ln\omega$, corresponding to the local entropy change.

In the isotropic phase, substituting Eq. (9.52) into Eq. (9.21) with $\langle S \rangle = 0$ and $D_m = 1$, the total volume fraction of reactive molecules is expressed as

$$\phi = \frac{\phi_1}{(1-\phi_1 K_a)^2}. \tag{9.54}$$

From this, the monomer volume fraction ϕ_1 can be written explicitly as a function of the total solute volume fraction ϕ and the association constant K_a:

$$\phi_1 = \frac{1}{2K_a^2\phi}\left(1 + 2K_a\phi - \sqrt{1+4K_a\phi}\right). \tag{9.55}$$

The association constant K_a is a function of temperature. The mean aggregation number in the isotropic phase, defined in Eq. (9.38), is then given by

$$\langle m \rangle_I = \frac{\phi(1 - K_a\phi_1)}{\phi_1} = \frac{1}{2}\left(1 + \sqrt{1 + 4K_a\phi}\right). \tag{9.56}$$

Eqs. (9.54)-(9.56) provide an analytical description of the aggregation state of reactive molecules in the isotropic phase as a function of concentration and temperature.

In the nematic phase ($\langle S \rangle \neq 0$), the m-function D_m depends on both the aggregation number m and the average orientational order parameter $\langle S \rangle$. The equilibrium values of the monomer fraction ϕ_1 and the order parameter $\langle S \rangle$ are determined numerically by solving the coupled equations Eqs. (9.21) and (9.31), together with Eq. (9.52), for a given total volume fraction ϕ and temperature. Once ϕ_1 and $\langle S \rangle$ are obtained, the volume fraction of the m-mers is calculated from Eq. (9.52), and the mean aggregation number is evaluated using Eq. (9.38).

9.2.3 Mixing free energy of associating systems

Due to the Gibbs-Duhem relations for multi-component systems, the free energy (Eq. (9.22)) can be expressed as

$$\begin{aligned} F &= N_0\mu_0 + \sum_{m=1}^{\infty} N_m\mu_m \\ &= N_t\left[\phi_0\mu_0 + \sum_{m=1}^{\infty} \frac{\phi_m}{n_r m}\mu_m\right], \end{aligned} \tag{9.57}$$

and using the chemical equilibrium condition Eq. (9.44), it is given by

$$F = N_t\left[(1-\phi)\mu_0 + \frac{\phi}{n_r}\mu_1\right]. \tag{9.58}$$

Then the free energy of mixing is given by

$$\Delta F = F - F_{ref} = N_t\left[(1-\phi)\Delta\mu_0 + \frac{\phi}{n_r}\Delta\mu_1\right], \tag{9.59}$$

where $\Delta\mu_0 = \mu_0 - \mu_0^\circ$ and $\Delta\mu_1 = \mu_1 - \mu_1^\circ$.

Substituting Eqs. (9.40) and (9.41) into Eq. (9.59), the free energy can be expressed as

$$\Delta F = \Delta F_{mix} + F_{nem}. \tag{9.60}$$

The first term ΔF_{mix} shows the free energy of mixing in the isotropic associating systems and is given by

$$\beta\Delta F_{mix} = \beta\left(\Delta E_{mix} - T\Delta S_{mix}\right), \tag{9.61}$$

where the energy term is given by

$$\beta\Delta E_{mix}/N_t = \chi\phi(1-\phi), \tag{9.62}$$

and the entropy of mixing:

$$\Delta S_{mix}/(N_t k_B) = -\left[(1-\phi)\ln(1-\phi) + \frac{\phi}{n_r}\ln\phi_1 + \frac{\phi}{n_r}\left(1 - \frac{1}{\langle m \rangle}\right)\right]. \tag{9.63}$$

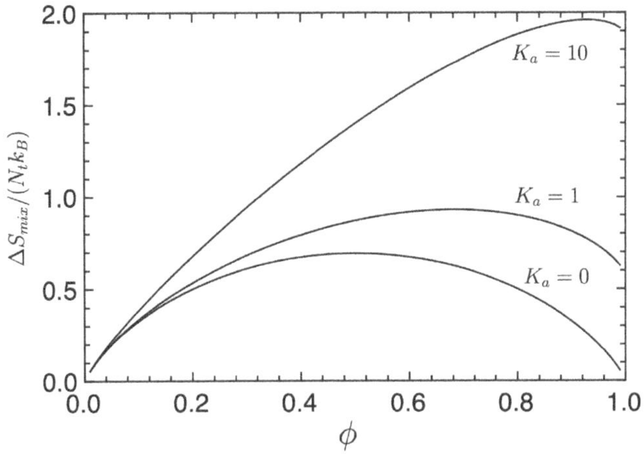

FIGURE 9.5
Entropy of mixing in the isotropic phase plotted against ϕ for various values of the association constant K_a with $n_r = 1$.

When $\langle m \rangle = 1$: $\phi = \phi_1$, the free energy of mixing results in that of the binary mixtures (Eq. (7.14)). The second term F_{nem} in Eq. (9.60) is given by Eq. (9.36).

Figure 9.5 shows the entropy of mixing (Eq. (9.63)) in the isotropic phase plotted against ϕ for various values of the association constant $K_a = 0, 1, 10$ with $n_r = 1$. The values of ϕ_1 and $\langle m \rangle$ are given by Eqs. (9.55) and (9.56), respectively. For the non-associating system $K_a = 0$, the mixing entropy is symmetric with respect to $\phi = 0.5$, which corresponds to the binary mixtures of low molecular weight molecules. As the association constant increases, the mixing entropy term becomes increasingly asymmetric and develops a maximum at higher concentrations. Furthermore, as molecular association becomes more pronounced, for example, $K_a = 1$ and $K_a = 10$, the entropy increases.

In the case of polymer solutions discussed in Eq. (7.14), the mixing entropy decreases with increasing molecular weight n_p. In contrast, in associating systems, molecular association increases entropy. This enhancement originates from the second term in Eq. (9.24), which accounts for the entropy associated with the polydispersity of the molecular aggregates.

The conditions for phase separation between the isotropic and nematic phases on the temperature-concentration plane are given by $\mu_0(\phi_I) = \mu_0(\phi_N)$ and $\mu_m(\phi_I) = \mu_m(\phi_N)$, where ϕ_I and ϕ_N denote the volume fractions of rods in the lower- and higher-concentration phases, respectively. Using Eq. (9.44), the coexistence (binodal) curves at thermal equilibrium are obtained from the coupled equations $\mu_0(\phi_I) = \mu_0(\phi_N)$ and $\mu_1(\phi_I) = \mu_1(\phi_N)$.

9.2.4 Nematic ordering of rod-like aggregates

We introduce the dimensionless reduced temperature

$$\tau = k_B T / U_a, \tag{9.64}$$

where U_a is a characteristic energy scale of the system. The bonding energy parameter is defined as

$$\gamma = \varepsilon_b / U_a. \tag{9.65}$$

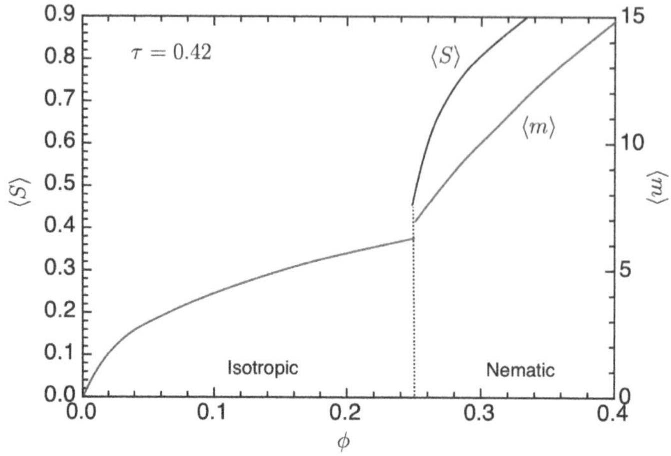

FIGURE 9.6
Mean value $\langle S \rangle$ of the order parameter and the mean aggregation number $\langle m \rangle$ plotted against the volume fraction ϕ at the temperature $\tau = 0.42$.

The association constant (Eq. (9.53)) can then be expressed as a function of temperature τ:

$$K_a = \omega \exp\left(\frac{\gamma}{\tau}\right). \tag{9.66}$$

For illustrative calculations, we adopt the parameter values $\chi = 0$ for a good solvent, $\omega = 0.01$, $\gamma = 4.0$, and $n_r = 1$.

Figure 9.6 shows the mean orientational order parameter $\langle S \rangle$ and the mean aggregation number $\langle m \rangle$ as a function of the solute volume fraction ϕ at $\tau = 0.42$. As ϕ increases, the system undergoes a first-order NIT, characterized by a discontinuous jump in $\langle S \rangle$. The mean aggregation number $\langle m \rangle$ also increases with ϕ and exhibits a discontinuous jump at the NIT concentration. In the isotropic phase, $\langle m \rangle$ follows Eq. (9.56) and is proportional to $\phi^{1/2}$. Upon entering the nematic phase, orientational ordering promotes the formation of longer rod-like aggregates, as evidenced by the sharp increase in $\langle m \rangle$.

Figure 9.7(a) shows the volume fraction ϕ_m as a function of aggregation number m at $\tau = 0.42$. In the isotropic phase ($\phi = 0.2$), ϕ_m exhibits a peak around $m \approx 5$, and large aggregates are absent ($\phi_m \sim 0$ for $m > 30$). In contrast, in the nematic phase ($\phi = 0.3$), the peak of ϕ_m shifts slightly to $m \simeq 3$, and the distribution becomes broader, indicating the emergence of larger aggregates. Figure 9.7(b) presents the orientational order parameter S_m (Eq. (9.33)) as a function of m for $\phi = 0.3$. As shown in Fig.9.6, the average order parameter is $\langle S \rangle \simeq 0.8$ at $\tau = 0.42$ and $\phi = 0.3$. The value of S_m increases monotonically with m, indicating that larger rods are more orientationally ordered than smaller ones. For small m, S_m increases approximately linearly with m, whereas for longer rods, S_m asymptotically approaches unity as m becomes large [139].

For small $\langle S \rangle$, the m-function D_m in Eq. (9.47) can be expanded as

$$D_m \simeq 1 + \frac{\Gamma_1^2}{10} m^2 \langle S \rangle^2 + \cdots . \tag{9.67}$$

Using Eqs. (9.52) and (9.67), we obtain

$$\sum_{m=1}^{\infty} \frac{\phi_m}{m} = M_0(k) + \frac{\Gamma_1^2}{10} \langle S \rangle^2 M_2(k), \tag{9.68}$$

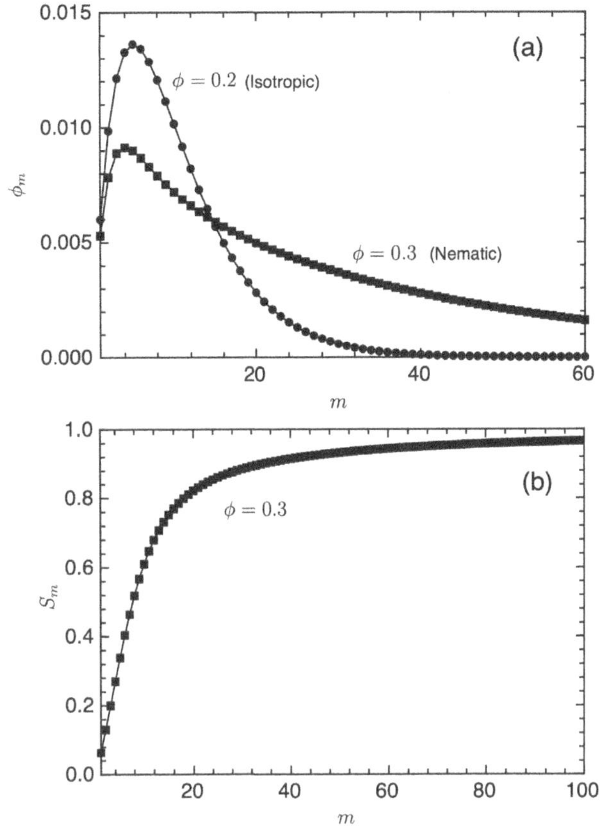

FIGURE 9.7
Volume fraction ϕ_m(a) and orientational order parameter S_m(b) plotted against the aggregation number m at the temperature $\tau = 0.42$.

where the moment functions are defined by

$$M_p(k) = \sum_{m=1}^{\infty} m^p k^m, \tag{9.69}$$

with $k = K_a\phi_1$ and $p = 0, 1, 2, 3, \cdots$. These functions satisfy the recurrence relation

$$M_{p+1}(k) = k\frac{dM_p(k)}{dk}, \quad |k| < 1. \tag{9.70}$$

Then, the mean aggregation number (Eq. (9.38)) in the nematic phase near the NIT can be approximated as

$$\langle m \rangle_N \simeq \frac{K_a\phi}{M_0(k) + (\Gamma_1^2/10)M_2(k)\langle S\rangle^2}, \tag{9.71}$$

where $M_0(k) = k/(1-k)$ and $M_2(k) = k(1+k)/(1-k)^3$. When $\langle S\rangle = 0$, Eq. (9.71) results in the mean aggregation number in the isotropic phase (Eq. (9.56)). The values of $M_0(k)$ and $M_2(k)$ decrease with decreasing k and then the mean aggregation number $\langle m\rangle_N$ increases, resulting in a discontinuous jump at the NIT concentration, as shown in Fig. 9.6.

Similarly, the orientational order parameter S_m (Eq. (9.33)) of the m-mer can be expanded as

$$S_m \simeq \frac{\Gamma_1}{10}\langle S\rangle m, \tag{9.72}$$

for a small value of m. At large m, it can be approximated by

$$S_m \simeq 1 - \frac{1}{(\Gamma_1/2)\langle S\rangle m}, \tag{9.73}$$

indicating that S_m asymptotically approaches unity as m increases. These expressions indicate that longer aggregates are more strongly aligned with the nematic director, consistent with the numerical results shown in Fig. 9.7. Thus, both the mean aggregation number and the orientational order parameter increase sharply at the NIT, highlighting the feedback between aggregation and orientational ordering in solutions of rod-like aggregates.

9.2.5 Landau expansion of the nematic free energy of associating systems

We now derive the Landau-de Gennes expansion of the nematic free energy to analyze isotropic-nematic (I+N) phase separation on the temperature-concentration plane. Expanding the exponent of the function $I_{[0,m]}$ (Eq. (9.34)) in powers of the orientational order parameter $\langle S\rangle$ and performing the angular integration, the nematic free energy (Eq. (9.36)) can be expressed in the Landau form: (see Eq. (4.69))

$$f_{nem} = \beta F_{nem}/N_t = \Gamma_0\left(C_1\langle S\rangle^2 - C_2\langle S\rangle^3 + C_3\langle S\rangle^4\right), \tag{9.74}$$

where the coefficients depend explicitly on the size distribution as

$$C_1 = \frac{1}{2}\left(\phi - \frac{\Gamma_0}{5}\sum_{m=1}^{\infty} m\phi_m\right), \tag{9.75}$$

$$C_2 = \frac{\Gamma_0^2}{105}\sum_{m=1}^{\infty} m^2\phi_m, \tag{9.76}$$

and

$$C_3 = \frac{\Gamma_0^3}{700}\sum_{m=1}^{\infty} m^3\phi_m. \tag{9.77}$$

This expansion resembles the conventional Landau theory for first-order phase transitions, but here the coefficients C_1, C_2, and C_3 are determined self-consistently through the aggregate distribution $\{\phi_m\}$. In particular, C_1 changes sign as the concentration or temperature varies, thereby controlling the stability of the isotropic phase. The cubic term with coefficient C_2 ensures that the NIT is first order, while $C_3 > 0$ provides stabilization at large $\langle S\rangle$. It should be emphasized that this Landau free energy is valid only in the vicinity of the NIT ($C_1 > 0$), where the expansion in $\langle S\rangle$ is justified.

To analyze the NIT more explicitly, we assume that near the NIT, the size distribution ϕ_m can be approximated by the isotropic form. Thus, we set $D_m = 1$ in Eq. (9.52), yielding

$$\phi_m = \frac{m}{K_a}k^m, \tag{9.78}$$

with $k = K_a\phi_1$. The monomer fraction ϕ_1 is then determined by Eq. (9.55). With this simplification, the summations in the coefficients of the Landau expansion (Eqs. (9.75)–(9.77)) can be evaluated analytically. We obtain

$$C_1 = \frac{1}{2}\left(\phi - \frac{\Gamma_0}{5}\frac{k(1+k)}{K_a(1-k)^3}\right), \tag{9.79}$$

$$C_2 = \frac{\Gamma_0^2}{105}\frac{k(k^2+4k+1)}{K_a(1-k)^4}, \tag{9.80}$$

and

$$C_3 = \frac{\Gamma_0^3}{700}\frac{k(k^3+11k^2+11k+1)}{K_a(1-k)^5}. \tag{9.81}$$

The equilibrium value of the order parameter $\langle S\rangle$ is given by minimizing the nematic free energy $(\partial f_{nem}/\partial\langle S\rangle)_T = 0$:

$$\langle S\rangle = \frac{1}{8C_3}\left[3C_2 + \sqrt{9C_2^2 - 32C_1C_3}\right]. \tag{9.82}$$

At the NIT, we have $f_{\text{nem}}(\langle S\rangle = 0) = f_{\text{nem}}(\langle S\rangle \neq 0) = 0$, as shown in Fig. 4.4, and thus the first-order NIT occurs at $4C_1C_3 = C_2$. The order parameter $\langle S\rangle_{\text{NI}}$ at the NIT is given by $\langle S\rangle_{\text{NI}} = C_2/(2C_3)$.

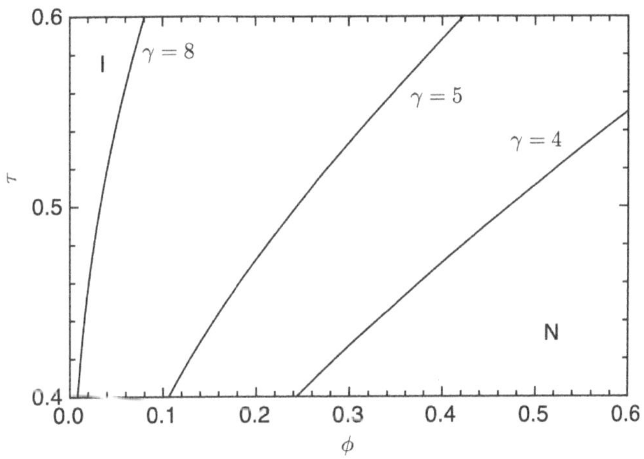

FIGURE 9.8
Nematic-isotropic phase transition (NIT) curves, calculated from $4C_1C_3 = C_2$, on the temperature-concentration plane for $\gamma = 4, 5, 8$.

Figure 9.8 shows the NIT curves, defined by $4C_1C_3 = C_2$, on the temperature-concentration plane for various values of the bonding energy parameter $\gamma(= 4, 5, 8)$. The right-hand side (left-hand side) of each curve corresponds to the nematic (isotropic) phase. The NIT curves depend on temperature because the mean aggregation number varies with temperature, as shown in Fig. 9.6. As the temperature decreases, the rod-like aggregates grow longer, and the NIT curves display a pronounced temperature dependence. With increasing bonding energy parameter γ, the NIT curves shift toward lower concentrations. At the NIT point, the order parameter takes the value $\langle S\rangle_{\text{NI}} \simeq 0.38$. This result indicates that stronger intermolecular bonding stabilizes the nematic phase, allowing its formation at lower concentrations. The chimney-type (I+N) phase separation appears on both sides of the NIT curve, as shown in Fig. 7.4. These phase transitions are qualitatively consistent with experimental observations in chromonic liquid crystals (Fig. 1.11) and nematic micellar systems [147].

9.3 Hydrogen bonds between a rod-like polymer and solvent molecules

Water-dispersible liquid crystalline polymers have attracted considerable attention as smart materials owing to their potential applications in next-generation display technologies, sensors, actuators, food science, and biomaterials [137, 148]. Representative examples include self-assembled liquid crystals [136], DNA solutions [10], chromonic liquid crystals [24, 149], graphene oxide [150, 151], and surfactant-coated carbon nanotubes [152, 153], among others. A common feature of these systems is the presence of physical bonding interactions—particularly hydrogen bonding—between rod-like polymers and solvent molecules, which play a crucial role in stabilizing the dispersion. In this section, we develop a mean-field theoretical framework to describe aqueous liquid crystalline systems, incorporating the effects of hydrogen bonding [154]. The role of hydrogen bonding between polymer and solvent molecules has been demonstrated both theoretically and experimentally in aqueous polymer solutions [141,155–157], which exhibit LCST-type and closed-loop phase diagrams.

9.3.1 Free energy of rod-like polymer solutions with hydrogen bonding

We consider a binary mixture of long rod-like polymers (rods) and solvent (water) molecules. Each rod-like molecule, with length L and diameter D, carries f identical functional groups capable of forming physical (hydrogen) bonds with solvent molecules. The bonding energy, taken here as the hydrogen bonding energy, is assumed to be on the order of the thermal energy, such that the bond formation is thermoreversible. In thermal equilibrium, each rod is associated with a certain number of solvent molecules, a process referred to as solvation. If m out of the f functional groups on a rod are bonded to solvent molecules, we refer to the rod as being in an m-mer state. Figure 9.9 schematically illustrates the dispersion of m-mers together with the free (unbonded) solvent molecules. The case of $m = 0$ corresponds to a pure rod-like polymer without any bound solvent molecules. We neglect physical bonding between solvent molecules here.

Let N_{m+1} denote the number of m-mers ($m = 0, 1, 2, \ldots, f$), and let N_0 represent the number of free, unbonded solvent molecules. The volume of a rod-like polymer is given by $v_r = (\pi/4)D^2L = v_0 n_r$, where $n_r = L/D$ is the axial ratio of the rod and $v_0 = (\pi/4)D^3$ denotes the reference volume. We assume that the volume of a solvent molecule is $v_s = v_0$. The total volume of the system is $V = v_0 N_t$, where

$$N_t = N_0 + \sum_{m=0}^{f} (n_r + m) N_{m+1}, \tag{9.83}$$

is the total number of lattice cells in the system. The volume fraction of the m-mers is given by

$$\phi_{m+1} = \frac{v_0(n_r + m)N_{m+1}}{V} = \frac{(n_r + m)N_{m+1}}{N_t}, \tag{9.84}$$

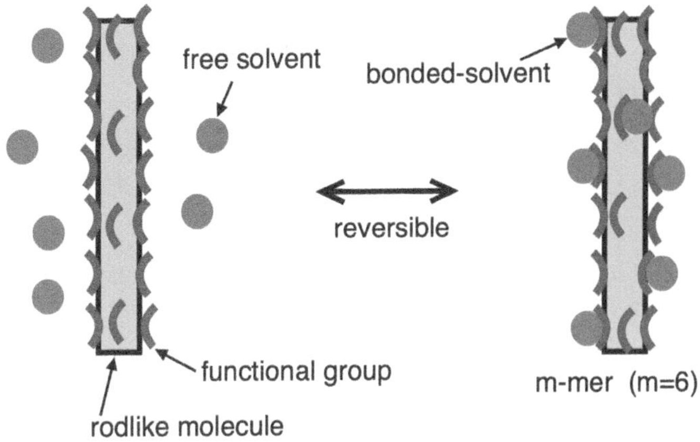

FIGURE 9.9
Schematic representation of m-mers formed via hydrogen bonding with solvent molecules. Each rod possesses f identical functional groups capable of forming hydrogen bonds with solvent molecules (closed circles), and m denotes the number of bonded solvent molecules per rod-like molecule.

while the volume fraction of free solvent molecules is

$$\phi_0 = \frac{v_0 N_0}{V} = \frac{N_0}{N_t}. \tag{9.85}$$

The total volume fraction of the m-mers satisfies

$$\sum_{m=0}^{f} \phi_{m+1} = 1 - \phi_0, \tag{9.86}$$

and the volume fraction of the rod-like polymer is given by

$$\phi = \sum_{m=0}^{f} \frac{n_r}{n_r + m} \phi_{m+1} = n_r \sum_{m=0}^{f} \frac{N_{m+1}}{N_t}. \tag{9.87}$$

The free energy of the system can be given by three contributions:

$$F = F_{ref} + \Delta F_{mix} + F_{nem}, \tag{9.88}$$

where F_{ref} denotes the free energy of the reference state of the m-mers and solvent molecules, ΔF_{mix} accounts for the free energy of mixing, and F_{nem} represents the free energy associated with orientational ordering of the rods. The reference free energy is written as

$$\begin{aligned} F_{ref} &= N_0 \mu_0^\circ + \sum_{m=0}^{f} N_{m+1} \mu_{m+1}^\circ \\ &= N_t \left[\phi_0 \mu_0^\circ + \sum_{m=0}^{f} \frac{\phi_{m+1}}{n_r + m} \mu_{m+1}^\circ \right], \end{aligned} \tag{9.89}$$

where μ_0° and μ_{m+1}° denote the standard chemical potentials of the pure solvent and the m-mers, respectively. The free energy of mixing is given, according to the Flory-Huggins theory for polydisperse polymer solutions [85], by

$$\begin{aligned}\beta\Delta F_{mix} &= N_t\left[\phi_0\ln\phi_0 + \sum_{m=0}^{f}\frac{\phi_{m+1}}{n_r+m}\ln\phi_{m+1} + \chi\phi(1-\phi)\right] \\ &= N_0\ln\phi_0 + \sum_{m=0}^{f} N_{m+1}\ln\phi_{m+1} + \chi(1-\phi)n_r\sum_{m=0}^{f} N_{m+1},\end{aligned} \tag{9.90}$$

where $\beta = 1/(k_BT)$, with k_B the Boltzmann constant and T the absolute temperature. The first two terms in Eq. (9.90) represent the entropy of mixing of the m-mers with the free solvent molecules. The last term contains the Flory-Huggins interaction parameter $\chi = U_0/(k_BT)$, which describes the nonspecific (dispersion-type) interaction between solute and solvent molecules, where U_0 is the corresponding interaction energy.

The last contribution in Eq. (9.88) corresponds to the free energy associated with nematic ordering. Let $f_d(\theta)$ denote the orientational distribution function of the rod-like polymers, subject to the normalization condition

$$\int f_d(\theta)d\mathbf{\Omega} = 1, \tag{9.91}$$

where $d\mathbf{\Omega} = 2\pi\sin\theta\, d\theta$ is the solid angle. Taking into account both the orientation-dependent attractive interactions of the Maier-Saupe type and the excluded volume interactions of the Onsager type (see Eq. (7.17)), the hybrid-free energy of polydisperse systems due to nematic ordering is given by

$$\begin{aligned}\beta F_{nem} &= N_t\Bigg[\sum_{m=0}^{f}\frac{\phi_{m+1}}{n_r+m}\int f_d(\theta)\ln(4\pi f_d(\theta))\, d\mathbf{\Omega} \\ &\quad -\frac{1}{2}\left(\nu_M+\frac{5}{4}\right)\sum_{m=0}^{f}\sum_{l=0}^{f}\phi_{m+1}\phi_{l+1}S^2\Bigg].\end{aligned} \tag{9.92}$$

The first term in Eq. (9.92) represents the entropy loss associated with the reduction of orientational degrees of freedom. The second term corresponds to the orientation-dependent mean field interaction: $\nu_M = U_M/(k_BT)$ denotes the Maier-Saupe parameter describing attractive interactions, and the constant 5/4 arises from the Onsager excluded volume contribution (Eq. (4.29)). The scalar nematic order parameter is defined by

$$S = \int P_2(\cos\theta)\, f_d(\theta)d\mathbf{\Omega}, \tag{9.93}$$

where $P_2(\cos\theta)$ is the second Legendre polynomial. In the present formulation, we assume that the orientational order parameter of an m-mer is independent of the number m of bonded solvent molecules, since the solvent molecules are small and nematic ordering is mainly governed by the axial ratio of the rods and their attractive interactions. If, however, the bonded moieties are large (e.g., surfactant molecules), it becomes necessary to introduce an m-dependent orientational order parameter S_{m+1} and the corresponding distribution function $f_{m+1}(\theta)$ for each m-mer.

Using Eqs. (9.86) and (9.87), the nematic free energy (Eq. (9.92)) can be rewritten as

$$\beta F_{nem} = N_t\left[\frac{\phi}{n_r}\int f_d(\theta)\ln(4\pi f_d(\theta))\, d\Omega - \frac{1}{2}\left(\nu_M+\frac{5}{4}\right)(1-\phi_0)^2S^2\right]. \tag{9.94}$$

In the absence of functional groups ($f = 0$), all rods are free from solvent bonding, and we have $1 - \phi_0 = \phi$. Equation (9.94) then results in the nematic free energy for a simple rod/solvent mixture (see Section 7.1.1), demonstrating that the present formulation extends the conventional theory by incorporating hydrogen bonding.

9.3.2 Chemical equilibrium conditions for a rod-solvent complex

The orientational distribution function $f_d(\theta)$ of the rod-like molecules is determined by the free energy (Eq. (9.92)) with respect to this function: $(\delta F_{\rm nem}/\delta f_d(\theta))_T = 0$, under the normalization condition Eq. (9.91). We then obtain

$$f_d(\theta) = \frac{1}{Z_0} \exp\left[n_r(\nu_M + \frac{5}{4}) g(\phi) S P_2(\cos\theta) \right], \tag{9.95}$$

where

$$g(\phi) = \frac{(1-\phi_0)^2}{\phi}. \tag{9.96}$$

The constants Z_0 in Eq. (9.95) is determined by the normalization condition (Eq. (9.91)) as $Z_0 = 4\pi I_0(S)$. The function $I_0(S)$ is defined as

$$I_q(S) = \int_0^1 [P_2(x)]^q \exp\left[n_r(\nu_M + \frac{5}{4}) g(\phi) S P_2(x) \right] dx, \tag{9.97}$$

where $q = 0, 1, 2 \cdots$. Substituting Eq. (9.95) into Eq. (9.93), the orientational order parameters S can be determined by the self-consistent equation

$$S = \frac{I_1(S)}{I_0(S)}. \tag{9.98}$$

Using the distribution function (Eq. (9.95)), the nematic free energy (Eq. (9.94)) is given by

$$\beta F_{nem} = N_t \left[\frac{1}{2}(\nu_M + \frac{5}{4})(1-\phi_0)^2 S^2 - \frac{\phi}{n_r} \ln I_0(S) \right]. \tag{9.99}$$

The total free energy (Eq. (9.88)) is given by the sum of Eqs. (9.89), (9.90), and (9.99).

The chemical potential of the free solvent molecule is given by

$$\begin{aligned} \beta\mu_0 &= \beta(\partial F/\partial N_0)_{N_{m+1}} \\ &= \beta\mu_0^\circ + \ln\phi_0 + 1 - (\phi_0 + \frac{\phi}{n_r}) + \chi\phi^2 \\ &\quad + \frac{1}{2}(\nu_M + \frac{5}{4}) S^2 \phi^2 g', \end{aligned} \tag{9.100}$$

and that of the m-mer is given by

$$\begin{aligned} \beta\mu_{m+1} &= \beta(\partial F/\partial N_{m+1})_{N_0} \\ &= \beta\mu_{m+1}^\circ + \ln\phi_{m+1} + 1 - (n_r + m)(\phi_0 + \frac{\phi}{n_r}) \\ &\quad + n_r\chi\left[(1-\phi)^2 + \frac{m}{n_r}\phi^2 \right] - \ln I_0(S) \\ &\quad + \frac{1}{2}(\nu_M + \frac{5}{4}) S^2 \left[n_r(g - \phi g') + (n_r + m)\phi^2 g' \right], \end{aligned} \tag{9.101}$$

for $m = 0, 1, 2, \ldots, f$, where we define $g' = dg/d\phi$.

In thermal equilibrium, each m-mer is in chemical equilibrium through bonding and unbonding processes. The chemical equilibrium conditions governing m-mer formation are given by

$$\mu_{m+1} = \mu_1 + m\mu_0, \quad (m = 0, 1, \ldots, f). \tag{9.102}$$

Substituting Eqs. (9.100) and (9.101) into Eq. (9.102), the volume fraction of the m-mer is given by

$$\phi_{m+1} = K_m \phi_1 \phi_0^m, \tag{9.103}$$

where

$$K_m = \exp(m - \Delta_m), \tag{9.104}$$

is the association constant for the m-mer formation and

$$\Delta_m = \beta(\mu_{m+1}^\circ - \mu_1^\circ - m\mu_0^\circ), \tag{9.105}$$

is the local free energy change due to the m-mer formation.

9.3.3 Derivation of the association constant

We derive here the free energy change Δ_m for the formation of the m-mer. The free energy change consists of the energy Δh and the entropy Δs terms:

$$\Delta_m = \beta(\Delta h - T\Delta s). \tag{9.106}$$

The energy change Δh due to the physical bonding between a functional group on the rod and a solvent molecule is proportional to the association number m, and is given by

$$\Delta h = -m\epsilon_b, \tag{9.107}$$

where $\epsilon_b (> 0)$ is the energy required for bond formation, such as hydrogen bonding energy. The entropy change consists of two terms: $\Delta s = \Delta s_{com} + \Delta s_h$. The first term Δs_{com} is the combinatorial entropy $\Delta s_{com} = k_B \ln W_m$, where W_m is the number of ways to select m functional groups out of f functional groups on a rod and is given by the binomial coefficient:

$$\Delta s_{com} = k_B \ln \left[\frac{f!}{m!(f-m)!}\right]. \tag{9.108}$$

The second term, Δs_h, is the entropy change arising from the hydrophobic interaction between the m-mer and surrounding water molecules. We here assume that the hydrophobic effect is proportional to the segment number $n_r + m$ of the m-mer: $s_h(n_r + m) = k_B \ln[\omega_m(n_r + m)]$, where ω_m is a constant. We then obtain $\Delta s_h = s_h(n_r + m) - s_h(n_r)$:

$$\Delta s_h = k_B \ln \left[\frac{w(n_r + m)}{n_r}\right], \tag{9.109}$$

where $w = \omega_m/\omega_0$ is a constant related to the hydrophobic effect due to m-mer formation, and $s_0 = k_B \ln w$ is the local entropy change associated with water-hydrocarbon interactions. We note that the free energy change (entropic component) at $T = 298$ K associated with transferring one molecule of methane from pure methane into pure water is $-k_B T \ln w = 2.5 \times 10^{-20}$ J [135]. From this, we can estimate $s_0/k_B = \ln w \simeq -6.0$, or equivalently, $w = 2.5 \times 10^{-3}$ as a typical example.

Substituting Eqs. (9.107)–(9.109) into Eq. (9.104), we obtain the association constant:

$$K_m = \frac{n_r + m}{n_r}\left[\frac{f!}{m!(f-m)!}\right] K_a^m, \tag{9.110}$$

where we define the association constant

$$K_a = w \exp(\beta \epsilon_b), \tag{9.111}$$

for a single bond formation, and the volume fraction (Eq. (9.103)) of the m-mer is given by

$$\phi_{m+1} = \frac{n_r + m}{n_r} \left[\frac{f!}{m!(f-m)!} \right] \phi_1 (K_a \phi_0)^m. \tag{9.112}$$

We then obtain

$$\sum_{m=0}^{f} \phi_{m+1} = \phi_1 (1 + K_a \phi_0)^f \left(1 + \frac{P K_a \phi_0}{1 + K_a \phi_0} \right), \tag{9.113}$$

where we define the fraction $P = f/n_r$ of the functional groups on a rod and Equation (9.86) yields

$$\phi_1 = \frac{1 - \phi_0}{(1 + K_a \phi_0)^f \left(1 + \frac{P K_a \phi_0}{1 + K_a \phi_0} \right)}. \tag{9.114}$$

The total volume fraction (Eq. (9.87)) of the rod is given by

$$\phi = \sum_{m=0}^{f} \frac{n_r}{n_r + m} \phi_{m+1} = \phi_1 (1 + K_a \phi_0)^f. \tag{9.115}$$

By combining Eqs. (9.114) and (9.115), we obtain the quadratic equation for ϕ_0:

$$K_a \phi_0^2 - \left[K_a - 1 - (1 + P) K_a \phi \right] \phi_0 - (1 - \phi) = 0, \tag{9.116}$$

and the solution is given by

$$\phi_0 = \left[K_a - 1 - (1 + P) K_a \phi + \sqrt{D(\phi)} \right] / 2K_a, \tag{9.117}$$

with

$$D(\phi) = \left[K_a - 1 - (1 + P) K_a \phi \right]^2 + 4 K_a (1 - \phi). \tag{9.118}$$

The volume fraction ϕ_0 of the free solvent molecules and the function $g(\phi)$ (Eq. (9.96)) are given as a function of temperature and concentration ϕ. We also obtain

$$g' = \left(\frac{dg(\phi)}{d\phi} \right) = - \left(\frac{1 - \phi_0}{\phi^2} \right) (1 - \phi_0 + 2 \phi \phi_0'), \tag{9.119}$$

in Eqs. (9.100) and (9.101), where

$$\phi_0' = \left(\frac{d\phi_0}{d\phi} \right) = - \frac{(1 + (1 + P) K_a \phi_0)}{\sqrt{D(\phi)}}. \tag{9.120}$$

9.3.4 Phase separation and nematic ordering

Due to the Gibbs-Duhem relations for multi-component systems, the free energy (Eq. (9.88)) can be expressed as

$$\begin{aligned} F &= \mu_0 N_0 + \sum_{m=0}^{f} \mu_{m+1} N_{m+1} \\ &= N_t \left[\mu_0 \phi_0 + \sum_{m=0}^{f} \frac{\mu_{m+1}}{n_r + m} \phi_{m+1} \right]. \end{aligned} \tag{9.121}$$

Using the chemical equilibrium condition (Eq. (9.102)) and Eq. (9.87), the free energy (Eq. (9.121)) can be written by

$$F = N_t\left[\mu_0(1-\phi) + \frac{\mu_1}{n_r}\phi\right], \tag{9.122}$$

and results in the free energy of binary mixtures. Then, the free energy for mixing is given by

$$\begin{aligned} \Delta F &= (F - F_{ref}) \\ &= N_t\left[\Delta\mu_0(1-\phi) + \frac{\Delta\mu_1}{n_r}\phi\right], \end{aligned} \tag{9.123}$$

where $\Delta\mu_0 = \mu_0 - \mu_0^\circ$ and $\Delta\mu_1 = \mu_1 - \mu_1^\circ$. Substituting Eqs. (9.100) and (9.101) into Eq. (9.123), the dimensionless free energy per unit lattice is given by

$$\begin{aligned} f_{LC} &= \beta\Delta F/N_t \\ &= \frac{\phi}{n_r}\ln\phi_1 + \phi_0\ln\phi_0 + (1+\ln\phi_0)(1-\phi_0-\phi) + \chi\phi(1-\phi) \\ &\quad + \frac{1}{2}(\nu_M + \frac{5}{4})S^2(1-\phi_0)^2 - \frac{\phi}{n_r}\ln I_0(S). \end{aligned} \tag{9.124}$$

The last two terms result in the free energy (Eq. (9.99)) of the nematic phase. Using the free energy f_{LC}, the chemical potentials can be expressed as

$$\beta\Delta\mu_0 = \beta\left(\frac{\partial\Delta F}{\partial N_0}\right)_{N_{m+1}} = f_{LC} - \phi\left(\frac{\partial f_{LC}}{\partial\phi}\right), \tag{9.125}$$

$$\beta\Delta\mu_{m+1} = \beta\left(\frac{\partial\Delta F}{\partial N_{m+1}}\right)_{N_0} = (n_r+m)\left[f_{LC} + (\frac{n_r}{n_r+m} - \phi)\left(\frac{\partial f_{LC}}{\partial\phi}\right)\right], \tag{9.126}$$

and result in Eqs. (9.100) and (9.101).

The conditions for phase separation on the temperature-concentration (T-ϕ) plane are given by

$$\mu_0(\phi') = \mu_0(\phi''), \tag{9.127}$$

$$\mu_{m+1}(\phi') = \mu_{m+1}(\phi''), \tag{9.128}$$

where ϕ' and ϕ'' are the volume fractions of the rods in the lower and higher concentration phases, respectively. Using Eq. (9.102), the coexistence (binodal) curves of phase equilibrium are determined by the coupled equations

$$\mu_0(\phi') = \mu_0(\phi''), \tag{9.129}$$

$$\mu_1(\phi') = \mu_1(\phi''). \tag{9.130}$$

The coexistence curves can also be derived by the common tangent method of the free energy (Eq. (9.124)). The spinodal curve, which separates stable and unstable regions, is given by $(\partial^2 f_{LC}/\partial\phi^2)_T = 0$, or equivalently by $(\partial\mu_0/\partial\phi)_T = 0$. This leads to

$$\frac{1}{n_r\phi} - \frac{1-\phi_0}{\phi_0\phi}\left(\frac{d\phi_0}{d\phi}\right) - 2\chi + \frac{1}{2}(\nu_M + \frac{5}{4})S^2(2\phi g' + \phi^2 g'') = 0, \tag{9.131}$$

where we define $g'' = (d^2g/d\phi^2)$ and $\phi_0'' = (d^2\phi_0/d\phi^2)$.

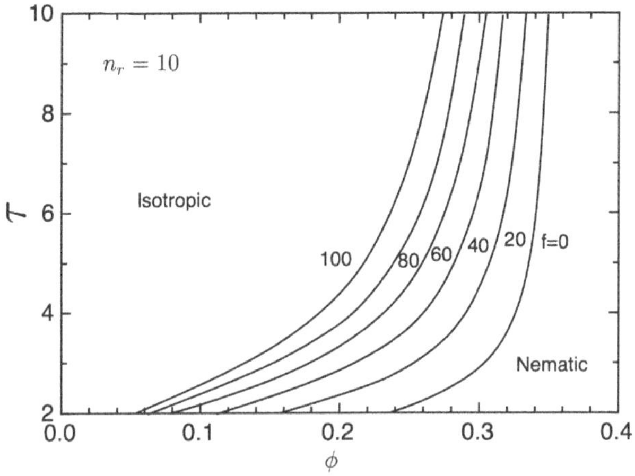

FIGURE 9.10
NIT curve on the $\tau - \phi$ plane for various values of f with $n_r = 10$. Reprinted with permission from Ref. [154].

The average number of solvent molecules bonded to the rod-like molecule is given by

$$\begin{aligned} \langle m \rangle &= \sum_{m=0}^{f} m N_{m+1} / \sum_{m=0}^{f} N_{m+1} \\ &= \sum_{m=0}^{f} \frac{m}{n_r + m} \phi_{m+1} / \sum_{m=0}^{f} \frac{1}{n_r + m} \phi_{m+1} \\ &= \frac{f K_a \phi_0}{1 + K_a \phi_0}. \end{aligned} \tag{9.132}$$

We here use the reduced-temperature $\tau = 1/\nu_M (= k_B T / U_M)$ and define the dimensionless numerical parameters: the nematic interaction parameter $\alpha = \nu_M / \chi (= 3)$ [84] and the bonding energy parameter $\gamma = \beta \epsilon_b / \nu_M$. Then the association constant is given by

$$K_a = w \exp(\gamma / \tau), \tag{9.133}$$

as a function of the temperature τ.

From Eq. (9.95), the first-order nematic-isotropic phase transition (NIT) occurs at $n_r(\nu_M + 5/4) g(\phi) = 4.55$, where the orientational order parameter S jumps. The NIT temperature is given by

$$\tau_{NI} = \frac{n_r g(\phi)}{4.55 - 1.25 n_r g(\phi)}. \tag{9.134}$$

Note that $g(\phi)$ is given by Eq. (9.96) as a function of ϕ and temperature. When $n_r (= L/D) > 3.63$ and $f = 0$, the NIT temperature diverges at $\phi_{NI}^* = 4.55/(1.25 n_r)$ at high temperatures due to the excluded volume interactions between rods (see Eq. (7.66)).

Figure 9.10 shows the NIT curves on the $\tau - \phi$ plane for $w = 0.01$ and $\gamma = 6$. The number of functional groups f is varied from 0 to 100, with the rod axial ratio fixed at $n_r = 10$. The region below (above) each NIT curve corresponds to the nematic (N) (isotropic (I)) phase. For $f = 0$, the NIT temperature increases with increasing rod concentration ϕ and diverges at $\phi_{NI}^* = 0.36$. As the number of functional groups f increases, the nematic phase region

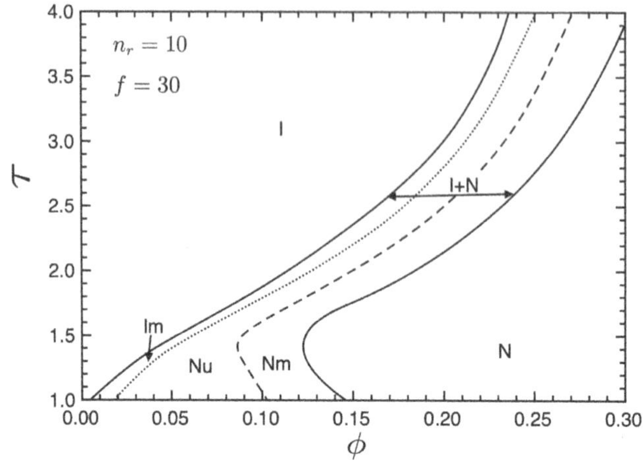

FIGURE 9.11
Phase diagram for $n_r = 10$ and $f = 30$. The solid (dashed) curves show the binodal (spinodal) curves, and the dotted curve corresponds to the NIT. Reprinted with permission from Ref. [154].

shifts toward lower concentrations, since solvent molecules bound to the rods contribute to the stabilization of the nematic phase. In other words, the formation of rod-solvent complexes enhances nematic ordering.

Figure 9.11 shows the phase diagram for $n_r = 10$ and $f = 30$. The solid and dashed curves represent the binodal and spinodal lines, respectively, while the dotted curve corresponds to the NIT. The region between the NIT and the spinodal line shows the unstable nematic region (Nu), where $(\partial^2 f_{LC}/\partial\phi^2)_T < 0$. Between the two binodal lines, we identify the isotropic metastable region (Im) and the nematic metastable region (Nm), where

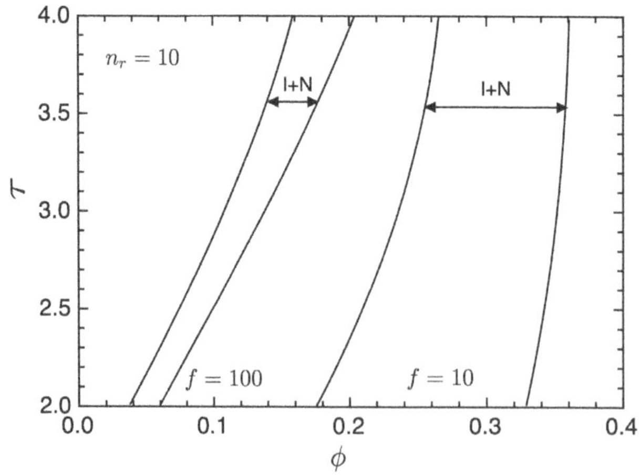

FIGURE 9.12
Chimney-type I+N phase separation on the $\tau - \phi$ plane for various values of the number f of functional groups with $n_r = 10$. Reprinted with permission from Ref. [154].

$(\partial^2 f_{LC}/\partial\phi^2)_T > 0$ [44]. The tie lines indicate phase separation (I + N) between the isotropic (I) and nematic (N) phases, and the coexistence curves exhibit a clear temperature dependence. As the temperature decreases, the N phase shifts to lower concentrations, consistent with observations in aqueous liquid crystalline polymer systems [10,158,159]. Upon further cooling, the binodal line of the N phase moves to higher concentrations. Near $\phi \simeq 0.13$, as the temperature decreases, the system undergoes the sequence $\text{I} \rightarrow \text{I} + \text{N} \rightarrow \text{N} \rightarrow \text{I} + \text{N}$, resulting in reentrant I+N phase separation.

Figure 9.12 shows the binodal curves for $f = 10$ and $f = 100$ with $n_r = 10$. Phase separation (I + N) occurs between the two binodal curves, and the N phase shifts to lower concentrations as the temperature decreases. The width of the biphasic region decreases with increasing f, indicating that the number of functional groups on the rods is an important parameter for controlling phase behavior, including I + N phase separation.

Phase diagrams show how the number of functional groups on the rods, and temperature influence the NIT and I + N phase separation. These results highlight the crucial role of rod-solvent complexes in promoting nematic ordering and provide insight into controlling phase behavior in water-dispersible liquid crystalline polymer systems. The theories introduced in this chapter will serve as a basis for further developments in other associating liquid crystalline systems.

Bibliography

[1] De Gennes P. G. and Prost J. *The Physics of Liquid Crystals.* 2nd ed., Oxford University Press, Oxford, 1993.

[2] Li Q. editor. *Liquid Crystals Beyond Displays.* John Wiley & Sons, New Jersey, 2012.

[3] Demus D., Goodby J., Gray G. W., Spiess H. W., and Vill V. editors. *Physical Properties of Liquid Crystals.* Wiley-VCH, Weinheim, 1999.

[4] Chandrasekhar S. *Liquid Crystals.* Cambridge University Press, Cambridge, 1992.

[5] Takezoe H. and Eremin A. *Bent-Shaped Liquid Crystals.* CRC Press, Boca Raton, 2017.

[6] Sied M. B., Diez S., Salud J., Lopez D. O., Cusmin P., Tamarit J. Ll., and Barrio M. Liquid Crystal Binary Mixtures of 8OCB + 10OCB. Evidence for a Smectic A-to-Nematic Tricritical Point. *J. Phys. Chem. B*, 109:16284–16289, 2005.

[7] Donald A. M., Windle A. H., and Hanna S. *Liquid Crystalline Polymers.* 2nd ed., Cambridge University Press, Cambridge, 2006.

[8] Wieczorek S. A., Freyssingeas E., and Holyst R. Relaxation Processes in Semidilute Solutions of Polymers in Liquid Crystal Solvents. *J. Phys. Chem. B*, 109:16252–16262, 2005.

[9] Miller W. G. and Wee E. L. Liquid Crystal-Isotropic Phase Equilibriums in the System Poly (gamma.-benzyl. alpha.-L-glutamate)-methylformamide. *J. Phys. Chem.*, 75:1446–1452, 1971.

[10] Yevdokimov Y. M., Salyanov V. I., Semenov S. V., and Skuridin S. G. *DNA Liquid-Crystalline Dispersions and Nanoconstructions.* Chapter 2, CRC Press, Boca Raton, 2012.

[11] Drzaic P. S. *Liquid Crystal Dispersions.* World Scientific, Singapore, 1995.

[12] Crawford G. P. and Zumer S. editors. *Liquid Crystals in Complex Geometries.* Taylor and Francis, London, 1996.

[13] Maschke U., Benmouna F., Roussel F., Daoudi A., Gyselinck F., Buisine J. M., Coqueret X., and Benmouna M. Relaxation Processes in Semidilute Solutions of Polymers in Liquid Crystal Solvents. *Macromolecules*, 32:8866–8871, 1999.

[14] Kyu T. and Chiu H. W. Phase Equilibria of a Polymer-Smectic-Liquid-Crystal Mixture. *Phys. Rev. E*, 53:3618–3622, 1996.

[15] Chiu H. W., Zhou Z. L., Kyu T., Cada L. G., and Chien L. C. Phase Diagrams and Phase Separation Dynamics in Mixtures of Side-Chain Liquid Crystalline Polymers and Low Molar Mass Liquid Crystals. *Macromolecules*, 29:1051–1058, 1996.

[16] Urayama K., Okuno Y., Nakao T., and Kohjiya S. Volume Phase Transition of Liquid Crystalline Gels in a Nematic Solvent. *Macromolecules*, 29:4567–4569, 2002.

[17] Kim S. T. and Finkelmann H. Cholesteric Liquid Single Crystal Elastomers (LSCE) Obtained by the Anisotropic Deswelling Method. *Macromol. Rapid Commun.*, 22:429–433, 2001.

[18] Urayama K., Okuno Y., Nakao T., and Kohjiya S. Volume Transition of Nematic Gels in Nematogenic Solvents. *J. Chem. Phys.*, 118:2903–2910, 2003.

[19] Warner M. and Terentjev E. M. *Liquid Crystal Elastomers.* Oxford University Press, Oxford, 2007.

[20] Meyer R. B. Effects of Electric and Magnetic Fields on the Structure of Cholesteric Liquid Crystals. *Appl. Phys. Lett.*, 12:281–282, 1968.

[21] Lombardo D., Kiselev M. A., Magazu S., and Calandra P. Amphiphiles Self-Assembly: Basic Concepts and Future Perspectives of Supramolecular Approaches. *Adv. in Cond. Matt. Phys.*, 2015:151683, 2015.

[22] Dierking I., Martins A., and Neto F. *Lyotropic Liquid Crystals.* Oxford University Press, Oxford, 2024.

[23] Park H. S., Kang S. W., Tortora L., Nastishin Y., Finotello D., Kumar S., and Lavrentovich O. D. Self-Assembly of Lyotropic Chromonic Liquid Crystal Sunset Yellow and Effects of Ionic Additives. *J. Phys. Chem. B*, 112:16307–16319, 2008.

[24] Lydon J. Chromonic Liquid Crystalline Phases. *Liq. Cryst.*, 38:1663–1681, 2011.

[25] Ichikawa T., Kuwana M., and Suda K. Chromonic Ionic Liquid Crystals Forming Nematic and Hexagonal Columnar Phases. *Crystals*, 12:1548, 2022.

[26] Schenning A. P. H. J., Crawford G. P., and Broer D. J. *Liquid Crystal Sensors.* CRC Press, Boca Raton, 2018.

[27] Rey A. D. Liquid Crystal Models of Biological Materials and Processes. *Soft Matter*, 6:3402–3429, 2010.

[28] Wang Y. and Li Q. Light-Driven Chiral Molecular Switches or Motors in Liquid Crystals. *Adv. Mater.*, 24:1926–1945, 2012.

[29] Oswald P. and Pieranski P. *Smectic and Columnar Liquid Crystals.* CRC Press, Boca Raton, 2019.

[30] Chandrasekhar S. and Ranganath G. S. Discotic Liquid Crystals. *Rep. Prog. Phys.*, 53:57–84, 1990.

[31] Kirkwood J. G. and Monroe E. Statistical Mechanics of Fusion. *J. Chem. Phys.*, 9:514–526, 1941.

[32] Matsuyama A. Mean Field Theory of Crystalline Ordering in Colloidal Solutions. *J. Phys. Soc. Jpn.*, 75:034604, 2006.

[33] Matsuyama A., Arikawa S., Wada M., and Fukutomi N. Uniaxial and Biaxial Nematic Phases of Banana-Shaped Molecules and the Effects of an External Field. *Liq. Cryst.*, 46:1672–1685, 2019.

[34] Luckhurst G. R. Biaxial Nematic Liquid Crystals: Fact or Fiction? *Thin Solid Films*, 393:40–52, 2001.

[35] Lubensky T. C. and Radzihovsky L. Theory of Bent-Core Liquid-Crystal Phases and Phase Transitions. *Phys. Rev. E*, 66:031704, 2002.

[36] Straley J. P. Ordered Phases of a Liquid of Biaxial Particles. *Phys. Rev. A*, 10:1881–1887, 1974.

[37] Priestley E. B., Wojtowicz P. J., and Sheng P. editors. *Introduction to Liquid Crystals.* Plenum Press, New York, 1976.

[38] Demus D. and Richter L. *Textures of Liquid Crystals.* Verlag Chemie, Weinheim, 1978.

[39] Gray G. W. and Goodby J. W. *Smectic Liquid Crystals-Textures and Structures.* Leonard Hill, Glasgow, 1984.

[40] Dierking I. Liquid Crystal Textures: an Overview. *Liq. Cryst.*, https://doi.org/10.1080/02678292.2025.2517865:1–33, 2025.

[41] Onuki A. *Phase Transition Dynamics.* Cambridge University Press., Cambridge, 2002.

[42] Onsager L. The Effects of Shape on the Interaction of Colloidal Particles. *Ann. N. Y. Acad. Sci.*, 51:627–659, 1949.

[43] Maier W. and Saupe A. Eine Einfache Molekular-Statische Theorie der Nematischen Kristallinflussigen Phase. *Z. Naturforsch*, 14a:882–889, 1959.

[44] Matsuyama A. *Thermodynamics of Flexible and Rigid Rod Polymer Blends.* In: Isayev A. I., editor. Encyclopedia of Polymer Blends. Vol. 1., Chapter 2, WILEY-VCH, Weinheim, 2010.

[45] Matsuyama A. Spinodal in a Liquid-Face Centered Cubic Phase Separation. *J. Phys. Soc. Jpn.*, 75:084602, 2006.

[46] Gast A. P. and Russel W. B. Simple Ordering in Complex Fluids. *Physics Today*, 51:24–30, 1998.

[47] Lekkerkerker H. M. W., Coulon P., van der Haegen R., and Deblieck R. On the Isotropic-Liquid Crystal Phase Separation in a Solution of Rodlike Particles of Different Lengths. *J. Chem. Phys.*, 80:3427–3433, 1984.

[48] Palffy-Muhoray P., De Bruyn J. J., and Dunmur D. A. Mean Field Theory of Binary Mixtures of Nematic Liquid Crystals. *Liq. Cryst.*, 127:301–319, 1985.

[49] Palffy-Muhoray P., Berlinsky A. J., De Bruyn J. J., and Dunmur D. A. Coexisting Nematic Phases in Binary Mixtures of Liquid Crystals. *Phys. Lett. A*, 104:159–162, 1984.

[50] Doi M. and Edwards S. F. *The Theory of Polymer Dynamics.* Chapter 10, Oxford University Press, Oxford, 1986.

[51] Lebwohl P. A. and Lasher G. Nematic Liquid Crystal Order. A Monte Carlo Calculation. *Phys. Rev. A*, 6:426–429, 1972.

[52] Bates M. A. and Luckhurst G. R. Biaxial Nematic Phases and V-Shaped Molecules: A Monte Carlo Simulation Study. *Phys. Rev. E*, 72:051702, 2005.

[53] McMillan W. L. Simple Molecular Model for the Smectic A Phase of Liquid Crystals. *Phys. Rev. A*, 4:1238–1246, 1971.

[54] Kobayashi K. K. On the Theory of Translational and Orientational Melting with Application to Liquid Crystals. *Phys. Lett.*, 31:125–126, 1970.

[55] Kventsel G. F., Luckhurst G. R., and Zewdie H. B. A Molecular Field Theory of Smectic A Liquid Crystals. *Mol. Phys.*, 56:589–610, 1985.

[56] Meyer R. J. and McMillan W. L. Simple Molecular Theory of the Smectic C, B, and H Phases. *Phys. Rev. A*, 9:899–906, 1974.

[57] Lee F. T., Tan H. T., Shin Y. M., and Woo C. W. Phase Diagram for Liquid Crystals. *Phys. Rev. Lett.*, 31:1117–1120, 1973.

[58] Nakagawa M. and Akahane T. A Molecular Statistical Theory of Smectic A Liquid Crystals. *J. Phys. Soc. Jpn.*, 53:1951–1960, 1984.

[59] Govind A. S. and Madhusudana N. V. A Simple Molecular Theory of Smectic-C Liquid Crystals. *Europhys. Lett.*, 55:505–511, 2001.

[60] Gorkunov M. V. and Osipov M. A. Molecular Models for the Smectic A–Smectic C Phase Transition in a System of Biaxial Molecules. *J. Phys. A: Math. Theor.*, 41:295001, 2008.

[61] Erkoreka A., Huang M., Aya S., and Martinez-Perdiguero J. Mean-Field Theory of the Uniaxial Ferroelectric Smectic A Liquid Crystal Phase. *Phys. Rev. Lett.*, 133:208101, 2024.

[62] Mandle R. J., Cowling S. J., and Goodby J. W. A Nematic to Nematic Transformation Exhibited by a Rod-Like Liquid Crystal. *Phys. Chem. Chem. Phys.*, 19:11429–11435, 2017.

[63] Nishikawa H., Shiroshita K., Higuchi H., Okumura Y., Haseba Y., Yamamoto S., Sago K., and Kikuchi H. A Fluid Liquid-Crystal Material with Highly Polar Order. *Adv. Mater.*, 29:1702354, 2017.

[64] Goossens W. J. A. A Molecular Theory of the Cholesteric Phase and of the Twisting Power of Optically Active Molecules in a Nematic Liqud Crystal. *Mol. Cryst. Liq. Cryst.*, 12:237–244, 1971.

[65] Lin-Liu Y. R., Shih Y. M., and Woo C. W. Molecular Theory of Cholesteric Liquid Crystals and Cholesteric Mixtures. *Phys. Rev. A*, 15:2550–2557, 1977.

[66] Lin-Liu Y. R., Shih Y. M., Woo C. W., and Tan H. T. Molecular Model for Cholesteric Liquid Crystals. *Phys. Rev. A*, 14:445–450, 1976.

[67] Sato T. and Teramoto A. Concentrated Solutions of Liquid-Crystalline Polymers. *Adv. Polym. Sci.*, 126:85–161, 1996.

[68] Dierking I. Chiral Liquid Crystals: Structures, Phases, Effects. *Symmetry*, 6:444–472, 2014.

[69] Matsuyama A. Theory of Polymer-Dispersed Cholesteric Liquid Crystals. *J. Chem. Phys.*, 139:174906, 2013.

[70] Matsuyama A. and Kan T. Helical Inversions and Phase Separations in Binary Mixtures of Cholesteric Liquid Crystalline Molecules. *Liq.* **Cryst.**, 46:45–58, 2019.

[71] Kimura H., Hoshino M., and Nakano H. Temperature **Dependent** Pitch in Cholesteric Phase. *J. Phys. Colloq.*, 139:C3–174, 1979.

[72] Tschierske C. Mirror Symmetry Breaking in Liquids and Liquid Crystals. *Liq. Crtst.*, 45:2221–2252, 2018.

[73] Ostapenko T., Wiant D. B., Sprunt S. N., Jakli A., and Gleeson J. T. Magnetic-Field Induced Isotropic to Nematic Liquid Crystal Phase Transition. *Phys. Rev. Lett.*, 101:247801, 2008.

[74] De Gennes P. G. Calcul de la Distorsion d'une Structure Cholesterique par un Champ Magnetique. *Solid State Commun.*, 6:163–165, 1968.

[75] Meyer R. B. Distortion of a Cholesteric Structure by a Magnetic Field. *Appl. Phys. Lett.*, 14:208–209, 1969.

[76] Matsuyama A. Theory of a Helicoidal Cholesteric Phase Induced by an External Field. *Liq. Cryst.*, 43:783–795, 2016.

[77] Xiang J., Shiyanovskii S. V., Imrie C., and Lavrentovich O. D. Electrooptic Response of Chiral Nematic Liquid Crystals with Oblique Helicoidal Director. *Phys. Rev. Lett.*, 112:217801, 2014.

[78] Dozov I. On the Spontaneous Symmetry Breaking in the Mesophases of Achiral Banana-Shaped Molecules. *Europhys. Lett.*, 56:247–253, 2001.

[79] Borshch V., Kim Y. K., Xiang J., Gao M., Jakli A., Panov V. P., Vij J. K., Imrie C. T., Tamba M. G., Mehl G. H., and Lavrentovich O. D. Nematic Twist-Bend Phase with Nanoscale Modulation of Molecular Orientation. *Nat. Comm.*, 4:2635, 2013.

[80] Mandle R. J. The Dependency of Twist-Bend Nematic Liquid Crystals on Molecular Structure: a Progression from Dimers to Trimers, Oligomers and Polymers. *Soft Matter*, 12:7883–7901, 2016.

[81] Barbero G., Evangelista L. R., Rosseto M. P., Zola R. S., and Lelidis I. Elastic Continuum Theory: Towards Understanding of the Twist-Bend Nematic Phases. *Phys. Rev. E*, 92:030501(R), 2015.

[82] Greco C., Luckhurst G. R., and Ferrarini A. Molecular Geometry, Twist-Bend Nematic Phase and Unconventional Elasticity: a Generalised Maier-Saupe Theory. *Soft Matter*, 10:9318–9323, 2014.

[83] Lekkerkerker H. N. W., Coulon P., Haegen R. V. D., and Deblieck R. On the Isotropic-Liquid Crystal Phase Separation in a Solution of Rodlike Particles of Different Lengths. *J. Chem. Phys.*, 80:3427–3433, 1984.

[84] Matsuyama A. and Kato T. Theory of Binary Mixtures of a Flexible Polymer and a Liquid Crystal. *J. Chem. Phys.*, 105:1654–1660, 1996.

[85] Flory P. J. *Principles of Polymer Chemistry.* Cornell University Press, Ithaca, 1953.

[86] Flory P. J. Phase Equilibria in Solutions of Rod-Like Particles. *Proc. R. Soc. London Ser. A*, 234:73–89, 1956.

[87] Cahn J. W. Spinodal Decomposition. *Trans. Metal. Soc. AIME*, 242:166–180, 1968.

[88] Matsuyama A., Evans R. M. L., and Cates M. E. Orientation Fluctuation-Induced Spinodal Decompositions in Polymer-Liquid Crystal Mixtures. *Phys. Rev. E*, 61:2977–2986, 2000.

[89] Langer J. S. editor. *Solid Far from Equilibrium.* Cambridge University Press, New York, 1992.

[90] Holyst R. and Schick M. Mixtures of Rigid and Flexible Nematogenic Polymers. *J. Chem. Phys.*, 96:721–729, 1992.

[91] Liu A. J. and Fredrickson G. H. Free Energy Functionals for Semiflexible Polymer Solutions and Blend. *Macromolecules*, 26:2817–2824, 1993.

[92] Shen C. and Kyu T. Spinodals in a Polymer Dispersed Liquid Crystal. *J. Chem. Phys.*, 102:556–562, 1995.

[93] Orendi H. and Ballauf M. Complete Phase Diagrams of Mixtures of a Nematic Liquid Crystal with n-alkanes. *Liq. Cryst.*, 6:497–500, 1989.

[94] Ahn W., Kim C. Y., Kim H. S., and Kim C. Phase Behavior of Polymer/Liquid Crystal Blends. *Macromolecules*, 25:5002–5007, 1992.

[95] Kronberg B., Bassignana I., and Patterson D. Phase Diagrams of Liquid Crystal + Polymer Systems. *J. Phys. Chem.*, 82:1714–1719, 1978.

[96] Dubault A., Casagrande C., Veyssie M., and Deloche B. Pseudo Clearing Temperature in Binary Polymer-Nematic Solutions. *Phys. Rev. Lett.*, 45:1645–1648, 1980.

[97] Dormoy Y., Gallani J. L., and Martinoty P. Behavior of a Flexible Polymer in Solution in the Isotropic Phase of a Nematic Liquid Crystal. *Mol. Cryst. Liq. Cryst.*, 170:135–149, 1989.

[98] Russel F., Canlet C., and Fung B. M. Morphology and Orientational Order of Nematic Liquid Crystal Droplets Confined in a Polymer Matrix. *Phys. Rev. E*, 65:021701, 2002.

[99] Lepena A. M., Glotzer S. C., Langer S. A., and Liu A. J. Effect of Ordering on Spinodal Decomposition of Liquid-Crystal/Polymer Mixtures. *Phys. Rev. E*, 60:R29–R32, 1999.

[100] Araki T. and Tanaka H. Stripe Pattern Formation in Phase Separation Accompanying Orientational Ordering under an External Field. *J. Phys.: Condens Matter*, 18:L305–L314, 2006.

[101] Fukuda J. Phase Separation Kinetics of Liquid Crystalline Polymers: Effect of Orientational Order. *Phys. Rev. E*, 59:3275–3288, 1999.

[102] Flory P. J. Statistical Thermodynamics of Semi-Flexible Chain Molecules. *Proc. R. Soc. London A*, 234:60–73, 1956.

[103] Matsuyama A., Sumikawa Y., and Kato T. Phase Separations Induced by Partial Stiffening of Semiflexible Polymers. *J. Chem. Phys.*, 107:4711–4718, 1997.

[104] Chiu H. W., Zhou Z. L., Kyu T., Cada L. G., and **Chien** L. C. Phase Diagrams and Phase Separation Dynamics in Mixtures of Side-Chain Liquid Crystalline Polymers and Low Molar Mass Liquid Crystals. *Macromolecules*, 29:1051–1058, 1996.

[105] Flory P. J. Theory of Systems of Rodlike Particles: I. Athermal Systems. *Mol. Cryst. Liq. Cryst.*, 54:289–309, 1979.

[106] Matsuyama A. Theory of Binary Mixtures of a Rodlike Polymer and a Liquid Crystal. *J. Chem. Phys.*, 132:214902, 2010.

[107] Casagrande C., Veyssie M., and Knobler C. M. Anisotropic Features in the Dynamics of Nematic-Nematic Phase Separation. *Phys. Rev. Lett.*, 58:2079–2082, 1987.

[108] Ratto J. A., Volino F., and Blumstein R. B. Phase Behavior and Order in Mixtures of Main-Chain Nematic Polyesters with Small Molecules: A Combined Proton and Deuterium NMR Study. *Macromolecules*, 24:2862–2867, 1991.

[109] Brochard F., Jouffroy J., and Levinson P. Phase Diagrams of Mesomorphic Mixtures. *J. de Phys.*, 45:1125–1136, 1984.

[110] Matsuyama A. Phase Separations in Mixtures of a Liquid Crystal and a Nanocolloidal Particle. *J. Chem. Phys.*, 131:204904, 2009.

[111] Soule E. R. and Rey A. D. Modelling Complex Liquid Crystal Mixtures: from Polymer Dispersed Mesophase to Nematic Nanocolloids. *Molecular Simulation*, 38:735–750, 2012.

[112] Gorkunov M. V. and Osipov M. A. Mean-Field Theory of a Nematic Liquid Crystal Doped with Anisotropic Nanoparticles. *Soft Matter*, 7:4348–4356, 2011.

[113] Adams M., Dozic Z., Kelle S. L., and Fraden S. Entropically Driven Microphase Transitions in Mixtures of Colloidal Rods and Spheres. *Nature*, 393:349–352, 1998.

[114] Yamamoto J. and Tanaka H. Transparent Nematic Phase in a Liquid-Crystal-Based Microemulsion. *Nature*, 409:321–325, 2001.

[115] Lagerwall J. P. F. and Scalia G. A. A New Era for Liquid Crystal Research: Applications of Liquid Crystals in Soft Matter Nano-, Bio- and Microtechnology. *Curr. App. Phys.*, 12:1387–1412, 2012.

[116] Musevic I. Nematic Liquid-Crystal Colloids. *Materials*, 11(1):24, 2018.

[117] Chen C. H. and Dierking I. Nanoparticles in Thermotropic and Lyotropic Liquid Crystals. *Front. Soft Matter*, 4:1518796, 2025.

[118] Terentjev E. M. Liquid Crystal Elastomers: 30 Years After. *Macromolecules*, 58:2792–2806, 2025.

[119] Tanaka T. Collapse of Gels and the Critical Endpoint. *Phys. Rev. Lett.*, 20:820–823, 1978.

[120] Tanaka T., Fillmore D., Sun S-T., Nishio I., Swislow G., and Shah A. Phase Transitions in Ionic Gels. *Phys. Rev. Lett.*, 45:1636–1639, 1980.

[121] Shibayama M. and Tanaka T. Volume Phase Transition and Related Phenomena of Polymer Gels. *Adv. in Polym. Sci.*, 109:1–62, 1993.

[122] Suzuki A. and Tanaka T. Phase Transition in Polymer Gels Induced by Visible Light. *Nature*, 346:345–347, 1990.

[123] Dusek K. and Duskova-Smrckova M. Volume Phase Transition in Gels: Its Discovery and Development. *Gels*, 6:22, 2020.

[124] Swislow G., Sun S. T., Nishio I., and Tanaka T. Coil-Globule Phase Transition in a Single Polystyrene Chain in Cyclohexane. *Phys. Rev. Lett.*, 44:796–799, 1980.

[125] Baysal B. M. and Karasz F. E. Coil-Globule Collapse in Flexible Macromolecules. *Macromol. Theory Simul.*, 12:627–646, 2003.

[126] Warner M. and Wang X. J. Phase Equilibria of Swollen Nematic Elastomers. *Macromolecules*, 25:445–449, 1992.

[127] Wang X. J. and Warner M. The Swelling of Nematic Elastomers by Nematogenic Solvents. *Macromol. Theory Simul.*, 6:37–52, 1997.

[128] Matsuyama A. and Kato T. Nematic Ordering-Induced Volume Phase Transitions of a Liquid Crystalline Gel. *J. Chem. Phys.*, 116:8175–8182, 2002.

[129] Matsuyama A. and Kato T. Discontinuous Elongation of Nematic Gels by a Magnetic Field. *Phys. Rev. E*, 64:010701(R), 2001.

[130] Hoppe C. E., Galante M. J., Oyanguren P. A., and Williams R. J. J. Polymer-Dispersed Liquid Crystals Based on Polystyrene and EBBA: Analysis of Phase Diagrams and Morphologies Generated. *Macromol. Chem. Phys.*, 204:928–935, 2003.

[131] Urayama K. Selected Issues in Liquid Crystal Elastomers and Gels. *Macromolecules*, 40:2277–2288, 2007.

[132] Matsuyama A. Volume Phase Transitions of Cholesteric Liquid Crystalline Gels. *J. Chem. Phys.*, 142:174907, 2015.

[133] Popa-Nita V. and Kralj S. Transformation of Phase Transitions Driven by an Anisotropic Random Field. *Phys. Rev. E*, 71:042701, 2005.

[134] Rey A. D. Liquid Crystal Models of Biological Materials and Silk Spinning. *Biopolymers*, 97:374–396, 2012.

[135] Israelachvili J. N. *Intermolecular and Surface Forces.* Academic Press, San Diego, 1985.

[136] Kato T. Self-Assembly of Phase-Segregated Liquid Crystal Structures. *Science*, 295:2414–2418, 2002.

[137] Rey A. D. Theoretical Platform for Liquid-Crystalline Self-Assembly of Collagen-Based Biomaterials. *Front. Phys.*, 7:1–15, 2019.

[138] Odijk T. Effect of Micellar Flexibility on the Isotropic-Nematic Phase Transition in Solutions of Linear Aggregates. *J. Phys.*, 48:125–129, 1987.

[139] McMullen W. E., Gelbart W. M., and Ben-Shaul A. Isotropic-Nematic Transition in Micellized Solutions. *J. Chem. Phys.*, 82:5616–5623, 1985.

[140] Herzfeld J. Liquid Crystalline Order in Self-Assembling Systems: Orientation Dependence of the Particle Size Distribution. *J. Chem. Phys.*, 88:2776–2779, 1988.

[141] Tanaka F. *Polymer Physics.* Cambridge University Press, Cambridge, 2011.

[142] Matsuyama A. Theory of Nematic-Isotropic Phase Transitions in Solutions of Rodlike Aggregates. *Liq. Cryst.*, 52:195–204, 2025.

[143] Chen Z. Y. Effect of Polydispersity on the Isotropic-Nematic Phase Transition of Rigid Rods. *Phys. Rev. E*, 50:2849–2855, 1994.

[144] Odijk T. and Lekkerkerker H. N. W. Theory of the Isotropic-Liquid Crystal Phase Separation for a Solution of Bidisperse Rodlike Macromolecules. *J. Phys. Chem.*, 89:2090–2096, 1985.

[145] De Michele C. Theory of Self-Assembly Driven Nematic Liquid Crystals Revised. *Liq. Cryst.*, 49:2003–2012, 2019.

[146] Van der Schoot P. and Cates M. E. Growth, Static Light Scattering, and Spontaneous Ordering of Rodlike Micelles. **Langmuir**, 10:670–679, 1994.

[147] Johannesson H., Furo I., and Halle B. Orientational Order and Micelle Size in the Nematic Phase of the Cesium Pentadecafluorooctanoate-Water System from the Anisotropic Self-Diffusion of Water. *Phys. Rev. E*, 53:4904–4917, 1996.

[148] Mezzenga R., Schurtenberger P., Burbidge A., and Michel M. Understanding Foods as Soft Materials. *Nat. Mater.*, 4:729–740, 2005.

[149] Binnemans K. Ionic Liquid Crystals. *Chem. Rev.*, 105:4148–4204, 2005.

[150] Geim A. K. and Novoselov K. S. The Rise of Graphene. *Nat. Mater.*, 6:4148–4204, 2007.

[151] Medhekar N. V., Ramasubramaniam A., Ruoff R. S., and Shenoy V. B. Hydrogen Bond Networks in Graphene Oxide Composite Paper: Structure and Mechanical Properties. *ACS Nano*, 4:2300–2306, 2010.

[152] Hirsch A. The Era of Carbon Allotropes. *Nat. Mater.*, 9:868–871, 2010.

[153] Badaire S., Zakri C., Maugey M., Derre A., Barisci J. N., Wallace G., and Poulin P. Liquid Crystals of DNA-Stabilised Carbon Nanotubes. *Adv.* **Mater.**, 17:1673–1676, 2005.

[154] Matsuyama A. Nematic-Isotropic Phase Separations Induced by Hydrogen Bonding Between a Rodlike Polymer and Solvent Molecules. *Liq.* **Cryst.**, 50:1679–1687, 2023.

[155] Matsuyama A. and Tanaka F. Solvation-Induced Reentrant Phase Separation in Polymer Solutions. *Phys. Rev. Lett.*, 65:341–344, 1990.

[156] Matsuyama A. and Tanaka F. Theory of Solvation-Induced Reentrant Coil-Globule Transition of an Isolated Polymer Chain. *J. Chem. Phys.*, 94:781–786, 1991.

[157] Coleman M. M. and Painter P. C. Hydrogen Bonded Polymer Blends. *Prog. Polym. Sci.*, 20:1–59, 1995.

[158] Kyu T. and Mukherjee P. Kinetics of Phase Separation by Spinodal Decomposition in a Liquid-Crystalline Polymer Solution. *Liq.* **Cryst.**, 3:631–644, 1988.

[159] Viale S., Best A. S., Mendes E., and Picken S. J. Formation of Aqueous Molecular Nematic Liquid Crystal Phase in Poly(p-sulfophenylene sulfoterephthalamide). *Chem. Commun.*, 12:1528–1530, 2005.

Index